Advancements and Innovations in Wireless Communications and Network Technologies

Michael R. Bartolacci
Pennsylvania State University – Berks, USA

Steven R. Powell
California State Polytechnic University – Pomona, USA

Managing Director:	Lindsay Johnston
Editorial Director:	Joel Gamon
Book Production Manager:	Jennifer Romanchak
Publishing Systems Analyst:	Adrienne Freeland
Development Editor:	Heather Probst
Assistant Acquisitions Editor:	Kayla Wolfe
Typesetter:	Lisandro Gonzalez
Cover Design:	Nick Newcomer

Published in the United States of America by
Information Science Reference (an imprint of IGI Global)
701 E. Chocolate Avenue
Hershey PA 17033
Tel: 717-533-8845
Fax: 717-533-8661
E-mail: cust@igi-global.com
Web site: http://www.igi-global.com

Library of Congress Cataloging-in-Publication Data

Advancements and innovations in wireless communications and network technologies / Michael R. Bartolacci and Steven R. Powell, editors.
    pages cm
  Summary: "This book is a collection of research and case studies that tackle the issues, advancements, and techniques on wireless communications and network technologies, offering expansive knowledge and different perspectives useful for researchers and students alike"-- Provided by publisher.
  Includes bibliographical references and index.
  ISBN 978-1-4666-2154-1 (hardcover) -- ISBN (invalid) 978-1-4666-2155-8 (ebook) -- ISBN (invalid) 978-1-4666-2156-5 (print & perpetual access) 1. Wireless communication systems--Case studies. I. Bartolacci, Michael R., 1954- editor of compilation. II. Powell, Steven, editor of compilation.
  TK5103.2.A385 2013
  621.382--dc23
                                2012019568

British Cataloguing in Publication Data
A Cataloguing in Publication record for this book is available from the British Library.

The views expressed in this book are those of the authors, but not necessarily of the publisher.

Table of Contents

Detailed Table of Contents

Chapter 1

Mohammad Anbar, Jawaharlal Nehru University, India
Deo Prakash Vidyarthi, Jawaharlal Nehru University, India

A Cellular IP (CIP) network involves a bulk of data transmission. It is highly reliable and guarantees the safe delivery of the packets required in such systems. Reliable traffic performance leads to efficient and reliable connectivity in Cellular IP network. CIP network, which consists of mobile hosts, base stations, and links, are often vulnerable and prone to failure. During the routing operation in the network, the base station, which works as router for the transmitted packets, may fail to perform. Reliable transmission is desirable, in terms of services of the base stations in the network, reliable routing, and processing the data. In this paper, the authors design a reliability model to increase the reliability of a flow, consisting of packets, passing through routers in a Cellular IP network. Particle Swarm Optimization (PSO) is able to solve a class of complex optimization problems. PSO is used to improve the reliability of the flow in CIP network. The proposed model studies the effects of packet processing rate (μ), packet arrival rate (λ), and the number of packets per flow on the reliability of the system. A simulation experiment is conducted and results reveal the effectiveness of the model.

Chapter 2

Jamal Haque, University of South Florida, USA
M. Cenk Erturk, University of South Florida, USA
Huseyin Arslan, University of South Florida, USA
Wilfrido Moreno, University of South Florida, USA

The paper explores the system and architecture requirements for cognitive driven reconfigurable hardware for an aeronautical platform, such as commercial aircraft or high altitude platforms. With advances in components and processing hardware, mobile platforms are ideal candidates to have configurable hardware that can morph itself, given the location and available wireless service. This paper proposes a system for an intelligent self-configurable software and hardware solution for an aeronautical system.

Chapter 3

This study evaluates video codec performance over VoIP using a campus wireless network. Today, the deployment of VoIP occurs in various platforms, including VoIP over LAN, VoIP over WAN and VoIP over VPN. Therefore, this study defines which video codec provides good video quality over VoIP transmission. The soft phone is used as a medium for communication between two parties. A network management system is used to evaluate and capture the video quality performance over VoIP. The quality of video codec is based on MOS, jitter, delay and packet loss. The experimental scope is limited to G.722 with MP4V-ES, G.726 (16) with H.261 and G.726 (24) with H.264. The results show that audio codec G.722 with MP4V-ES generates good video quality over VoIP using wireless local area network. Whereas audio codec G.726 (16) with H.261 generates low rate video and voice quality performance. Therefore, using the appropriate video and audio, the codec selection increases video quality over VoIP transmission.

Chapter 4

Quality-of-Service (QoS) routing protocol is developed for mobile Ad Hoc Networks. MANET is a self configuring network of mobile devices connected by wireless links. Each device in the MANET is free to move independently in any direction; therefore, it changes links to other devices frequently. The proposed QoS-based routing in the Optimized Link State Routing (OLSR) protocol relates bandwidth and delay using a fuzzy logic algorithm. The path computations are examined and the reason behind the selection of bandwidth and delay metrics is discussed. The performance of the protocol is investigated by simulation. The results in FQOLSR indicate an improvement in mobile wireless networks compared with the existing QOLSR system.

Chapter 5

Today's National Airspace System (NAS) is managed using an aging surveillance radar system. The radar technology is not adequate to sustain the support of aviation growth and cannot be adapted to use 21st century technologies Therefore, FAA has begun to implement the Next Generation Air Transportation System (NextGen) that would transform today's aviation and ensure increased safety and capacity. The first building block of the NextGen system is the implementation of the Automatic Dependent Surveillance-Broadcast (ADS-B) system. One of the most important design aspects of the ADS-B program is the design of the terrestrial radio station infrastructure. This design determines the layout of the terrestrial radio stations throughout the United States and is optimized to meet system performance, safety and security. Designing this infrastructure to meet system requirements is at the core of Service Volume (SV) Engineering. In this paper, the authors present a complex Geospatial/RF design ER Development model-based ADS-B SV Engineering design that captures radio sites layout and configuration parameters. CORE software is selected to implement the model-based SV Engineering environment.

Hesham A. Ali, Mansoura University, Egypt
Ahmed I. Saleh, Mansoura University, Egypt
Mohammed H. Ali, Mansoura University, Egypt

The main objective of PCS networks is to provide "anytime-anywhere" cellular services. Accordingly, lost calls as well as the network slow response have become the major problems that hardly degrade the network reliability. Those problems can be overcome by perfectly managing the Mobile Terminals (MTs) locations. In the existing location management (LM) scheme, Location Area (LA) is the smallest unit for registration. A MT must register itself when passing through its LA boundary to a neighboring one. Moreover, such registration takes place at the MTs' master HLR (even though currently managed by another HLR), which increases communication costs. As a result, existing LM scheme suffers from; (1) excessive location registrations by MTs located around LA boundaries (ping-pong effect) and (2) requiring the network to poll all LA cells to locate the callee MT. In this paper, a novel LM strategy is introduced by restructuring LAs into smaller areas called Base Areas (BAs), which impacts the paging cost. The proposed LM strategy uses caching to reduce unwanted updates and 2LP to reduce paging cost. Experimental results show that the proposed scheme introduces a distinct improvement in network response and tracing process.

Izabella Lokshina, SUNY Oneonta, USA

This paper examines turbo codes that are currently introduced in many international standards, including the UMTS standard for third generation personal communications and the ETSI DVB-T standard for Terrestrial Digital Video Broadcasting. The convergence properties of the iterative decoding process associated with a given turbo-coding scheme are estimated using the analysis technique based on so-called extrinsic information transfer (EXIT) chart. This approach provides a possibility to anticipate the bit error rate (BER) of a turbo code system using only the EXIT chart. It is shown that EXIT charts are powerful tools to analyze and optimize the convergence behavior of iterative systems utilizing the turbo principle. The idea is to consider the associated SISO stages as information processors that map input a priori LLR's onto output extrinsic LLR's, the information content being obviously assumed to increase from input to output, and introduce them to the design of turbo systems without the reliance on extensive simulation. Compared with the other methods for generating EXIT functions, the suggested approach provides insight into the iterative behavior of linear turbo systems with substantial reduction in numerical complexity.

Dario Di Zenobio, Fondazione Ugo Bordoni, Italy
Massimo Celidonio, Fondazione Ugo Bordoni, Italy
Lorenzo Pulcini, Fondazione Ugo Bordoni, Italy
Arianna Rufini, Fondazione Ugo Bordoni, Italy

Broadband Wireless Access is a strategic opportunity for mobile operators which aim to provide connectivity in digital divide areas, in order to accelerate speed of deployment and save in installation costs. This paper presents an innovative approach to access the end user, relying on infrastructural integration of femtocellular technology with existing cabled network. Usually, the adoption of Femtocell Access

Points, operating in the licensed cellular bands typically designed to be used in SOHO, improves the radio coverage and the building penetration of the existing mobile networks, based on macrocells. In the proposed solution, the peculiar functionality of femtocells is further improved using a MATV/SMATV cabled infrastructure which facilitates the signal connection inside the building. The potentiality of the solution is even more evident, taking into account the growing interest towards the possible deployment of new mobile technologies, like LTE in both the last portion of the UHF band V and the GSM frequency band, resulting from the re-farming process.

Chapter 9

Mike Sabelkin, École de Technologie Supérieure, Canada
François Gagnon, École de Technologie Supérieure, Canada

The proposed communication system architecture is called TOMAS, which stands for data Transmission oriented on the Object, communication Media, Application, and state of communication Systems. TOMAS could be considered a Cross-Layer Interface (CLI) proposal, since it refers to multiple layers of the Open Systems Interconnection Basic Reference Model (OSI). Given particular scenarios of image transmission over a wireless LOS channel, the wireless TOMAS system demonstrates superior performance compared to a JPEG2000+OFDM system in restored image quality parameters over a wide range of wireless channel parameters. A wireless TOMAS system provides progressive lossless image transmission under influence of moderate fading without any kind of channel coding and estimation. The TOMAS system employs a patent pending fast analysis/synthesis algorithm, which does not use any multiplications, and it uses three times less real additions than the one of JPEG2000+OFDM.

Chapter 10

Mahdy Saedy, University of Texas at San Antonio, USA
Vahideh Mojtahed, Purdue University, USA

This paper introduces an efficient machine-to-machine (M2M) communication model based on 4G cellular systems. M2M terminals are capable of establishing Ad Hoc clusters wherever they are close enough. It is also possible to extend the cellular coverage for M2M terminals through multi-hop Ad Hoc connections. The M2M terminal structure is proposed accordingly to meet the mass production and security requirements. The security becomes more critical in Ad Hoc mode as new nodes attach to the cluster. A simplified protocol stack is considered here, while key components are introduced to provide secure communications between M2M and the network and also amongst M2M terminals.

Chapter 11

Elizabeth Avery Gomez, New Jersey Institute of Technology, USA

Sensing technologies by design are calibrated for accuracy against an expected measurement scale. Sensor calibration and signal processing criteria are one type of sensor data, while the sensor readings are another. Ensuring data accuracy and precision from sensors is an essential, ongoing challenge, but these issues haven't stopped the potential for pervasive application use. Technological advances afford an opportunity for sensor data integration as a vehicle for societal well-being and the focus of ongoing research. A lean and flexible architecture is needed to acquire sensor data for societal well-being. As such, this research places emphasis on the acquisition of environmental sensor data through lean application programming protocols (APIs) through services such as SMS, where scant literature is presented. The contribution of this research is to advance the research that integrates sensor data with pervasive applications.

This paper is devoted to modeling and simulation of traffic with integrated services at media gateway nodes in the next generation networks, based on Markov reward models (MRM). The bandwidth sharing policy with the partial overlapped transmission link is considered. Calls arriving to the link that belong to VBR and ABR traffic classes, are presented as independent Poisson processes and Markov processes with constant intensity and/or random input stream, and exponential service delay time that is defined according to MRM. Traffic compression is calculated using clustering and learning vector quantification (e.g., self-organizing neural map). Numerical examples and simulation results are provided for communication networks of various sizes. Compared with the other methods for traffic compression calculations, the suggested approach shows substantial reduction in numerical complexity.

This paper presents the Cognitive Radio framework for wireless Ad Hoc networks. The proposed Cognitive Radio framework is a complete model for Cognitive Radio that describes the sensing and sharing procedures in wireless networks by introducing Queued Markov Chain method in spectrum sensing and Competitive Indexing Algorithm in spectrum sharing part. Queued Markov Chain method is capable of considering waiting time and is well generalized for an unlimited number of secondary users. It includes the sharing aspect of Cognitive Radio. Power-law distribution of node degree in scale-free networks is important for considering the traffic distribution and resource management thus we consider the effect of the topology on sensing and sharing performances. The authors demonstrate that CIF outperforms Uniform Indexing (UI) algorithm in Scale-Free networks while in Random networks UI performs as well as CIF.

This paper presents the results of mobile application which helps in preventing mobile phone accidents to the great extent. An electronic circuit (Transmitter and Receiver block) also designed to detect the driver's mobile phone automatically once he or she starts the vehicle and the circuit will switch OFF and then ON the mobile phone without human intervention with the help of 5 pin relay in order to start the application automatically. The authors further extend the research by comparing the obtained results after installing this application with a recent study of the US National Safety Council, conducted on 2010. The authors also show how far this application helps in reducing economic losses in India.

The flexibility of movement for the wireless ad hoc devices, referred to as node mobility, introduces challenges such as dynamic topological changes, increased frequency of route disconnections and high packet loss rate in Mobile Ad hoc Wireless Network (MANET) routing. This research proposes a novel on-demand routing protocol, Speed-Aware Routing Protocol (SARP) to mitigate the effects of high node mobility by reducing the frequency of route disconnections in a MANET. SARP identifies a highly mobile node which forms an unstable link by predicting the link expiration time (LET) for a transmitter and receiver pair. NS2 was used to implement the SARP with ad hoc on-demand vector (AODV) as the underlying routing algorithm. Extensive simulations were then conducted using Random Waypoint Mobility model to analyze the performance of SARP. The results from these simulations demonstrated that SARP reduced the overall control traffic of the underlying protocol AODV significantly in situations of high mobility and dense networks; in addition, it showed only a marginal difference as compared to AODV, in all aspects of quality-of-service (QOS) in situations of low mobility and sparse networks.

Chapter 16

Computing and communication technologies have merged to produce an environment where many applications and their associated data reside in remote locations, often unknown to the users. The adoption of cloud computing promises many benefits to users and service providers, as it shifts users' concerns away from the physical location of system components and toward the accessibility of the system's services. While this adoption of cloud computing may be beneficial to users and service providers, it increases areas of concern for computer forensic examiners that need to obtain data from cloud computing environments for evidence in legal matters. The authors present an overview of cloud computing, discuss the challenges it raises from a digital forensics perspective, describe suitable tools for forensic analysis of cloud computing environments, and consider the future of cloud computing.

Chapter 17

Many strategies have been developed to manage supply chain operations effectively. Vendor Managed Inventory (VMI) system is one of the prevalent strategic tools of the supply side logistics based on the electronic data exchange and business process automation among the suppliers and customers to enhance the competitive advantage. VMI is widely used in different industries including automotive sector. The VMI concept is a continuous replenishment program where suppliers are given access to demand and inventory level of customers and they are fully responsible for managing and replenishing the customer's stock. VMI's extension on customer satisfaction cannot be perceived sufficiently by decision-makers who are responsible to develop and invest in the customer-supplier relationship. This paper presents a path model using the method of Partial Least Squares (PLS) regression to give insight to decision-makers to understand effect of the VMI adoption on customer satisfaction. This paper investigates both determinants of relative factors of successful VMI adoption and the relationship in the supply chain with an empirical automotive industry case. The results show that the collaboration and coordination between customer and supplier and infrastructure of the information-sharing are the important dimensions to add value to the supply chain and to enhance customer satisfaction.

Chapter 18

Hierarchical Mobile IP (HMIP) reduces the signaling delay and number of registration messages to home agent (HA) by restricting them to travel up to a local gateway only. It uses centralized gateways that may disrupt the communications, in the event of a gateway failure, between a gateway and the mobile users residing with underlying foreign agents (FAs) in a regional network. Dynamic mobility management schemes, using distributed gateways, proposed in literature, tend to circumvent the problems in HMIP. These schemes employ varying regional network sizes or hierarchy levels that are dynamically selected according to call-to-mobility ratio (CMR) of individual user. In reality, this information cannot be readily available in practice. Also, any unusual alterations in CMR values may hamper the system performance. This paper proposes a new mobility management strategy for IP-based mobile networks, which is independent of individual user history. The proposed scheme uses subnet-specific registration areas and is fully distributed so that the signaling overheads are evenly shared at each FA. The scheme provides a viable alternative to dynamic mobility management schemes for its simplicity, performance, and ease of implementation.

Chapter 19

This paper explores the influence of digital technologies on media networks, in particular how they affect the traditional gatekeeping model. Wireless communications are the hot point of all digital technologies, and their application to the transmission of the Olympic Games is a milestone for the global creative industries every two/four years. The authors argue that the research and innovation (R&I) industries' involvement with the media industries needs to be reconsidered within the framework of an updated media gatekeeping model. To investigate this research question, results are reported from a case study examining the gatekeeping processes in the 2008 Olympic Games in Beijing, and the subsequent Olympics up to 2016. Results show the need for a new gatekeeping model that takes into consideration the impact of digital technologies, especially wireless communications. Additionally, new decision models regarding innovation investment in the global media industry are suggested by the impact of R&I on the media gatekeeping model itself.

Preface

This book contains articles from the four issues of Volume 3 of the *International Journal of Interdisciplinary Telecommunications and Networking* (IJITN). As has been the case with our previous two books containing articles from IJITN, this book reflects the journal's mission of publishing high-quality original interdisciplinary academic and practitioner research, surveys, and case studies that address telecommunications and networking issues, answer telecommunications and networking questions, or solve telecommunications and networking problems. The articles also reflect the journal's objective of covering a wide variety of topics related to telecommunications and networking technology, management, policy, economics, and social impact from a diversity of disciplinary viewpoints, including electrical engineering, computer science, operations research, business, and law.

The first article of Volume 3, Issue 1, "Maximizing the Flow Reliability in Cellular IP Network Using PSO," is by Mohammad Anbar and Deo Prakash Vidyarthi. In cellular networks the geographical area covered by the network is divided into cells serviced by base stations at the center of the cells. In cellular IP networks the base stations serve as access points and routers for the IP packets in the network. Since two critical factors in a network's performance are its response time and reliability, the performance in a cellular IP network can be enhanced by minimizing the time required for the base stations to process the packets and route them as well as decreasing the probability that the network will fail. Particle Swarm Optimization (PSO), a tool based on the behavior of social insects such as bird flocks and fish schools in which individuals mimic the behavior of their neighbors, can be used to solve difficult optimization problems. In their paper the authors develop a model using PSO to reduce the flow processing time in the base stations' router to improve the reliability of the network. Simulation results indicate that PSO can increase the cellular network's reliability.

Cognitive systems can learn. The second article of this issue, "Cognitive Aeronautical Communication System," by Jamal Haque, M. Cenk Ertuk, Huseyin Arslan, and Wilfrido Moreno, investigates cognitive aeronautical radio systems, which can learn the aeronautical environment for various locations and altitudes. Such systems face a number of difficult issues, including channel impairments, Doppler effects, and spectrum bandwidth use and frequency band allocation in changing geographical, political, and regulatory environments. The paper proposes an intelligent self-configurable hardware and software system, whose heart is a data base updated after each aircraft's flight. The data, which is collected by the aircraft's navigation and radar system during the flight and downloaded to the data base afterwards, includes the route traversed, wireless links available, frequency band, bandwidth, data rate, wireless standard and signal quality. The authors believe that such a system will result in an efficient use of the spectrum and a high data rate for global connectivity.

One of the engines behind the proliferation of video services is digital compression. The third article of the issue, "Best Approach for Video Codec Selection Over VoIP Conversation Using Wireless Local Area Network" by Mohd Nazri Ismail, evaluates the performance over VoIP on a campus WLAN of three audio/video codec pairs: G.722/MP4V-ES; G276 (16)/H.261; and G.726 (24)/H.264. Four factors are considered in the analysis: Mean Opinion Score (MOS); packet delay; jitter; and packet loss. In the experiment a soft phone is used to transmit audio and video; performance is monitored using network management software. The results indicate that the G.722/MP4V-ES pair possesses the least delay, the best video quality, and the least jitter.

Accounting for Quality of Service (QoS) in a mobile ad hoc network (MANET) is more difficult than in other networks since the network topology changes as the nodes move. In the fourth article of this issue, "Fuzzy QoS Based OLSR Network" by G. Uma Mahewswari, a QoS routing protocol is developed based on Optimized Link State Routing (OLSR) with a fuzzy logic algorithm relating bandwidth and delay. Simulation results indicate the performance of the OLSR routing protocol with this QoS particular enhancement is superior to the existing OLSR protocol using QoS.

The fifth and final article of Volume 2, Issue 2 is "Development of a Complex Geospatial/RF Design Model in Support of Service Volume Engineering Design" by Erton Boci, Shahram Sarkani, and Thomas Mazzuchi. The article deals with the design in the U.S. of the terrestrial radio station infrastructure of the Automatic Dependent Surveillance – Broadcast (ADS-B) system, the first building block of the FAA's Next Generation Air Transportation System. The design must optimize system performance, safety, and security at a minimum cost. In particular, the article examines the model-centric approach adopted by the Service Volume (SV) Engineering design team to capture, manage, and distribute its design and configuration data. The CORE software package that the team employed provides a customizable, collaborative entity-relationship-attribute environment with a single database repository. The article describes the steps the team took in formulating the data model and implementing it with CORE software.

In order to efficiently establish any service in a PCS network, the location of the mobile terminal (MT) must be clearly identified. A strategy for accomplishing this can be found in the IS-41 and GSM standards. In the first article of Volume 2, Issue 2, "Location Management in PCS Networks Using Base Areas (BAs) and 2 Level Paging (2LP) Schemes" by Hesham Ali, Ahmed Saleh, and Mohammed Ali, present a different location management strategy than that found in the IS-41 and GSM standards. Their novel strategy is based on restructuring the cells into smaller areas, moving MT registration from the master home location register (HLR) to the MT's HLR, using caching to reduce unwanted database updates, and two-level paging. Experimental results indicate that location costs decrease and network response improves when the proposed strategy is introduced.

In turbo processes the channel decoder and receiver demodulator exchange extrinsic information. Turbo codes have excellent performance, especially at low and medium signal-to-noise ratios, and can be found in many international standards, including the 3G standard UMTS and the DVB-T digital broadcasting standard. The second article in Volume 3, Issue 2, "Application of Extrinsic Information Transfer Charts to Anticipate Turbo Code Behavior" by Izabella Lokshina, presents a method based on the extrinsic information transfer (EXIT) chart for estimating the convergence properties of the iterative decoding process associated with a given turbo coding scheme. Compared with other methods for generating EXIT functions, the suggested approach provides insight into the iterative behavior of linear turbo systems while substantially reducing numerical complexity.

The advent of femtocells provides mobile operators with a new architectural solution for improving signal quality in critical areas. Compelling business cases can be made for replacing landlines entirely with femtocellular technology as well as for using it in fixed/mobile network infrastructural solutions. The third article of Volume 3, Issue 2, "A Femtocellular-Cabled Solution for Broadband Wireless Access: A Qualitative and Comparative Analysis" by Dario Di Zenobio, Massimo Celidonio, Lorenzo Pulcini, and Arianna Rufini, presents a solution to broadband wireless access based on integrating an "enhanced" femtocell with a cable distribution network. Simulation results indicate that the solution can be cost-effective with enhanced QOS performance when providing IMT-Advanced services to users in digital divide areas.

In the fourth article in Volume 3, Issue 2, "Data Transmission Oriented on the Object, Communication Media, Application, and State of Communication Systems," Mike Sabelkin and Francois Gagnon propose a communication system architecture based on the object, communications media, application, and state of the communication system. In particular, the authors focus their architecture on wireless communications media and consider a scenario where a gray scale image is transmitted losslessly over a wireless line-of-sight channel. Simulation results validate use of the architecture and compare the proposed system with other alternatives.

The next phase of mobile growth is expected to occur in machine-to-machine (M2M) communication. The fifth article in Volume 3, Issue 2, "Machine-to-Machine Communications and Security Solution in Cellular Systems" by Mahdy Saedy and Vahideh Mojtahed, introduces a model of M2M based on 4G cellular networks. The efficient model assumes that M2M terminals will use 4G cellular network resources if they are located in the 4G coverage area, but far from other terminals, and ad hoc mode if they are close to other terminals. The model incorporates a simplified protocol stack and various measures to secure communications, a critical concern for M2M networks.

In order to discharge their responsibilities, agencies charged with preparing and protecting the public require an ongoing flow of accurate and precise environmental data from human and non-human sensors. Advances in sensing technologies along with a proliferation of sensor data have increased the need for data integration and a flexible architecture for acquiring the data. In the sixth and final article of Volume 3, Issue 2, "Sensing Technologies for Societal Well-Being: A Needs Analysis," Elizabeth Avery Gomez suggests how this might be done using lean application programming protocols through services such as SMS. While the author views her research as a vehicle for discussion and feedback on the alignment of sensor data acquired through information and communications technologies and sustainable business processes needed for response readiness and crisis management, she believes that it can have applicability in day-to-day sustainable systems for overall social well-being and in global initiatives to tackle environmental issues.

The third issue of the calendar year 2011 included four papers, three of which were based on research conducted at U.S. universities. The first of these articles, "Modeling and Simulation of Traffic with Integrated Services at Media Gateway Nodes in Next Generation Networks" by Izabella Lokshina of SUNY College at Oneonta, represents a mathematical approach to the problem of dealing with varying types of traffic in network nodes. The approach utilizes the characterization of traffic represented as independent Poisson processed and Markov processes with constant intensity and random input streams.

The future of telecommunications and networking lies in the growth of mobile devices and their capabilities. One only has to remember the large awkward mobile phones of the early 1990's and their limited analog voice – only communications capabilities to realize that the world of telecommunications and networking has come a long way. It is not coincidence that the growth of the Internet has corresponded with the growth of mobile device capabilities and usage. One of the key issues addressed in this work is dealing with the Quality of Service (QoS) aspects of various types of network traffic. One of the problems of today's networks that is becoming more acute is the variation in the types of traffic that share a given network. If the broader network can be considered a collection of heterogeneous networks and sources of traffic, then the amalgamation of such in the next generation of overall networks will only be faced with greater variations in traffic, demand, and heterogeneity of its infrastructure. Dr. Lokshina's work in this article focuses on how this variety can be dealt with "media gateway nodes" in such networks. Such nodes require the sharing of bandwidth between multiple types of multimedia traffic from various sources and destined for various nodes. Each individual piece of the traffic "pie" must be given some sort of priority with respect to the other pieces while seeking to still provide overall efficiency and minimizing delays in the network. The notion of QoS grew out of work in the 1990's on Asynchronous Transfer Mode (ATM) and has evolved to define the nature of all broadband networks where some time time-sensitive traffic such as real time video must share bandwidth with less time-sensitive traffic such as email; yet all must be served in an efficient manner. Lokshina's approach in this work takes a probabilistic traffic modeling point of view stressing traffic compression in order to achieve optimal throughput and reduce the blocking probability for traffic arrivals.

The second article of this issue is "A Complete Spectrum Sensing and Sharing Model for Cognitive Radio Ad Hoc Wireless Networks Using Markov Chain State Machine" by three researchers at the University of Texas at San Antonio. This work also takes a probabilistic approach to modeling network traffic, but with strictly wireless ad hoc networks in mind. Wireless ad hoc networks, having no fixed infrastructure to rely on for services, have their own set of unique and difficult problems to deal with to ensure network efficiency. In this work, the authors deal with ad hoc networks that rely on cognitive radio. Cognitive radio allows more general mobile devices to be programmed to be used on a particular network without "hardwiring" them to work only on a given network or set of frequencies. It allows for adaptive capabilities for devices to work across networks, which becomes important for mobile devices in certain situations. One scenario in which cognitive radio would be useful includes so-called battlefield networks where soldiers and their associated means for transport and weaponry form their own ad hoc networks in a given area and as others move from group to group, their devices should be able to adapt to the network being used in a given area. Cognitive radio, with its ability to program devices to sets of frequencies, also offers the ability to more efficiently utilize bandwidth in a given region. Typically, there is a limited amount of bandwidth that must be shared by mobile devices operating in half duplex or full duplex mode. The ability to utilize more of the available bandwidth or spectrum can allow for greater numbers of nodes to operate in a region or allow segmentation of traffic. This segmentation may even result in the creation of several independent ad hoc networks operating simultaneously.

The next article of this issue is "A Novel Approach to Avoid Mobile Phone Accidents While Driving and Cost-Effective Fatalities" by two authors from AMa University in India. This work truly displays the applied and interdisciplinary nature of the work published in the journal. Distractions while driving have been proven to increase accidents and ultimately fatalities and costly injuries. Although governments are starting to intervene with restrictions on mobile device use while driving, developments from the mobile network providers and their associated application developers have also come to the forefront in trying to limit these unwanted consequences of device use while in an automobile or similar moving vehicle.

The authors discuss two types of "events" for mobile phones while driving: incoming calls and outgoing calls; both of which involve slightly different kinds of distractions for drivers. The authors then proceeded to develop an application that could handle both types of events while the vehicle is in motion. Consequently, the proposed application includes a mechanism for locating a mobile user and measuring their speed in the vehicle. The motivation for their approach, while obviously including the notion of saving lives through accident reduction, was the mitigation of risk involved and economic impact associated with driver-distracted mobile-phone related accidents.

The final article of this issue is entitled "A Study of Speed Aware Routing for Mobile Ad Hoc Networks" written by Kirthana Akunuri, Ritesh Arora and Ivan G. Guardiola, all of the Missouri University of Science and Technology in the U.S. This article examines how speed of movement and general mobility affects the topology for a Mobile Ad Hoc Network (MANET). One of the motivating factors for this research, as explained in the article, is that the effects of mobility on routing have been studied less than other aspects of MANET network performance. The authors propose a Speed-Aware Routing Protocol (SARP) that essentially ignores nodes traveling at too great a speed from initial route discovery. Essentially, packets from a high speed node are dropped due to the fact that maintaining a route related to that node is not a worthwhile endeavor. Part of the protocol involves the determination of a link expiration time (LET) which essentially predicts the period of time that a route will be feasible. Thus, some minimum threshold level is set for a LET and ones below a threshold force the exclusion of nodes generating these values to be excluded from the routing process in the immediate future. The key to this approach is the identification and exclusion of "fast moving" nodes. Thus the approach is sacrificing the ability to deal with such nodes in an equal fashion for greater stability in the route determination process. The authors used simulation to display the advantages of their routing approach.

The fourth issue of the calendar year 2011 included four articles on a wide variety of telecommunications-related topics. The first of these, "A Survey of Cloud Computing Challenges from a Digital Forensics Perspective" by Gregory H. Carlton and Hill Zhou, both of the California State Polytechnic University, focused on the problem of cloud computing-stored information and obtaining digital forensic information on it. Cloud computing describes the process by which information is stored on servers on the greater Internet and not locally. Since information is not stored in-house, gathering forensics that pass legal muster is a challenge since the information is somewhat controlled externally, yet was placed there by an internal source. The authors begin by providing an overview of cloud computing that leads into a discussion of the implications for digital forensics. In today's digital society, the ability to document the sources of digitally stored information and other key metadata provides the foundation for modern criminal law enforcement and also some civil litigation such as copyright infringement.

The second article of this last issue is entitled "Value Creation in Electronic Supply Chains by Adoption of a Vendor Managed Inventory System." This article, written by Yasanur Kayikci of the Vienna University of Economics and Business, is a very "applied" one in that it focused on a specific application of Internet-based technologies and telecommunication networks, utilizing them for electronic supply chains in which a supplier manages a retailer's inventory. This is a very unique application of networks and one that has grown in notoriety over the past few years as companies focus on increasing efficiency and reducing inventory costs. The author looks at factors that influence the adoption of such technologies. In particular, the work looked at an industry that has made such technology a strategic priority, the automotive industry.

The third article, written by Paramesh C. Upadhyay and Sudarshan Tiwari of India, details a distributed and fixed mobility management system for IP-based mobile networks. This work bases mobility management on mobile device user history and not other more traditional metrics. Signaling cost in their approach revolves around an analytical model utilizing Markov chains and random walk modeling. The approach utilizes both a micro-registration model as well as a macro-registration model (in other words mobile devices register at two different layers). Their approach was test for varying distances from a mobile host and performed in a robust fashion.

The final article of this issue, "The Media Gatekeeping Model Updated by R and I in ICTs: The Case of Wireless Communications in Media Coverage of the Olympic Games" is a unique look at the impact of telecommunication and networking technologies on media coverage for the olympic games. Since 2012 was an olympic year with the summer games being hosted in London, this work is ever more topical and interesting. In contrast to most of the published work in the journal, this paper takes a broader perspective that focuses on the impact of the technologies rather than the technologies themselves. In particular, the globalization of mass media has transformed the way most populaces view world events. This globalization has taken place through the evolution of telecommunication and networking technologies and has accordingly transformed the "gatekeeping" model described by Cossiavelou, Bantimaroudis, Kavakli, and Illia in this article.

I would like to now elaborate a bit on an important aspect of telecommunication and networking technologies that is growing in its importance and one in which I, Michael Bartolacci, am engaged. This is the application of wireless technologies to the field of disaster planning and management for China. Whether or not you personally believe in global warming or feel that the earth is undergoing a rapid change in climate, one cannot debate the tremendous impact of natural disasters such as hurricanes and tornado have had in recent history. Add to these the unpredictable nature of other types of natural disasters such as earthquakes and volcano eruptions and the need to both prepare for such events and to have systems in place for managing the human and natural damage that occurs during them is apparent. China, due to its large land area size (third in terms of land area trailing only the Russian Federation and Canada and just ahead of the United States) and population (currently over 1.3 billion by some conservative estimates) presents an enormous challenge in preparing for and dealing with natural and manmade disasters. In this context, Bartolacci and fellow researchers explore the applications of telecommunications technologies for disaster planning and management there.

Disaster management systems in China still have some problems to overcome in order to improve their effectiveness. In general, these problems lie primarily in their ability to disseminate information, in other words, in their ability to communicate. The following statistic discourages improvement and greater development of such systems; 57.7% of China's population and about 60% poorer population live in rural areas which has a tremendous impact on the nature of emergency management. Such people are more vulnerable due to their limited resources and these vulnerable groups are become more fragile when confronting disasters. Disaster planners should use this critical piece of information as they identify relevant preparedness actions. However, the management of natural disaster emergencies is different in various regions. The level of study and management in regions with relatively developed economies is higher than that in regions with relatively undeveloped economies. Poor and remote rural areas must receive the same level of attention in the process of disaster management to be able to reduce their vulnerability. China has no real "teeth" to its goals with respect to wireless rural deployment or even universal service access and therefore has a mish mash of wireless infrastructure that could support disaster management and to assist the rural disadvantaged in times of need and even in day to day existence.

Wireless communications in China, much like its landlines, lagged behind more industrialized countries until the early 1990's. The Chinese government, in an attempt to introduce free market competition into what was traditionally a state run service (under the auspices of the Ministry of Posts and Telecommunications), created the incentives for the development of a duopoly in the mobile telecommunications market for China. Although state-controlled for the most part, both companies sought global partners and had ambitious goals for growth and profitability. Both China Unicom and China Mobile focused their efforts on serving the growing industrialized urban markets of Eastern China. Very large cities and their surrounding areas such as Shanghai, Beijing, and Shenzhen were profitable markets where the companies made extensive investments in infrastructure in order to satisfy the ever-growing demand for wireless voice and data communications with China's rapidly expanding economy. This focus on the profitable Eastern markets was in direct contrast to the overall goals set forth by the Chinese government of providing universal telephony service for its populace. By 2004, China had enacted the "Village Access Project" (VAP) with the goal of connecting all rural villages to the rest of the country with some form of telephone service, be it via landlines, satellites, or mobile telephony. While this effort met with some success in that many rural villages were given access to some form of telephony, there was no universal service regulatory regime associated with the overall project. Under the auspices of the VAP, companies that sought to maximize their profit were faced with rather expensive network expansions to reach small and seemingly unprofitable market regions. No direct monetary or regulatory incentives beyond the mere existence of the project spurred carriers into investing in such connections for the long term. Local, regional and national governmental aspects and control of the overall project sometimes conflicted in their tone, focus, and commitment. This mandate without a clear overall regulatory unit policing it soon led to unwanted behavior by China's telecommunications companies including the wireless carriers. It is clear that China is not really prepared to implement wireless technologies to their best advantage in the context of disaster planning and management. This discussion of ongoing research highlights the ever-growing importance of telecommunications technologies play in our world.

Michael R. Bartolacci
Pennsylvania State University – Berks, USA

Steven R. Powell
California State Polytechnic University – Pomona, USA

Chapter 1
Maximizing the Flow Reliability in Cellular IP Network Using PSO

Mohammad Anbar
Jawaharlal Nehru University, India

Deo Prakash Vidyarthi
Jawaharlal Nehru University, India

ABSTRACT

A Cellular IP (CIP) network involves a bulk of data transmission. It is highly reliable and guarantees the safe delivery of the packets required in such systems. Reliable traffic performance leads to efficient and reliable connectivity in Cellular IP network. CIP network, which consists of mobile hosts, base stations, and links, are often vulnerable and prone to failure. During the routing operation in the network, the base station, which works as router for the transmitted packets, may fail to perform. Reliable transmission is desirable, in terms of services of the base stations in the network, reliable routing, and processing the data. In this paper, the authors design a reliability model to increase the reliability of a flow, consisting of packets, passing through routers in a Cellular IP network. Particle Swarm Optimization (PSO) is able to solve a class of complex optimization problems. PSO is used to improve the reliability of the flow in CIP network. The proposed model studies the effects of packet processing rate (μ), packet arrival rate (λ), and the number of packets per flow on the reliability of the system. A simulation experiment is conducted and results reveal the effectiveness of the model.

INTRODUCTION

Cellular IP is a protocol for mobility management at the micro level and inherits many features of cellular systems. CIP Network is divided into geographical areas called cells and each cell is serviced by a Base Station (BS) located at the centre of the cell. Base station plays the role of router

during routing operation in Cellular IP networks. The BS provides a connection end point for the roaming Mobile Hosts (MHs) (Campbell, Gomez, Kim, Valko, Wan, & Turanyi, 2000).

In communication networks, especially in Cellular IP network, router CPU cycle is an important resource. CPU at router takes some amount of time to process a packet of the flow. This time is important for a flow as the routers should not

DOI: 10.4018/978-1-4666-2154-1.ch001

fail while processing the packet. Any router in CIP network (including base stations as routers) has its processing capability and can process a certain number of packets per second. Router CPU is a scarce resource (Tanenbaum, 2004) and it is desired that the reliability of the flow passing through the CIP network should be maximized.

The growing importance of mobile network has motivated an essential research into how data packets passing through the mobile communication network can be transmitted and processed reliably. This suggests reducing the processing time of the flow of packets while they are being transmitted in presence of failures in form of uncertainties.

Reliability is the ability of a system to perform its functions successfully in routine as well as in hostile or unexpected circumstances. Reliability is the probability that the network, consisting of various components, performs its intended function for a given time period when operated under normal (or stated) environmental conditions. The unreliability of a connection is the probability that the experienced outage probability for the connection is larger than a predefined maximum tolerable value. The connection reliability is related to the traffic parameters (Zhao, Shen, & Mark, 2006). The design of reliable resource management algorithms for CIP networks is an important issue. Reliability studies for mobile computing are still under extensive research (Liao, Ke, & Lai, 2000). The design of a reliable bandwidth management for cellular IP networks is also an important issue deliberated in (Olivena, Kim, & Suda, 1998; Jayaram, Sen, Kakani, & Das, 2000; Tipper, & Dahlberg, 2002; Dahlberg & Jung, 2001). Thus, the reliability issue is to be addressed and studied well in the field of wireless communications.

In a wireless cellular network environment, base stations are prone to failure (Prakash, Shivaratri, & Singhal, 1999). A BS may either crash or fail to send, receive or process flows of data packets during routing operation. Due to the failure of a BS, all the connections in the failed cell area

get terminated and all the services are interrupted until the failed BS is restored. BS failure significantly degrades the performance and bandwidth utilization of the Cellular IP networks. Specifically, services for high priority ongoing calls such as real-time traffic could be interrupted, which is usually not acceptable.

In recent years the applications of Particle Swarm Optimization (PSO), which is a useful search procedure for optimization problems, have attracted the attention of researchers of various disciplines as a problem solving tool. PSO is a search procedure based on the natural evolution. PSO has successfully been applied for various optimization problems for which no straightforward solution exists (Anbar & Vidyarthi, 2009). This paper discusses the effects of the base station (router) failure in CIP networks, with emphasis on improving the reliability being affected by the wireless environment. A PSO based reliability model for router CPU time management is being proposed here to facilitate CIP network design that meets users' demand in terms of reliable services.

The rest of the paper is organized as follows. First section contains a review of literature. In the second section, routing operation in Cellular IP networks has been explained. The third section addresses the problem studied in this paper that includes reliability and the PSO. The proposed model has been explained in details along with the algorithm in the fourth section. In the fifth section, the evaluation of the proposed model through experiments is done. Final section contains observations on the conducted experiments and conclusions of the study.

REVIEW OF LITERATURE

A very little work on the reliability based CPU router time management model in CIP Networks have been done. However, some of the models that address the other reliability issues in cellular networks have been briefed here.

Three cost functions associated with the retransmission-based partially reliable transport service were introduced by (Marasli, Amer, & Conrad, 1996).

An algorithm for computing low-latency recovery strategy in a reliable network was proposed by (Zhang, Ray, Kannan, & Iyengar, 2003).

An optimal forward link power allocation model for data transmission was proposed by (Sun, Krzymien, & Jalai, 1998).

A soft handoff/power distribution scheme had been proposed for cellular CDMA downlinks, and its effect on connection reliability had been studied by (Zhao, Shen, & Mark, 2006).

A neural-network-based multicast routing algorithm for constructing a reliable multicast tree that connects the participants of a multicast group was proposed by (Kumar, & Venkataram, 2003).

A protocol called Reliable Mobile Multicast Protocol (RMMP) was proposed in (Liao, Ke, & Lai, 2000) to provide reliable multicast services for mobile IP networks. The mobile agent in mobile IP was extended to assist reliable multicasting for mobile devices.

ROUTING IN CELLULAR IP NETWORK

Cellular IP network is divided into cells. These cells are controlled by the base stations considered as access points and wireless routers. Gateway connects the network to the Internet. CIP protocol is designed for mobility management at the micro level. Mobile hosts in CIP network implement Cellular IP protocol. Many operations such as handoff, paging and routing take place in this network. Structure of the Cellular IP networks is shown in Figure 1.

Routing in Cellular IP network is one of the important operations by which the packets of a flow are forwarded from source to destination. Two types of routing are often used in communication systems; end-to-end routing and hop-by-hop routing. In general, hop-by-hop routing is used in mobile communications (Heimlicher, & Karaliopoulos, 2007) especially in Cellular IP (Campbell, Gomez, Kim, Valko, Wan, & Turanyi, 2000). Hop-by-hop routing may help in reducing energy consumption and packet processing time

Figure 1. Cellular IP Network

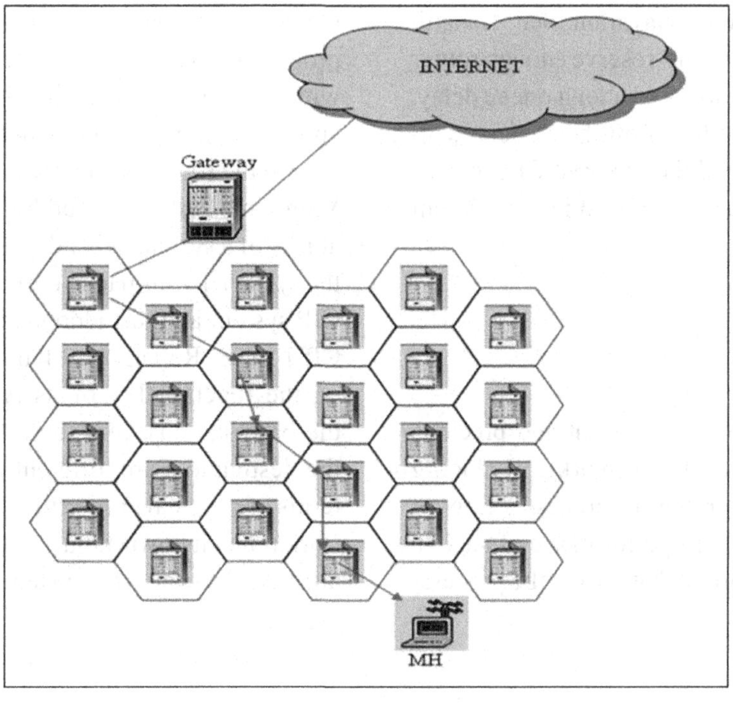

(Heimlicher, & Karaliopoulos, 2007). In hop-by-hop routing, packets are forwarded (routed) according to an independent decision taken in the router (base station in case of Cellular IP networks) based on the destination addresses for the incoming flow. Each BS, in CIP network, maintains a routing cache. Routing cache stores two types of information; the source IP address and the previous neighbor from which the packet reached the current base station. In hop-by-hop routing, one packet carrying the path information is enough as an entry to the routing table and there is no need for carrying the full header to the destination by the packets to be followed (Heimlicher, & Karaliopoulos, 2007). It requires the route information to be updated through the data packets being sent. As long as the mobile host is sending packets through this route regularly, the routing cache will keep valid routing information. It is to be observed that route in Cellular IP network stays valid for a specific period of time known as route-time-out (Sobrinho, 2002).

Each router in the CIP network has a packet processing capacity i.e. each router has a policy to accept or reject a flow if the load exceeds its capacity. Any router cannot accept a flow if it is preloaded by the maximum allowable limit as per its policy (capacity) (Jayaram, Sen, Kakani, & Das, 2000). A flow must reserve enough router CPU cycles in order to avoid a long queue delay. Out of the flow a real-time flow, being delay sensitive, must be offered the required CPU cycles to ensure that they are processed in a minimum time at the router.

THE PROBLEM

Router CPU time is an important and precious resource in Cellular IP networks. Reducing the time, taken by a router to process a flow of packets, can improve the performance of the CIP networks. It is important that during the process-

ing of the flow the router CPU should perform reliably (error free) so that the communication is guaranteed. The reliability of such systems is introduced as follows.

Reliability

Reliability is an important characteristic parameter in any system. It reflects the ability of a system to perform its tasks and functions in a proper manner. Reliability is a general term and is not clear unless the associated aspect with reliability is specified. In some systems, reliability might be associated with the quality of production; in other systems it may be associated with the validity of the system. It is associated with many aspects in wireless communications such as Quality of Service, Connection Completion, Packet Loss, etc. For example, a wireless system can be reliable in terms of connection completion in which it is able to preserve an important connection from being dropped. Any designed model for improving the performance of any communication system should be reliable depending on the purpose for which the model has been developed. In CIP network, many models have been designed and developed. One of them is by (Anbar & Vidyarthi, 2009) in which a better bandwidth management has been proposed. A reliability aspect may be associated with this in terms of Connection Dropping Probability. Another model in which packet processing time at the router for a flow is reduced (Anbar & Vidyarthi, 2009) may address reliability. Reliability of a system depends on the components of the system. In this paper, the considered system is CIP system and the addressed resource is router CPU time. Reliability of this system depends on the functionality of its base stations. Since the base stations, in a route from the source to the destination, are different form each other in terms of physical properties, load, etc., each base station has its own failure rate and therefore its attribution towards the system's reliability. So the reliability of the flow is determined according to the performance of the base stations, present in the route, during the routing operation.

Particle Swarm Optimization

Swarm intelligence is an intelligent paradigm to solve the hard optimization problems. It is based on the behavior of the social insects such as bird flocks, fish school, ant colony etc., in which individual species change its position and velocity depending on its neighbor's movement. This is done by mimicking the behavior of the creatures within their swarm or colony. Particle Swarm Optimization (PSO) is an algorithm based on the swarm intelligence. It is a population based optimization tool. Main strength of the PSO is its fast convergence attributed to its well organized logic and procedures. PSO starts with a group of randomly generated populations. It has a fitness function for the population evaluation. It updates the population and searches for the optimal solution with random techniques (Abdelhalim, Salama, & Habib, 2006; Ying, Yang, & Zeng, 2006; Franken, & Engelbrecht, 2005). A problem to be solved by PSO (e.g. bandwidth reservation problem) can be transformed to the

function optimization problem. PSO model is a swarm of individuals called particles. Particles are initialized with random solutions. These particles move through many iterations to search a new and better solution. Each particle is represented by two factors; the position x and the velocity v. Each particle has a specific position (x) and in the beginning it is initialized. The other factor, velocity determines the movement of each particle in the space with specific velocity. During the iteration time t, particles update their position and velocity (x and v) (Anbar & Vidyarthi, 2009). PSO simulates the behavior of the bird flocking.

In Figure 2, a flock of birds are flying randomly and searching for food in an area. There is only one piece of food (optimal result) in the area being searched. All the birds in the flock are aware that there is some food in this area but none of them know the location. Nearest bird may locate the food and the strategy for others to find the food is to follow the nearest bird to the food.

In PSO, two types of best values are used. One is Pbest, which is the local (personal) best position for each particle in the swarm and must be updated depending on the fitness value for each particle. The second best value is Gbest which is the global best value for all the swarm in general.

Figure 2. An example of Swarm (Bird Flocking)

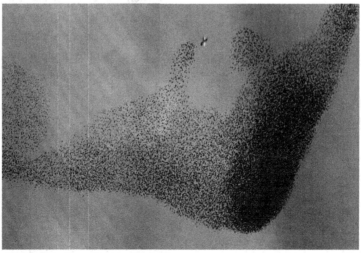

This value must be checked, in each iteration, and be changed by Pbest of any individual in the swarm if the Pbest value at the current iteration is better than the fitness value of Gbest in the previous iteration.

During the search procedure, each individual in the swarm adjusts its velocity (V) and position (X) depending on its past experience. Velocity update is done as per Equation 1.

$$V_j^{k+1} = w.V_j^k + c1.r1.(Pbest_j^k - X_j^k) \\ + c2.r2.(Gbest^k - X_j^k) \tag{1}$$

where

- w is inertia weight and is varied from 0.4 till 0.9.
- r1, r2 are random numbers between 0 and 1
- c1, c2 are acceleration factors that determine the relative pull for each particle toward Pbest and Gbest and usually c1, c2 = 2.
- V_j^{k+1} is the velocity of particle j in iteration K+1.
- V_j^k is the velocity of particle j in iteration K.
- $Pbest_j^k$ is the best position of particle j so far.
- X_j^k is the current position of particle j in iteration k.
- $Gbest^k$ is the global best position of the swarm so far.

Position update is done as given in Equation 2.

$$X_j^{K+1} = X_j^K + V_j^K \tag{2}$$

where

- V_j^k is the velocity of particle j in iteration k.

Router CPU Time Management Using PSO

CPU Time Management model takes into account a flow consisting of packets passing through a route having number of routers. Packet processing at a router starts when a packet reaches to this router. The time taken by the router to complete processing a packet is T_i. Therefore the time taken to process a flow consists of N packets is ($N * T_i$). Total time taken to process a complete flow in the route, having M routers, is:

$$T = \sum_{i=1}^{M} N.T_i \tag{3}$$

Few assumptions, considered in this model, have been listed as below.

Flow rather than a packet is studied in this model; therefore the flow is assumed to be consisting of N packets. Flows are assumed to be real-time flows and thus the admission control is not discussed because they all are considered to be accepted flows by the routers. The main concern of this model is the processing time at the router for each flow.

Each wireless router (base station) in the Cellular IP network has its own capacity (packet processing rate referred as μ_i). Packet arrival rate at each router is different and is referred as λ_i. Queue is built up until the processing time of each flow consisting of N packets is completed.

Based on the queuing system theory the time taken to process a packet by a router (Robertazzi, 2000) is:

$$T_i = \frac{1}{\mu_i} \times \frac{1}{1 - \frac{\lambda_i}{\mu_i}} \tag{4}$$

The model considers routing operation in Cellular IP networks. In Figure 1, an example of routing operation is depicted in which a flow (group of packets) passes through the route starting from the Gateway which connects the network to Internet until it reaches to the final destination. Each base station forwards the flow based on an independent decision, the main principle of Hop-by-Hop routing as discussed earlier. At each router (base station) CPU takes time to process the packets and forward them to the next hop. This model reduces the total processing time taken in this specific route to process the flow. To achieve this objective the model applies Particle Swarm Optimization. Two main parameters that affect the processing time at each router (evident from Equation 4) are processing rate of the routers and packet arrival rate. Due to these two parameters, it results in different processing times at the routers. The total time taken to process a flow on a route is different for every new flow. A system should reliably perform for this time period and thus reliability becomes an optimization problem which is desired to be as high as possible. PSO is applied as one of the powerful tested optimization tools to solve such optimization problems. PSO, in this model, optimizes reliability of a flow that passes through number of routers in the way.

Each individual of the swarm in PSO is represented by the two parameters (velocity and position). Each router is considered as an individual in the swarm that consists of number of routers (individuals). Therefore, at an individual in the swarm two main parameters are considered; packets arrival rate (λ_i) and packet processing rate (μ_i). Packet arrival rate is the velocity of each individual in the swarm; packet processing rate is the position of each individual in the swarm.

THE PROPOSED MODEL

The work, proposed here, considers router CPU time and the failures of base stations for reliability estimation. Failure of the base station means the failure of its CPU in processing the packets

passing through it for that time period. Each cell has one base station responsible for processing the incoming flow of packets and passing them to the next hop. The proposed model uses PSO to improve the reliability of the communication network based on reducing the flow processing time in the router. The computation of the reliability depends on the failure rates of the base stations (routers) and the flow processing time.

Explanation of the Model

The reliability of the data flow depends on the services of the base stations falling in the route. The availability of these services depends on the failure rates of the devices (base station). The failure of the base station is determined by various factors such as the transmission power, heat, signal to noise ratio between the terminal equipment and the base station etc. In this model, the reliability parameter is chosen to be represented by the exponential distribution during the processing of a flow of packets. This means that this entity (BS) which has been in use for some time (any number of hours) is as good as a new entity in regard to the amount of time remaining until the entities fail (Khanbary & Vidyarthi, 2009). Thus the reliability of the base station (router) over the time period T_i is

$$R_i = e^{-\alpha_i . T_i} \tag{5}$$

where, α is the failure rate of the base station (Figure 3) and T is the time taken by the base station to process the flow of packets in the router.

The model is designed such that the reliability of the system increases. The simulation study is conducted using router CPU time management strategy developed by (Anbar & Vidyarthi, 2010). To observe the effect of the number of routers and the number of packets per flow on the reliability of the designed network system, experiment has

Figure 3. Base stations with different failure rates

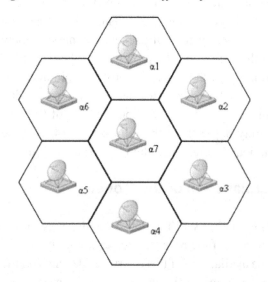

been conducted for different number of routers and different number of packets per flow.

The Fitness Function

As the data packet flow passes in a route consisting of m routers, the total reliability of the system is computed as

$$R = \prod_{i=1}^{m} e^{-\alpha_i T_i} \qquad (6)$$

To obtain the best reliability for the designed network system, the reliability R in Equation 6 is to be maximized.

The Algorithm

The algorithm used for the proposed model is as follows.

1. Input number of routers (hops) in the route (chromosome size).
2. Input number of packets per the considered flow.
3. Input number of Iteration.

4. Randomly generate the values of packet processing rate and packet arrival rate in given range.
5. Randomly generate the failure rates for each base station between 0 and 1 for the experiment.
6. Compute the initial values of flow processing times in each router using Equations 3 and 4.
7. Initialize the best processing time for each router (Pbest) by selecting the initial computed values as the best values for each router considering $\mu best_j$ with the individual which has the best processing time.
8. Initialize the global best value among the individuals in the swarm (Gbest) by selecting it as the smallest processing time among all the individuals in the swarm.
9. Compute the initial values of reliabilities for routers (base stations) using Equation 5.
10. Compute the initial (total) reliability of the system (fitness function) using Equation 6.
11. Score the initial population against the total reliability (fitness function).
12. For a given number of iteration repeat steps 13 to 18.
13. Update the velocity of each individual (packet arrival rate λ_i) as follows.
14. $\lambda_j^{k+1} = w.\lambda_j^k + c1.r1.(\mu best_j^k - \mu_j^k)$
    $+ c2.r2.(Gbest - \mu_j^k)$
15. Update the position of each individual (packet processing rate μ_i) as follows.
16. $\mu_j^{k+1} = \mu_j^k + \lambda_j^k$
17. Check the values of λ_i and μ_i, make sure that each value is within its allowable range.
18. Update (Pbest) for each individual as follows: if packet processing time for this router at iteration k+1 is less than packet processing time for the same router at iteration k then

make Pbest as the packet processing time at iteration k+1.

19. Update (Gbest) as follows: for all routers if packet processing time of any of them at iteration k+1 is less than Gbest then make Gbest as the packet processing time for this individual $\mu best_j^K$ (Gbest is the best value for the swarm found so far).

20. Compute the reliability of each router using Equation 5.

21. Compute the total reliability of the system using Equation 6.

22. Display the total reliability of the system computed as an output.

EXPERIMENTAL EVALUATION

In this section, the performance of the proposed algorithm is evaluated. It has been observed that the solution converges by 20 iterations. The experiments have been designed by writing programs in C++.

Simulation Parameters

The simulation parameters used in the experiment conforms to other similar study (Anbar & Vidyarthi, 2010) and is as follows.

- Number of routers varies (8, 12, 16, 20) in every experiment.
- Number of packets per flow varies (200, 400, 600, 800) for each number of routers.
- Packet processing rate (μ) and packet arrival rate (λ) in four different ranges of values but they are fixed for every router in each experiment.
- As the failure rate for the router is the property of the device, the values of failure rates are randomly generated between 0 and 1, 0 exclusive and are static for the same experiment.

Experiment 1: First Range of λ and μ

a. **Input Values:** $\lambda = 500 - 1000$ packets/sec, $\mu = 1000 - 1500$ packets/sec.
   - Number of packets per flow is 200.
   - Number of routers: 8, 12, 16, 20.
   - $0 <$ failure rate of router ≤ 1
b. **Input Values:** $\lambda = 500 - 1000$ packets/sec, $\mu = 1000 - 1500$ packets/sec.
   - Number of packets per flow is 400.
   - Number of routers: 8, 12, 16, 20.
c. **Input Values:** $\lambda = 500 - 1000$ packets/sec, $\mu = 1000 - 1500$ packets/sec.
   - Number of packets per flow is 600.
   - Number of routers: 8, 12, 16, 20.
d. **Input Values:** $\lambda = 500 - 1000$ packets/sec, $\mu = 1000 - 1500$ packets/sec.
   - Number of packets per flow is 800.
   - Number of routers: 8, 12, 16, 20.
e. Final Results

Experiment 2: Second Range of λ and μ

a. **Input Values:** $\lambda = 1000 - 1500$ packets/sec, $\mu = 1500 - 2000$ packets/sec.
   - Number of packets per flow is 200.
   - Number of routers: 8, 12, 16, 20.
b. **Input Values:** $\lambda = 1000 - 1500$ packets/sec, $\mu = 1500 - 2000$ packets/sec.
   - Number of packets per flow is 400.
   - Number of routers: 8, 12, 16, 20.
c. **Input Values:** $\lambda = 1000 - 1500$ packets/sec, $\mu = 1500 - 2000$ packets/sec.
   - Number of packets per flow is 600.
   - Number of routers: 8, 12, 16, 20.
d. **Input Values:** $\lambda = 1000 - 1500$ packets/sec, $\mu = 1500 - 2000$ packets/sec.
   - Number of packets per flow is 800.
   - Number of routers: 8, 12, 16, 20.
e. Final Results

Experiment 3: Third Range of λ and μ

a. **Input Values:** $\lambda = 1500 - 2000$ packets/sec, $\mu = 2000 - 2500$ packets/sec.
    - Number of packets per flow is 200.
    - Number of routers: 8, 12, 16, 20.
b. **Input Values:** $\lambda = 1500 - 2000$ packets/sec, $\mu = 2000 - 2500$ packets/sec.
    - Number of packets per flow is 400.
    - Number of routers: 8, 12, 16, 20
c. **Input Values:** $\lambda = 1500 - 2000$ packets/sec, $\mu = 2000 - 2500$ packets/sec.
    - Number of packets per flow is 600.
    - Number of routers: 8, 12, 16, 20.
d. **Input Values:** $\lambda = 1500 - 2000$ packets/sec, $\mu = 2000 - 2500$ packets/sec.
    - Number of packets per flow is 800.
    - Number of routers: 8, 12, 16, 20.
e. Final Results

Experiment 4: Fourth Range of λ and μ

a. **Input Values:** $\lambda = 2000 - 2500$ packets/sec, $\mu = 2500 - 3000$ packets/sec.
    - Number of packets per flow is 200.
    - Number of routers: 8, 12, 16, 20.
b. **Input Values:** $\lambda = 2000 - 2500$ packets/sec, $\mu = 2500 - 3000$ packets/sec.
    - Number of packets per flow is 400.
    - Number of routers: 8, 12, 16, 20.
c. **Input Values:** $\lambda = 2000 - 2500$ packets/sec, $\mu = 2500 - 3000$ packets/sec.
    - Number of packets per flow is 600.
    - Number of routers: 8, 12, 16, 20.
d. **Input Values:** $\lambda = 2000 - 2500$ packets/sec, $\mu = 2500 - 3000$ packets/sec.
    - Number of packets per flow is 800.
    - Number of routers: 8, 12, 16, 20.
e. Final Results

OBSERVATIONS AND CONCLUSIONS

The observations derived from the experiments are as follows.

The advantage of the model is clear from all the Figures 4 through 23 where the reliability of the flow processing is increased. A close observation of few experiment in which the number of packets is big (800 packets) and number of routers is big (20 routers) indicates that the reliability is increasing despite these big values. This is a conspicuous achievement.

As clear from the figures, initial values for reliability are random values based on the processing time values. Reliability keeps on increasing in successive iteration.

The purpose behind dividing the experiments into four sets is to experiment with all possible values of packet arrival rate and packet processing rate and to observe their effects on the reliability of the system. All the cases in four experiments indicate the enhancement in the reliability of the studied Cellular IP system.

As clear from Equation 5, there are two main parameters affecting the reliability of the router. The first one is the failure rate of the router. The value of failure rate for each router is randomly generated at the beginning of the experiment and fixed until the end because it is a physical property of the router and shouldn't change in the same experiment. Therefore, when the generated value is small the reliability is less; this can't be noticed explicitly in the plotted graphs, but it is clear from Equation 5. The other parameter affecting the reliability is the processing time for a flow of packets at each router. Value of the processing time is affected also by packet arrival rate and packet processing rate (Anbar & Vidyarthi, 2010). Figures 9, 14, 19 are indicative that the reliability is improving with the decrease in processing time.

Number of packets plays an important role in changing the reliability value. From Figures 10, 11, 12, and 13 (for example), when number of

Figure 4. 200 packets per flow

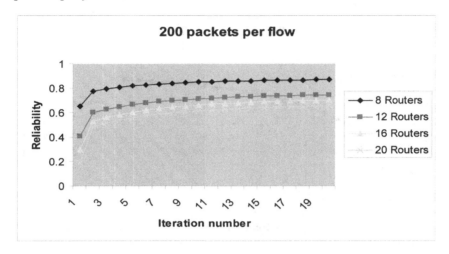

Figure 5. 400 packets per flow

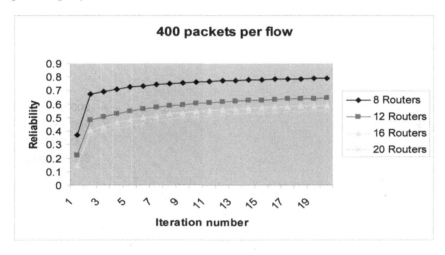

Figure 6. 600 packets per flow

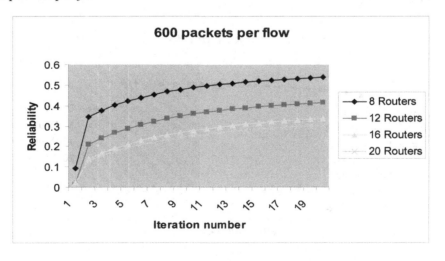

Figure 7. 800 packets per flow

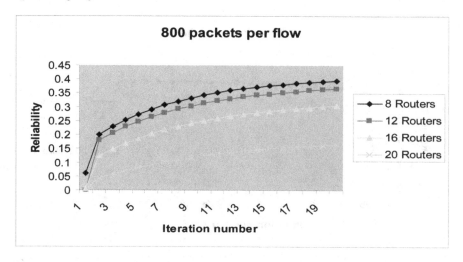

Figure 8. End values for the first range

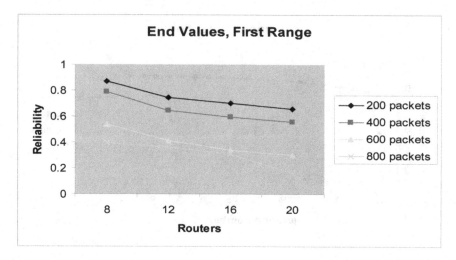

Figure 9. 200 packets per flow

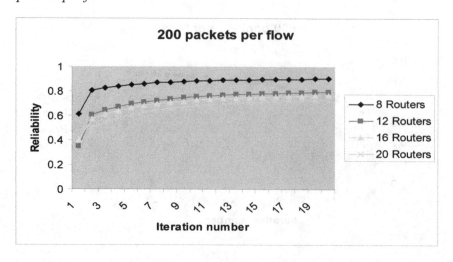

Figure 10. 400 packets per flow

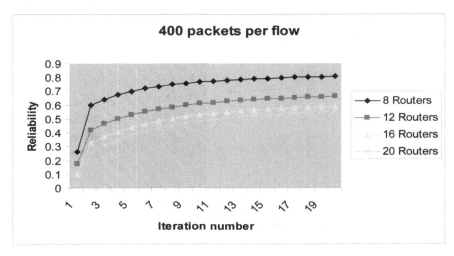

Figure 11. 600 packets per flow

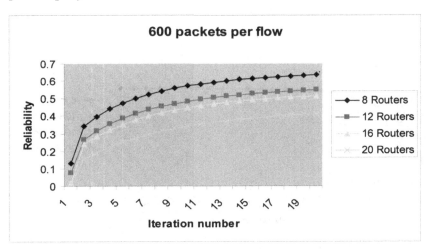

Figure 12. 800 packets per flow

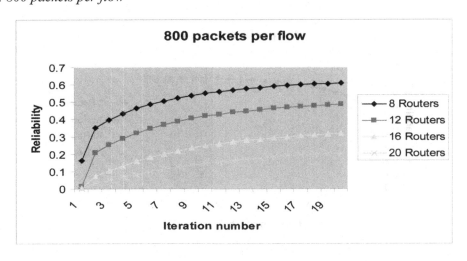

Figure 13. End values for the second range

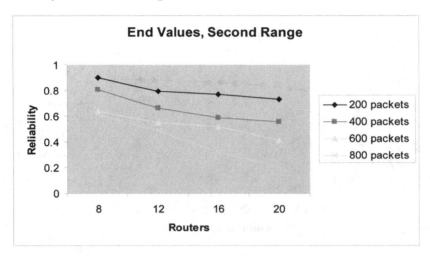

Figure 14. 200 packets per flow

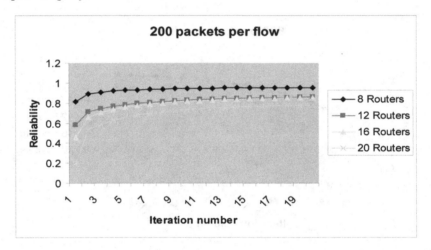

Figure 15. 400 packets per flow

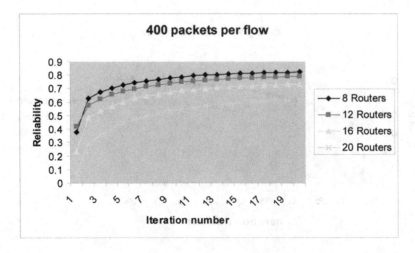

Figure 16. 600 packets per flow

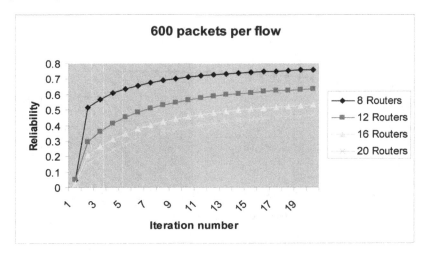

Figure 17. 800 packets per flow

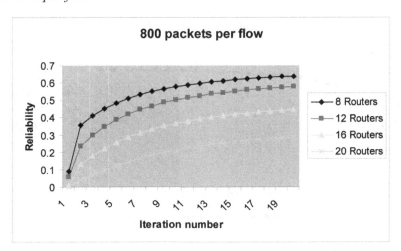

Figure 18. End values for the third range

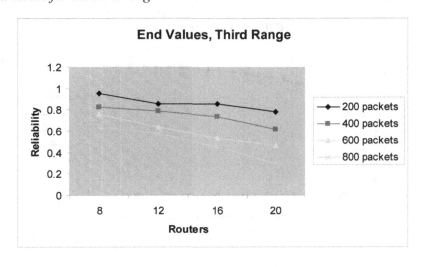

Figure 19. 200 packets per flow

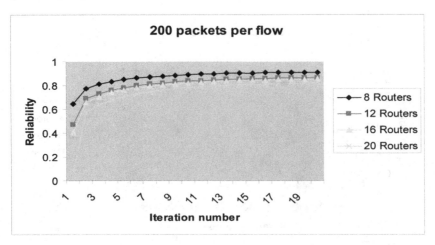

Figure 20. 400 packets per flow

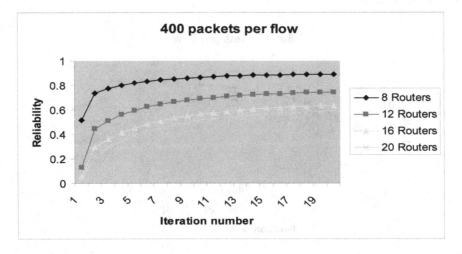

Figure 21. 600 packets per flow

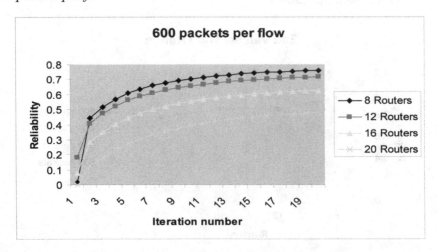

Figure 22. 800 packets per flow

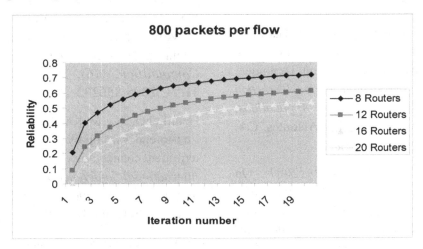

Figure 23. End values for the fourth range

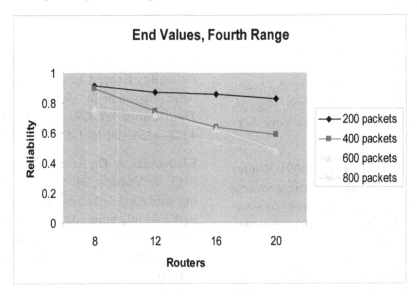

packets per flow is increasing from 200 packets to 800 packets the processing time is increasing resulting in decrease in reliability. This is logical because the proposed model is reliability model based on flow processing time.

Number of routers in the route of a flow also has its own effect on the reliability. The route that has 20 routers takes more time to process the flow passing through than the route that has 8 routers. It is clear from the graphs that when the number of routers is less, the reliability of the system is better.

Very little work has been done in the field of reliability in CIP networks and therefore the comparison with another model is not possible.

The proposed model doesn't discuss the routing operation in Cellular IP network. It proposes a model that ensures data packet flow reliability when this flow is processed in the path passing through routers. PSO is used to increase the reliability of the flow in the route.

REFERENCES

Abdelhalim, M. B., Salama, A. E., & Habib, S. E.-D. (2006). Hardware Software Partitioning using Particle Swarm Optimization Technique. In *Proceedings of the 6th International Workshop on System on Chip for Real Time Applications*, Cairo, Egypt (pp. 189-194). Los Alamitos, CA: IEEE Press.

Anbar, M., & Vidyarthi, D. P. (2009). On Demand Bandwidth Reservation for Real-Time Traffic in Cellular IP Network using Particle Swarm Optimization. *International Journal of Business Data Communications and Networking*, 5(3), 53–66. doi:10.4018/jbdcn.2009070104

Anbar, M., & Vidyarthi, D. P. (2009). Buffer Management in Cellular IP Network using GA. In *Proceedings of the 2nd International Conference on Advanced Computer Theory and Engineering (ICACTE 2009)*, Egypt, Cairo (Vol. 2, pp. 1163-1173). New York, NY: ASME Press.

Anbar, M., & Vidyarthi, D. P. (2009). Router CPU Time Management using Particle Swarm Optimization in Cellular IP Networks. *International Journal of Advancements in Computing Technology*, 1(2), 48–55.

Anbar, M., & Vidyarthi, D. P. (2010). Comparative study of two CPU router time management algorithms in cellular IP networks. *International Journal of Network Management*.

Campbell, A. T., Gomez, J., Kim, S., Valko, A. G., Wan, C.-Y., & Turanyi, Z. R. (2000). Design, implementation, and evaluation of cellular IP. *IEEE Personal Communications*, 7(4), 42–49. doi:10.1109/98.863995

Dahlberg, T., & Jung, J. (2001). Survivable Load Sharing Protocols: A Simulation Study. *ACM Wireless Networks Journal*, 7(3), 283–296. doi:10.1023/A:1016630206995

Franken, N., & Engelbrecht, A. P. (2005). Particle swarm optimization approaches to coevolve strategies for the iterated prisoner's dilemma. *IEEE Transactions on Evolutionary Computation*, 9(6), 562–579. doi:10.1109/TEVC.2005.856202

Heimlicher, S., & Karaliopoulos, M. (2007). End-to-end vs. hop-by-hop transport under intermittent connectivity. In *Proceedings of the 1st International Conference on Autonomic Computing and Communication Systems* (Article No. 20). Brussels, Belgium: ICST Press.

Jayaram, R., Sen, S. K., Kakani, N. K., & Das, S. K. (2000). Call Admission and Control for Quality-of-Service (QoS) Provisioning in Next Generation Wireless Networks. *ACM Wireless Networks Journal*, 6(1), 17–30. doi:10.1023/A:1019160708424

Khanbary, L. M. O., & Vidyarthi, D. P. (2009). Reliability Based Channel Allocation using Genetic Algorithm in Mobile Computing. *IEEE Transactions on Vehicular Technology*, 58(8), 4248–4256. doi:10.1109/TVT.2009.2019666

Klamargias, A. D., Parsopoulos, K. E., Alevizos, P. D., & Vrahatis, M. N. (2008). Particle filtering with particle swarm optimization in systems with multiplicative noise. In *Proceedings of 10th Annual Conference on Genetic and Evolutionary Computation*, Atlanta, GA (pp. 57-62). New York, NY: ACM Press.

Kumar, B. P. V., & Venkataram, P. (2003). Reliable multicast routing in mobile networks: a neural-network approach. *IEEE Proceedings Communications*, 150(5), 377–384. doi:10.1049/ip-com:20030649

Liao, W., Ke, C.-A., & Lai, J.-R. (2000). Reliable Multicast with Host Mobility. In *Proceedings of the IEEE Global Telecommunication Conference (GLOBECOM '00)*, San Francisco, CA (Vol. 3, pp. 1692-1696). Los Alamitos, CA: IEEE Press.

Marasli, R., Amer, P. D., & Conrad, P. T. (1996). Retransmission-Based Partially Reliable Transport Service: An Analytic Model. In *Proceedings of the 15th Annual Joint Conference: IEEE Computer Societies: Networking the next generation,* San Francisco, CA (Vol. 2, pp. 24-28). Los Alamitos, CA: IEEE Press.

Olivena, C., Kim, J. B., & Suda, T. (1998). An Adaptive Bandwidth Reservation Scheme for High-Speed Multimedia Wireless Networks. *IEEE Journal on Selected Areas in Communications, 16*(6), 858–874. doi:10.1109/49.709449

Prakash, R., Shivaratri, N. G., & Singhal, M. (1999). Distributed Dynamic Fault-Tolerant Channel Allocation for Cellular Networks. *IEEE Transactions on Vehicular Technology, 48*(6), 1874–1888. doi:10.1109/25.806780

Robertazzi, T. G. (2000). *Computer Networks and Systems, Queuing theory and performance evaluation.* New York, NY: Springer-Verlag.

Ross, S. M. (2007). *Introduction to Probability Models.* Amsterdam, The Netherlands: Elsevier.

Sobrinho, J. L. (2002). Algebra and algorithms for QoS path computation and hop-by-hop routing in the internet. *IEEE/ACM Transactions on Networking, 10*(4), 541–550. doi:10.1109/TNET.2002.801397

Sun, S., Krzymien, W. A., & Jalai, A. (1998). Optimal forward link power allocation for data transmission in CDMA systems. In *Proceedings of the IEEE International Symposium on Personal, Indoor and Mobile Radio Communication,* Boston, MA (Vol. 2, pp.848-852). Los Alamitos, CA: IEEE Press.

Tanenbaum, A. S. (2004). *Computer Networks.* New Delhi, India: Pearson Education.

Tipper, D., & Dahlberg, T. (2002). Providing fault tolerance in wireless access networks. *IEEE Communications Magazine, 40*(1), 17–30. doi:10.1109/35.978050

Ying, T., Yang, Y.-P., & Zeng, J.-C. (2006). An Enhanced Hybrid Quadratic Particle Swarm Optimization. In *Proceedings of the Sixth International Conference on Intelligent Systems Design and Applications,* Jinan, China (Vol. 2, pp. 980-985). Los Alamitos, CA: IEEE Press.

Zhang, D., Ray, S., Kannan, R., & Iyengar, S. S. (2003). A Recovery Algorithm for Reliable Multicasting in Reliable Networks. In *Proceedings of the IEEE 32nd International Conference on Parallel Processing,* Kaohsiung, Taiwan (pp. 493-500). Los Alamitos, CA: IEEE Press.

Zhao, D., Shen, X., & Mark, J. W. (2006). Soft Handoff and Connection Reliability in Cellular CDMA Downlinks. *IEEE Transactions on Wireless Communications, 5*(2), 354–365. doi:10.1109/TWC.2006.1611059

This work was previously published in the International Journal of Interdisciplinary Telecommunications and Networking, Volume 3, Issue 1, edited by Michael R. Bartolacci and Steven R. Powell, pp. 1-19, copyright 2011 by IGI Publishing (an imprint of IGI Global).

Chapter 2
Cognitive Aeronautical Communication System

Jamal Haque
University of South Florida, USA

Huseyin Arslan
University of South Florida, USA

M. Cenk Erturk
University of South Florida, USA

Wilfrido Moreno
University of South Florida, USA

ABSTRACT

The paper explores the system and architecture requirements for cognitive driven reconfigurable hardware for an aeronautical platform, such as commercial aircraft or high altitude platforms. With advances in components and processing hardware, mobile platforms are ideal candidates to have configurable hardware that can morph itself, given the location and available wireless service. This paper proposes a system for an intelligent self-configurable software and hardware solution for an aeronautical system.

INTRODUCTION AND OVERVIEW

Wireless connectivity has come a long ways in providing reliable and bandwidth efficient data connectivity. Given the limited resource in multi-dimensional electro-space, i.e. time, frequency, space, polarization, modulation/orthogonal signalization; an increase in data rate can be achieved through multi-dimensional resource efficiency. To achieve higher spectral density (Wertz & Larson, 1999), higher signal energy over noise (E_b/N_o) is required (Signal Processing Design Line, n.d.) to support the different links.

In the past few decades, the increased time spent in the air by higher numbers of users (Bureau of Transportation Statistics, n.d.) is creating a demand for data in in-flight services (Lai, 1998). In addition, the aircraft can be used as a relay. Therefore, aircraft based Aeronautical Data Networks (ADN) for future wireless communication structure is increasingly being discussed. All programs lead by National Aeronautics and Space Administration (NASA), Federal Aviation Administration (FAA), European Union (EU) and EUROCONTROL are including the aeronautical platform as part of the network (NASA, 2009; Gilbert, Jin, Berger, & Henriksen, 2008; Newsky, n.d.). A key enabler would be a robust physical layer. From the networking point of view, there

DOI: 10.4018/978-1-4666-2154-1.ch002

are a couple of studies, where in-flight internet with both aeronautical ad-hoc networking and centralized manner strategies are discussed (Sakhaee & Jamalipour, 2006; Medina, Hoffman, Ayaz, & Rokitansky, 2009). The global movement of the aeronautical system can take advantage of emerging wireless services and standards. This paper explores the concept for a Cognitive Aeronautical Software Defined Radio (CASDR). The organization of this paper is as follows: the following section discusses the driving motivation for CASDR. The CASDR system requirements are then established. In addition, the problem statement for Doppler and its effects to the physical layer in aeronautical scenario are investigated. The proposed hardware system definition for aeronautical software defined radio is given. In the last section, conclusions and a roadmap for future studies are provided.

MOTIVATION AND CHALLENGES

The ever-changing geographical environment of an aircraft and an increasing availability of differ-ent wireless services makes one wonder, what if such services can be accessed in real time?

This provided the motivation to develop a concept system and its hardware that would ac-commodate to the rapid changes, not just due to the aircraft location, but also to support the growth of services and industry evolution.

Figure 1 depicts the notional framework of opportunistic wireless data service that may be available for an aircraft in flight. At higher al-titude the services may be more traditional and fixed, however on ground, the growing WiMAX and local area network services may be available to be accessed from the aircraft. The high-speed mobility of an aircraft adds additional challenges to the design of system physical layer, such as path loss and multi-Doppler spread (Medina et al., 2008; Erturk, Haque, & Arslan, 2010).

REVIEW OF LITERATURE

The desire for a universal and a reconfigurable terminal first appeared in the military area. The need for mobility and accessibility was the driv-

Figure 1. Aeronautical system

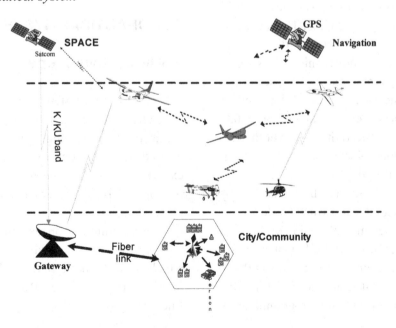

ing requirement. One of the early concept was a reconfigurable system appeared as an equipment called "SPEAKeasy" (Lackey & Upmal, 1995). The Software Communications Architecture (SCA) developed by the Joint Tactical Radio System (JTRS) program of the U.S. Department of Defense (DoD) further fueled the growth of SDR (JTRS Enterprise, n.d.). JTRS aims to provide a family of digital, programmable, multiband, multimode, modular radios to alleviate communications interoperability problems. Finally the work of Mitola (1995), there is now a growing interest in reconfigurable terminals.

The increase in air traffic is resulting in the surge of commercial airborne communication system (European Organization for the Safety of Air Navigation, 2008). Aircell and AeroSat have developed the ground based hardware and now offer in flight Internet service. Aircell uses a concept of air-to-ground link (Blumenstein, 2007) and provides the in-flight Internet service called 'gogo' on aircrafts. GOGO service works of cellular phone base stations in the continental US, which act as access points for an en route flight. A recent flight from Tampa, Florida to Detroit, Ohio USA, a user using GOGO service experienced an average upload speed of 0.27 Mbits/s and an average download speed of 0.33 Mbits/s with latency of 233ms. However, the ground based service is limited to flight coverage over land only. For the oceanic flight satellite based connectivity is required. AeroSat developed satellite communication (SATCOM) Ku band for commercial airliners (AeroSat Corporation, 2008). This offers broad connectivity, however the cost and data throughput of satellite based service is not conducive to user demand.

The growth in SDR has been enabled by advances in semiconductor, which has led to the development of programmable multi-core General Purpose Processor (GPP), Digital Signal Processor (DSP), Field Programmable Gate Array (FPGA) and Analog to Digital Converter (ADC). GPP, DSPs and FPGAs provide the programmability

and processing capability to realize such a system. Hence, the processing chain starting from digital intermediate frequency (IF) down to demodulation can be implemented in digital signal processing (Srikanteswara, Chembil Palat, Reed, & Athanas, 2003; Mohebbi, Filho, Maestre, Davies, & Kurdahi, 2003). Another key enabler is the high speed ADC that bridges the analog and digital world (Zanikopoulos, Hegt, & van Roermund, 2006; Salkintzis, Hong, & Mathiopoulos, 1999a). Advance algorithms that require intense processing can now be implemented in the combination of these moderate size, weight and power processing engines. FPGA's, with their ability to parallelize, can implement intense processing algorithms that may be difficult to implement in a DSP or GPP.

Therefore the maturity of; SDR algorithm's, high bandwidth processing engines, development of tunable antenna and availability of high speed ADC makes the implementation of CASDR a possibility. The global mobility of an Aeronautical platform is the ideal implementation of a CASDR concept. A CASDR will learn and configure itself in order to provide multi standard/service modem's as it traverses continents, countries and cities (Cummings & Haruyama, 1999).

AERONAUTICAL SYSTEM

ASDR System Scope

The scope of this system would be to provide an intelligent configurable radio system, provide connectivity for a changing geographical, political and regulatory environments that an aircraft experiences. Such a system will take advantage of opportunistic services available today and planned in future.

The communication design is beginning to converge on standard building blocks, or systems, which form the basic building block of a communication system, i.e., Read Solomon, Turbo Encoder, Modulations, Viterbi etc. Whether a communication link is being developed for short range, long range, line of sight (LOS) or non line of sight (NLOS) the basic building blocks of communication system are the same. If available in software they can be stitched together to build a radio transceiver. Aeronautical Networks (ANs) could be an important application of such systems, since different regions or countries assign different frequency bands based on their needs and spectrum allocation policies.

Aeronautical Network Geometry

Geometric relations are observed between an aircraft station (AS) or an aircraft's altitude (h1) with a Ground Station (GS). The LOS communication distance (without considering Fresnel and other parameters) from AS to GS can be calculated using the Pythagoras theorem as follows:

$$d_1 = (h_1 \times [2R + h_1])^{0.5} \cong (2Rh_1)^{0.5} \qquad (1)$$

where, R is the radius of the Earth which varies from 6336 km to 6399 km, but assumed 6370 km (for the purpose of calculations). For distances

between the two nodes above the sea level, the above formula needs additional steps for calculating the communication distance. The formula is calibrated by a statistically measured parameter by International Telecommunication Union (ITU), i.e., 'k'.

$$d_1 \cong (2Rkh_1)^{0.5} \qquad (2)$$

Figure 2 shows the maximum communication distances that can be achieved between AS and AS/GS. The jump in the first 2 km altitudes for GS communications can be considered a very low orbit AS which can reach a communication zone of D=120 km. Many commercial planes flying at the altitude of 9 km can potentially create communication zones about D=250 km with a very conservative approach (k=0.5). On the other hand, considering the communication distance between two ASs, it can be inferred that it could reach up to D=480 km with k=1/2.

Figure 2 shows that ASs could be used as a backhaul or relay for wireless infrastructures, since they have the capability of communicating long distances as compared to wireless ground

Figure 2. Communication zone of an AS

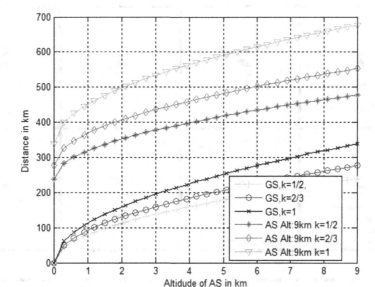

backhauls. Aeronautical Network (AN) will have a substantial lower round trip delay, which will allow for a low delay telephone and voice over IP services.

Aeronautical Network Scenarios and Data Access

Aeronautical Networks can provide critical services for various situations, such as; public safety communications, Denial of Service (DoS), disaster situations, in-flight Internet, as well as mobile communication on the ground, providing services for highways, trains, etc. The network structure that is being proposed in this paper is as follows: Given a particular region to be covered, initially Service Gateway Ground Stations (SGGS) should be built according to the communication distance (see Figure 3). Assuming that a GS can communicate to an AS within the distance of 200 km, roughly 8 SGGS will be able to provide service for an area of 1600 km by 800 km.

Data access in an AN can be defined as follows: When a GS or AS has data to send, the flow of the data should be from/to SGGS so the con-

nection with other networks such as public switched telephone network (PSTN), cellular networks and Internet Protocol (IP) could be established.

To provide in-flight services, a centralized configured network should be considered; SGGSs act like Base Stations (BS), covering a particular region where Subscriber Stations (SS) are simply ASs. Scheduling is done by the SGGS and in this structure, AS's are not communicating with each other, except with SSGS's. However, if an AS is not able to register to a SGGS, which could be a case of oceanic flights, then data of that particular AS should be routed to an AS which was already registered to a SGGS with ad-hoc networking strategies.

For an AN network the use of AS as a base station used for cellular network is also discussed. In this case, SS's are the GS's, which can be fixed or mobile. When a GS has data to send, it sends its data to an AS. This can be considered as a relay, reflecting the data to its associated SGGS to finalize the establishment. This structure is feasible to provide public safety services in disaster scenarios, provide backhaul option for terrestrial networks

Figure 3. Aeronautical network scenarios

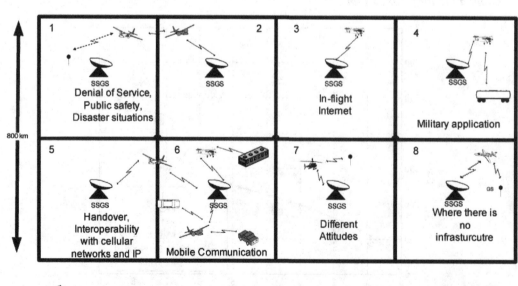

and military communication applications. Moreover, in this structure, since both AS's and GS's are not fixed, the handover of a GS between multiple AS is also another challenging issue. It is important to note that the handover process in this structure is somehow different, since GS are doing handovers not only because of their own mobility, but also due to mobility of AS's.

One of the main issues in AN's is the topology estimation. Since there are many mobile stations, in terms of GS and AS, the scheduling and routing of data would differentiate from time to time. In these cases the topology estimation of the network should be done properly, so that the data can be routed and scheduled in mesh and centralized networks strategies respectively.

Physical Layer

In a wireless system design, understanding the limits and bounds of a channel impairments theoretically and empirically are critical to the design of the system. An aeronautical environment poses a daunting task to cover a huge area for any system designer. Global channel characteristics need to be

understood to establish model parameters. However, this would lean toward statistical average and will result in inefficient system parameters. Current system based on 'gogo' service, uses a ground based link and provides a limited data rates. A data connectivity sample was taken for a Delta flight traversed between Tampa, Florida to Detroit, Ohio USA, using 'Speed test'(http://www.speedtest.net/). Different global servers were pinged periodically duration the flight to measure download, upload and latency. Figure 4 and Figure 5 are the global data rates and latency experienced during the flight.

Most of the current system, assumes a line of sight (LOS). This is also the case for the aeronautical platforms connectivity modeling. However, an intelligent CASDR will allow for the ability to configure the system and learn the channel condition over the flight route and establish history, hence establish accurate channel parameters for a given location, altitude and speed. Since the aircraft traverses pre-planned route, over time this channel parameters will provide accurate characteristics (International Air Transport Association, n.d.). This will allow higher order spectral efficient

Figure 4. Global in-flight data rates

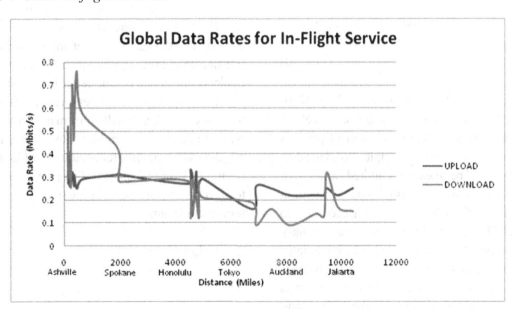

Figure 5. Global in-flight latency

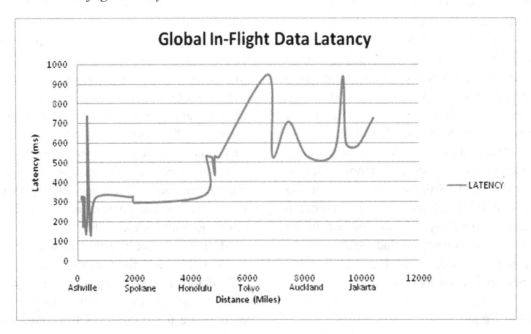

modulations and multi-carrier system to be used and provide higher data throughput. Details of this cognitive channel sensing behavior are discussed.

A time varying wireless impulse response is represented by equation (3), where the signal is impaired by amplitude, phase, Doppler and time delay.

$$h(n) = \frac{1}{\sqrt{N}} \sum a_k \exp\left(\frac{j2\pi\theta}{N}\right) \exp\left(\frac{j2\pi f_d n}{N}\right) \delta(t - \tau_k)$$

$$(3)$$

where a_k, θ, f_d and τ_k are amplitude, phase, Doppler shift and delay for each path respectively.

For a LOS, the effect of number of paths is significantly less, $\tau \approx 0$, f_d, Doppler shift based on platform would be fixed and a limited variation of phase will lead mostly to amplitude degradation due to path loss. For the diffused path, according to Bello (1973), it represents a wide-sense stationary uncorrelated (WSSUS) channel, which emulates a small area characterization for multipath channel. The effect of Doppler to the line of sight is mostly frequency shift; however the diffused and specular reflections will have a spread due to Doppler. This Doppler spread for an aeronautical communications depicts a bandwidth less than 360° (Haque, Erturk, & Arslan, 2010; Hoeher & Haas, 1999; Haas, 2002; Elnoubi, 1992). Most of the current research assumes a two-ray model as the channel model for flat surface areas. In an extremely mountainous terrain environment, the channel model results in an intermittent loss of LOS along with increasing angle spread that could match the Jakes Doppler spread. In the two limiting cases; the angular spread at the receiver depicts either a two ray model or Jakes spectrum. Therefore, a modified Doppler spread model needs to be developed, that will go from a narrow beam width to 360°, as the mobile moves from flat to rough environments. Hence, the use of D_f factor from 0 to 1 for the growing Doppler spread, due to beam width, represents going from flat to rough terrain environment:

$$D_f = \frac{\theta_H - \theta_L}{2\pi},$$

where

$$(\theta_H - \theta_L) \leq 2\pi \tag{4}$$

$$D = \begin{cases} 2D_f & , \theta_H > \theta > \theta_L \\ \infty & , else \end{cases}$$

where

$$\theta = \cos^{-1}\left(\frac{f_d \lambda}{v}\right)$$

$$|f_d| \leq f_{d\max} \tag{5}$$

Doppler density going from non-isotropic to isotropic:

$$p_{fd}(f_d) = \begin{cases} \dfrac{1}{\pi D f_{d\max}\sqrt{1 - (\frac{f_d}{f_{d\max}})^2}} & if \ |f_d| \leq f_{d\max} \\ 0 & else \end{cases} \tag{6}$$

Figure 6 shows the Air to Ground (A/G) and Ground to Air (G/A) aeronautical communications in an en-route scenario and their corresponding Doppler spectrum. The arrival/take-off, taxi and parking scenarios depicts different multipath and received angle spreads (Salkintzis, Hong, & Mathiopoulos, 1999). The spectrum in an en-route scenario depicts a Doppler shift with a narrow beam Doppler spread, where it can be assumed as another Doppler shift. Among carrier and modulation systems, Orthogonal frequency division multiplexing (OFDM) is the most sensitive to Doppler. OFDM based systems has been adopted/proposed for several current/future communication systems all over the world, i.e., asymmetric digital subscriber line (ADSL) services, IEEE 802.11a/g/n, IEEE 802.16, IEEE 802.20, digital audio broadcast (DAB), digital terrestrial television broadcast (DVD) in Europe, ISDB in Japan and fourth generation (4G) cellular systems. Therefore, it is reasonable to assume that any SDR application will also need to support OFDM in the ADN network. In an OFDM based systems, a serial symbol stream is converted into parallel streams and each symbol is modulated with different orthogonal sub-carriers. With the usage of cyclic prefix (CP), since OFDM based systems have already relatively longer symbol durations compared to single carrier systems, they are known for their robustness against frequency selectivity of the channel, i.e., delay spread. However, longer symbol durations lead to weakness of the OFDM systems to time variation of the channel, i.e., Doppler shift/spread which is a challenging issue in ADN.

Figure 6. Doppler Power Spectrums for ADNs

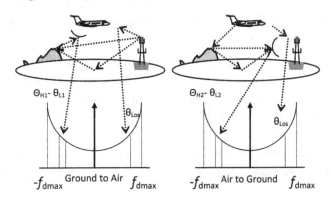

Figure 7. Doppler ICI vs. Sub-carrier BW

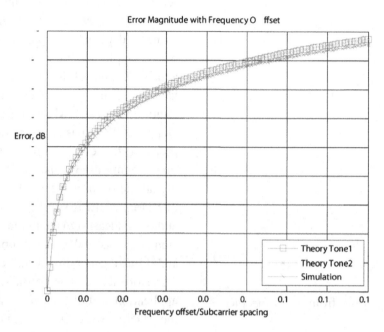

The two Doppler shifts affecting the system can be described as follows:

$$y(n) = x(n) * h(n) + w(n) \qquad (7)$$

where, $w(n)$ is noise and $h(n)$ is the channel impulse response defined as:

$$h(n) = \sum_{i=1}^{2} a_i \exp\left(\frac{j2\pi f_{\Delta i}(n - \tau_i)}{N}\right) \delta(n - \tau_i) \qquad (8)$$

where a_i is the attenuation value, N is the number of FFT bins, τ_i and $f_{\Delta i}$ are the delay and the normalized Doppler frequency shift (NDF) for the first and second ray respectively where

$f_{\Delta i} = \dfrac{f_{Di}}{\Delta_f}$. For the ADN, Figure 4 presents

the two-path channel model. In OFDM, as long as the carrier's orthogonality is maintained, then there is no bleeding of energy. Intercarrier interference (ICI) is related to sub-carrier bandwidth and their proportional interference due to Doppler

offset. As an example, an estimation of ICI interference for a system with 64 point FFT OFDM symbol, which has a 312 kHz subcarrier bandwidth, is plotted in Figure 7.

The Figure 7 shows the ICI error vector contribution due to frequency shift caused by Doppler in a two ray model. At 0.1 fraction of sub carrier frequency the ICI error contribution approaches -10dB. To support higher spectral efficiency generally ICI should remain within or less than 0.02 fraction of sub-carrier frequency. This will allow ICI interfering energy to remain well below -25 dB allowing higher spectral efficiency.

Cognitive Route Based Physical Layer Estimates

The aircraft routes driven by FAA for various segments are ideal to establish a history of wireless channel conditions for the route. Once a route is traversed, its history of channel impairments are stored with associated coordinates and aircraft attitude information. This data is downloaded to a central database to be shared with another aircraft.

For new routes, the cognitive channel estimator will try to understand the channel condition. Over time, the channel history collected from different aircraft will create a channel map for each route. The ASDR will then be able to download this data and prior to a flight adjust the physical layer parameters for the route. For a mobile platform that has a predetermined route, such as AN, the channel estimation is broken down to static and dynamic components. The static components effecting the channel would be large objects i.e., mountains, buildings, etc. The averaging over multiple routes will provide of stable static channel estimates. The dynamic components will be due to time varying objects.

AERONAUTICAL SOFTWARE DEFINED RADIO

The advances in components and signal processing techniques are the leading enabler for a configurable hardware and intelligent software. Software defined radio emerges from the desire of single radio hardware that molds its feature to different radio schemes (Salkintzis, Hong, & Mathiopoulos, 1999b). The artificial intelligence needed for the smarts of such configurable hardware is emerging into what is known as cognitive radio (Mitola & Maguire, 1999).

Cognitive algorithms combined with configurable hardware can take full advantage of varying location of an aircraft, whether that is in the air, en route and lends themselves to take advantage of opportunistic spectrum and network connectivity.

A system with the ability to morph to accommodate the aeronautical changing environments, channels conditions across domestics and international boundaries is required. Aeronautical software defined radio (ASDR) platform will also allow the flexibility to comply with countries regulations governing the spectrum usage and interference.

Spectrum Coverage

The spectrum bandwidth use and frequency band allocation for different systems is one of the challenges to overcome for truly building an ASDR capable of accommodating itself for different regions. For a given region or country, the standard may be the same but the frequency band used may be different. For example, the 802.16 specification applies across a wide range of radio frequency spectrum and WiMAX could function on any various frequencies i.e., 2.5 GHz is predominantly being used in the USA, elsewhere in the world 2.3 GHz used in Asia and some countries are using 3.5 GHz.

The Analog TV bands (700 MHz) may become available for WiMAX usage, but currently it is being used for digital TV, however different countries might choose to use the spectrum that best suits their needs. Table 1 lists opportunistic frequency data network available (Zhang & Ansari, 2010; Lemme, Glenister, & Miller, 1999).

Another feature that will be necessary in a SDR application is a tunable RF front end capable of locking on the various bands.

Frequency bands and bandwidths for future wireless communication studies in terms of aeronautical communications are discussed at the World Radio communication Conference (WRC) 2007. This international body maintains and agrees to abide by the use of spectrum by international treaty. Aeronautical Mobile (Route) Service (AM(R)S) communication is defined as a safety system requiring high reliability and rapid response. Safety and security applications together with, Air Traffic Control (ATC) and Air Traffic Management (ATM) communications are considered to be AM(R)S. To accommodate the future growth of aeronautical communication, new band allocations are being made in AM(R)S rather than VHF band in L and C. L band (960-1164 MHz) and C band (5091-5150 MHz) allocations are discussed in the meeting.

Table 1. Wireless standards

Band (GHz)	BW (MHz)	Standard	Region	Service
2.4	20	802.11b/g	US	Wi-Fi
5	20	802.11a	US	Wi-Fi
2.5	20	802.16	US	Fixed WiMAX
3.5,2.5	20	802.16a	Can	Fixed WiMAX
2.3	20	802.16e	Aus	Fixed WiMAX
1.616 -1.6265	10.5	Custom	Global	Iridium Down Link
19.4 -19.6	20	Custom	Global	Iridium Up Link
2,4		Sirius/XM	US	QPSK, OFDM
1.9, 0.85	1.23, 5	W/CDMA	US	3G Cellular
1.8, 0.9	1.23, 5	W/CDMA	EU	3G Cellular
0.5 – 0.8	n/a	n/a		Analog TV

L band is suggested as a suitable band for future aeronautical communication studies. C band is considered to be used in Airport surface network systems, since it is thought to be useful for short range, high data throughput.

Critical System Parameters

Cognitive radio system will require optimization of system performance. Algorithms capable of real time optimizing the system performance as well as pre/post flight will create pre-flight configuration (Table 2).

Aeronautical Cognitive Radio

The term cognitive comes from psychology meaning "brains" the ability to learn and understand. The aeronautical environment is ideal application for an intelligent radio, which is capable of

Table 2. Parameters

Aeronautical Optimization Parameters
Customer Usage
DQ: Quantity of data transferred at various flight segments. DT: Duration of data transfer per segmented route. TC: Traffic classes: multi-media, navigation, system health & safety. BER: Required Bit error rate per Traffic Classes.
Network & Data Access Layer
Protocol Selection, Routing configuration, Forward error selection given the customer driven BER, Available to provide relay service, Packet error rate
Physical Layer
T_M:Multipath delay spread: Characterizes channel smearing due to arrival of multi-signal reflection arrivals. f_{DS}: Doppler spread or Doppler bandwidth. f_d: Doppler Shift. A: Attenuation: power loss, function of frequency and distance. L: Impulse Response Length: length, in signal elements, of CIR. $Band$: Carrier frequency Band. BW: Available bandwidth. SWP: Standard waveform performance.

Figure 8. Route based channel sensing

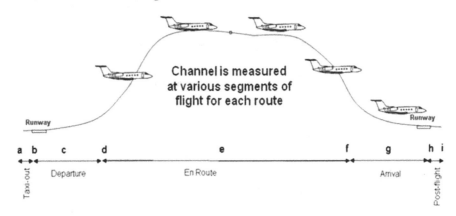

learning the environment for various locations and altitudes (see Figures 7 and 8). Over time, each aircraft flying over certain route will store the data on board storage devices. This data shall contain the route the airline/aircraft traversed, the opportunist wireless links available, frequency band, bandwidth, data rate, wireless standard, signal quality for the route, etc. Upon arrival at the destination, data is then downloaded to a centralized flight communication data bank. This data is then available for flights heading on the same route (Figure 9).

Cognitive Intelligence

The brain of the aeronautical cognitive would be to work of its constant awareness of aircraft

geographical location and RF channel. It will sense weather conditions that may affect the radio transmission and available services available (Figure 10).

Awareness

The aircraft navigation and radar systems will provide the sensing stimulation to the cognitive engine. The inertial measurement unit (IMU) used for flight navigation will provide aircraft speed, altitude, and attitude. Advance forward looking radar will provide the weather conditions that may affect the radio transmission performance. Global position system (GPS) will provide location of the aircraft with respect to global geography. Furthermore, the awareness engine will have the

Figure 9. Aeronautical SDR and CE

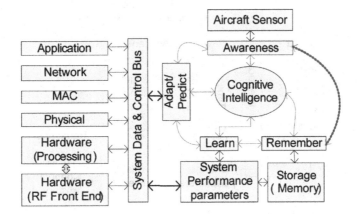

Figure 10. Cognitive engine

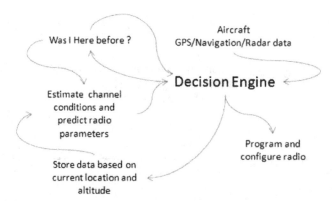

ability to estimate the data requirements based on past data use and flight profile, before accessing the spectrum for services.

Learn

The cognitive awareness provides an opportunity for CASDR to learn the spectrum usage, data demand and system throughput based flight route during day or night. Such statistics will allow a constant learning and developing statistics profile that is stored for each route. This allows cognitive radio of other airlines that have not travelled that particular route to have a priori knowledge and schedule services accordingly. The system

parameters available at particular location can be configured for that country or location.

The channel sensing and estimation for the flight route will serve to establish channel statistics, as shown in Figure 11. The CASDR cognitive channel awareness can configure the system to establish channel impairments for the flight route.

Remember

System performance data gathered for different flight routes through different airlines will serve as means to remember these flight parameters, exchanged through a centralized data archives. Such data will grow in time and averaging over

Figure 11. Aeronautical channel sensing

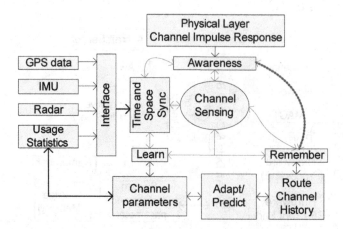

time will provide a reliable statistics for configuring the SDR radio parameters.

Adapt and Predict

The cognitive engine learning and sensing ability with an aircraft system will allow ability to predict system configuration parameters and adapt them to data gathered through flights travelled by other carriers.

Aeronautical Configurable Hardware

The key to a configurable system for an aeronautical system is to design hardware with minimum analog front-end, access different antenna system, digitize the signal and a scalable architecture. Figure 12 presents such a system. The RF front-end board will support multiple bands with varying gain amplifiers (Pilgrim, 2008). Closely coupled A/D boards with FPGAs are required for high-speed data connectivity and processing. A technique such as under sampling for demodulations is used to reduce the front-end components. The advances in ADC devices as well as non-compliance feature of Nyquist sampling theory is an enabler for an ASDR application. Violation of Nyquist theory will create signals aliased at integer multiple of sampling frequency (N^*f_s) (Susaki, 2002). This put the challenge on front-end processing system. The advances in programmable Digital Signal Processing (DSP) and Field Programmable Gate Array (FPGA) are ideal for such processing (Xilinx, n.d.). FPGA offers ability

to parallelize processing hence, allow a high-end processing throughput. The Virtex-6 FPGA family by Xilinx provides up to 2,016 DSP48 slices that deliver up to 1000 Giga MAC/s of DSP processing performance. Xilinx offers solutions for evolving standards such as WCDMA, WiMAX, TD_SCDMA and LTE. Texas Instruments DSP products are now offering six DSP processors in a single package with processing capability 4000 million MACS (16-Bits) at 500 MHz (Texas Instruments, 2010).

CONCLUSION

The advances in component technology, evolution in communication services, and an increase in data demand and aircraft mobility creates an ideal application for CASDR to support the aeronautical system. Current deployed systems are beginning to form shape (i.e., gogo); however, they are now adding another hardware box to provide connectivity. Since the system is hardwired for a particular modem, the evolution will require hardware modification to keep up with growth in the telecommunication growth. Accurate measurements of channel characteristics, such as Doppler, will allow spectral efficient modulation to be used for higher data rates. Advance algorithms along with processing capabilities can resolve the impact Doppler, due to aircraft high mobility. The novel cognitive channel measurement and estimation for each route will increase spectrum efficiency and in return provide high data throughput. An

Figure 12. Aeronautical software defined radio

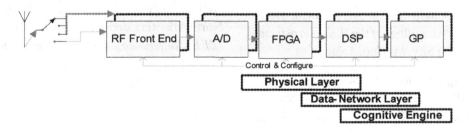

optimum combination of bandwidth, subcarrier bandwidth, acceptable Doppler frequency and multipath immunity system can be developed for ADN. This will result in an efficient use of the spectrum and provide a high data rate for the global connectivity.

REFERENCES

AeroSat Corporation. (2008). *About Airborne SatCom.* Retrieved from http://www.aerosat.com/about/about_airborne_satcom.asp

Bello, P. A. (1973). Aeronautical channel characterization. *IEEE Transactions on Communications, 21*(5), 548–563. doi:10.1109/TCOM.1973.1091707

Blumenstein, J. (2007). *Aircell: Inflight Wi-Fi Built for the Airline Business.* Retrieved from http://blog.aircell.com

Bureau of Transportation Statistics. (n.d.). *Airlines and Airports.* Retrieved from http://www.bts.gov/programs/airline_information

Cummings, M., & Haruyama, S. (1999). FPGA in the Software Radio. *IEEE Communications Magazine, 37*(2), 108–112. doi:10.1109/35.747258

Elnoubi, S. M. (1992, May 10-13). A simplified stochastic model for the aeronautical mobile radio channel. In *Proceedings of the Vehicular Technology Conference (VTC 1999),* Denver, CO (Vol. 2, pp. 960-963).

Enterprise, J. T. R. S. (n.d.). *Connecting the tactical edge.* Retrieved from http://jpeojtrs.mil

Erturk, M. C., Haque, J., & Arslan, H. (2010, March 6-13). Challenges in Aeronautical Data Networks. In *Proceedings of the IEEE Aerospace Conference,* Big Sky, MT (pp. 1-7).

European Organization for the Safety of Air Navigation (Eurocontrol). (2008). *STATFOR - Air Traffic Statistics and Forecasts.* Retrieved from http://www.eurocontrol.int/statfor/

Gilbert, T., Jin, J., Berger, J., & Henriksen, S. (2008). *Future Aeronautical Communication Infrastructure Technology Investigation.* Huntsville, AL: NASA.

Haas, E. (2002). Aeronautical channel modeling. *IEEE Transactions on Vehicular Technology, 51*(2), 254–264. doi:10.1109/25.994803

Haque, J., Erturk, M. C., & Arslan, H. (2010, April). Doppler Estimation for OFDM based Aeronautical Data Communication. In *Proceedings of the IEEE Wireless Telecommunications Symposium (WTS),* Tampa, FL (pp. 1-6).

Hoeher, P., & Haas, E. (1999, September). Aeronautical channel modeling at VHF-band. In *Proceedings of the Vehicular Technology Conference (VTC 1999),* Amsterdam, The Netherlands (Vol. 4, pp. 1961-1966).

International Air Transport Association (IATA). (n.d.). *Schedule Reference Service (SRS).* Retrieved from http://www.iata.org/ps/publications/srs/

Lackey, R. J., & Upmal, D. W. (1995). Speakeasy: The Military Software Radio. *IEEE Communications Magazine, 33*(5), 56–61. doi:10.1109/35.392998

Lai, J. (1998). *Broadband Wireless Communication Systems Provided by Commercial Airplanes.*

Lemme, P. W., Glenister, S. M., & Miller, A. W. (1999). Iridium(R) aeronautical satellite communications. *IEEE AES Systems Magazine, 14*(11), 11–16. doi:10.1109/62.809197

Medina, D., Hoffman, F., Ayaz, S., & Rokitansky, C. H. (2008, June 16-20). Feasibilty of an Aeronautical Mobile Ad-Hoc Network over the North Atlantic Corridor. In *Proceedings of the 5th IEEE Conference on Sensor, Mesh, and Ad hoc Communications and Networks,* San Francisco, CA (pp. 109-116).

Mitola, J. (1995). The software Radio architecture. *IEEE Communications Magazine, 33*(5), 26–38. doi:10.1109/35.393001

Mitola, J., & Maguire, G. Q. (1999). Cognitive Radio:bMaking Software Radios More Personal. *IEEE Personal Communications, 6*(4), 13–18. doi:10.1109/98.788210

Mohebbi, B., Filho, E. C., Maestre, R., Davies, M., & Kurdahi, F. J. (2003, October). A case study of mapping a software-defined radio (SDR) application on a reconfigurable DSP core. In *Proceedings of the 1st IEEE/ACM/IFIP International Conference on Hardware/Software Codesign and System Synthesis,* Newport Beach, CA (pp. 103-108).

NASA. (2009). *Advanced CNS Architectures and System Technologies.* Retrieved from http://acast.grc.nasa.gov/main/projects/

Newsky. (n.d.). NEWSKY – Networking the Sky for Aeronautical Communications. Retrieved from http://www.newsky-fp6.eu/

Pilgrim, D. (2008). *Simplifying RF front-end design in multiband handsets.* Retrieved from http://rfdesign.com/microwave_millimeter_tech/rf_front_end_mmic/radio_simplifying_rf_frontend/index1.html

Sakhaee, E., & Jamalipour, A. (2006). The Global In-Flight Internet. *IEEE Journal on Selected Areas in Communications, 24*(9), 1748–1757. doi:10.1109/JSAC.2006.875122

Salkintzis, A. K., Hong, N., & Mathiopoulos, P. T. (1999a). ADC and DSP challenges in the development of software radio base stations. *IEEE Personal Communications, 6*(4), 47–55. doi:10.1109/98.788215

Salkintzis, A. K., Hong, N., & Mathiopoulos, P. T. (1999b). ADC and DSP Challenges in the Development of Software Radio Base Stations. *IEEE Personal Communications, 6,* 47–55. doi:10.1109/98.788215

Signal Processing Design Line. (n.d.). *EE times design.* Retrieved from http://www.dspdesignline.com

Srikanteswara, R., Chembil Palat, R., Reed, J. H., & Athanas, P. (2003). An Overview of Configurable Computing Machines for Software Radio Handsets. *IEEE Communications Magazine, 41*(7), 134–141. doi:10.1109/MCOM.2003.1215650

Susaki, H. (2002). A Fast Algorithm for High-Accuracy Frequency Measurement: Application to Ultrasonic Doppler Sonar. *IEEE Journal of Oceanic Engineering, 27*(1). doi:10.1109/48.989878

Texas Instruments. (2010). *TMS320C6472.* http://focus.ti.com/docs/prod/folders/print/tms320c6472.html

Wertz, J. R., & Larson, W. J. (1999). *Space Mission Analysis and Design* (3rd ed.). Bloomington, IN: Microcosm Press.

Xilinx. (n.d.). *Xilinx DSP Platform.* Retrieved from http://www.xilinx.com/technology/dsp.htm

Zanikopoulos, A., Hegt, H., & van Roermund, A. (2006). Programmable/Reconfigurable ADCs for Multi-standard Wireless Terminals. In *Proceedings of the IEEE Conference on Communications, Circuits and Systems* (Vol. 2, pp. 1337-1341).

Zhang, Y., & Ansari, N. (2010). Wireless Telemedicine service over integrated IEEE802.11/Wlan and 802.16/Wimax networks. *IEEE Wireless Communications, 17*(1), 30–36. doi:10.1109/MWC.2010.5416347

This work was previously published in the International Journal of Interdisciplinary Telecommunications and Networking, Volume 3, Issue 1, edited by Michael R. Bartolacci and Steven R. Powell, pp. 20-35, copyright 2011 by IGI Publishing (an imprint of IGI Global).

Chapter 3
Best Approach for Video Codec Selection Over VoIP Conversation Using Wireless Local Area Network

Mohd Nazri Ismail
University of Kuala Lumpur, Malaysia

ABSTRACT

This study evaluates video codec performance over VoIP using a campus wireless network. Today, the deployment of VoIP occurs in various platforms, including VoIP over LAN, VoIP over WAN and VoIP over VPN. Therefore, this study defines which video codec provides good video quality over VoIP transmission. The soft phone is used as a medium for communication between two parties. A network management system is used to evaluate and capture the video quality performance over VoIP. The quality of video codec is based on MOS, jitter, delay and packet loss. The experimental scope is limited to G.722 with MP4V-ES, G.726 (16) with H.261 and G.726 (24) with H.264. The results show that audio codec G.722 with MP4V-ES generates good video quality over VoIP using wireless local area network. Whereas audio codec G.726 (16) with H.261 generates low rate video and voice quality performance. Therefore, using the appropriate video and audio, the codec selection increases video quality over VoIP transmission.

INTRODUCTION

The objective of this study is to: i) define the best approach for video codec selection over VoIP communication using campus wireless network environment. The experiment of this study is limited to the three types of video and audio codec such as G.722 with MP4V-ES, G.726 (16) with H.261 and G.726 (24) with H.264. This experiment will study the performance of video quality over VoIP communication. Codec is an algorithm used to encode and decode the voice conversation.

This paper presents the evaluation of audio and video codec performances over wireless local area networks (WLAN). This study will identify the problematic areas of audio and video codecs over a WLAN environment. The objectives of this research are:

DOI: 10.4018/978-1-4666-2154-1.ch003

1. To study the characteristics of V2oIP performance in WLAN environment
2. To study which audio and video codecs that are able to provide better performance in wireless environment

The contributions and signification values of this study are:

1. To produce a significant knowledgeable on V2oIP (video and audio) performance in campus wireless environment to social network especially to researchers and academic institutions
2. The results of the V2oIP performance in wireless environment are useful and can be used as a guidelines for next generation network

REVIEW OF LITERATURE

Today's communication networks are greatly affected by a number of technological changes resulting in the development of new and innovative end-user services. One of the elements for these new applications is video services that impact on the appearance of new multimedia services. Voice services are complemented with video and text (instant messaging and videoconference). Examples are multimedia conferences and collaborative applications that are now enhanced to support nomadic (traveling employees with handheld terminals) and IP access (workers with an SIP client on their PC and WLAN access) (Alfonso et al., 2008; Alcatel, 2005; Pérez & Fernández, 2006).

Since the early 1990s, when the technology was in its infancy, international video coding standards – chronologically, H.261 (ITU-T, 1993), MPEG-1 (ISO/IEC, 1993), MPEG-2 / H.262 (ITU-T and ISO/IEC, 1994), H.263 (ITU-T, 2000), and MPEG-4 (Part 2) (ISO/IEC, 1999) – have been the engines behind the commercial success of

digital video compression. H.264/MPEG-4 AVC is the latest international video coding standard. It was jointly developed by the Video Coding Experts Group (VCEG) of the ITU-T and the Moving Picture Experts Group (MPEG) of ISO/IEC (Gary et al., 2004), MP4V-ES is the video mpeg4 bitstream.

Today, many researchers concentrate on wireless technology implementations on VoIP service. In the digital era, the increase of network bandwidth and the ubiquitous wireless access facilitate the creation of more and more innovative network services. Among these services, Voice over Internet Protocol (VoIP) is surely one of the most popular and successful real-time multimedia services on the Internet (Wang & Wu, 2008).

Many organizations are using WLANs as a medium for communication, so it is important to investigate how VoIP over WLAN performs based on previous study (Stuedi & Alonso, 2007). Wireless VoIP applications make the very inefficient use of WLAN resources. Due to the large overhead involved in transmitting small packets in an 802.11 WLAN, the bandwidth available for VoIP traffic is far less than its maximal 11Mbps data rate it currently supports (Narbutt & Davis, 2006).

METHODOLOGY

In the experiment, campus wireless network environment is used as a communication medium between two parties using audio and video transmission. This study posits several research questions: i) what is the video quality performance using different types of video and audio codec and ii) what is the acceptance level for video and audio codec selection. VQ manager is used to capture the video quality performance over VoIP. Figure 1 shows the interface of VQ manager application.

Soft phone is used to transmit audio and video application between two parties in campus environment. Soft phone is a software program

Figure 1. VQ manager application interface

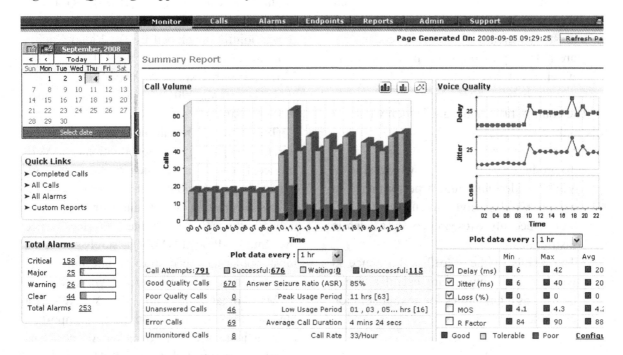

for making telephone calls over the Internet using a general purpose computer, rather than using dedicated hardware. Soft phone is designed to behave like a traditional telephone. A soft phone is usually used with a headset connected to the sound card of the PC. X-Lite is a proprietary freeware soft phone and uses the Session Initiation Protocol (SIP). Some of the features support by X-Lite soft phone as follows: i) Open Standards and Session Initiated Protocol (SIP); ii) IM and Presence Management; and iii) Multi-party Voice and Video Conferencing (refer to Figure 2).

Figure 2. X-Lite soft phone interface

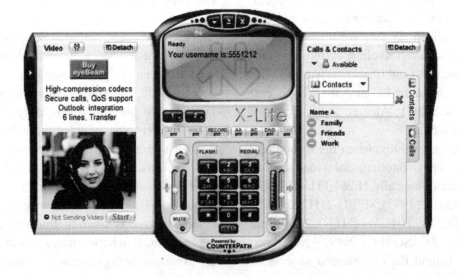

Figure 3, Figure 4 and Figure 5 show the configuration and flow of video packets over VoIP between the source and destination parties using different audio and video codec selections. The measurement of video quality is based on human perception (Mean Opinion Score (MOS)), delay, jitter, packet loss. Figure 6 shows the evaluation process of video quality over VoIP communication.

ANALYSIS AND RESULTS

This section is measures and compares video quality performance using differences audio and video codec selection through campus wireless network environment. Three experiments have been conducted to evaluate video quality over

VoIP such as G.722 with MP4V-ES, G.726 (16) with H.261 and G.726 (24) with H.264.

- **Audio and video codec experiment (G.722 with MP4V-ES):** VQ manager system generates delay on video approximately 250 to 300 ms (refer to Figure 7). Using this audio and video codec selection, jitter and packet loss are achieved approximately 4 to 6 ms and 0% (refer to Figure 8 and Figure 9).

Mean opinion score (MOS) generated by VQ manager system is rated approximately 4 to 4.5 (refer to Figure 10). MOS is expressed in one number, from 1 to 5, 1 being the worst and 5 the best (refer to Table 1) (Moura et al., 2007; Cole & Rosenbluth, 2001; Masuda & Ori, 2001).

Figure 3. Audio Codec G.722 with Video Codec MP4V-ES

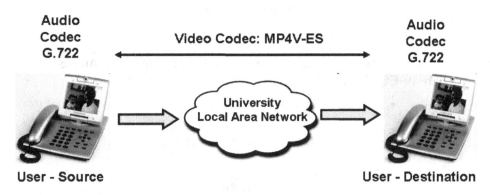

Figure 4. Audio Codec G.726 -16 with Video Codec H.261

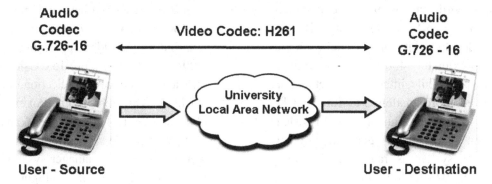

Figure 5. Audio Codec G.726 -24 with Video Codec H.264

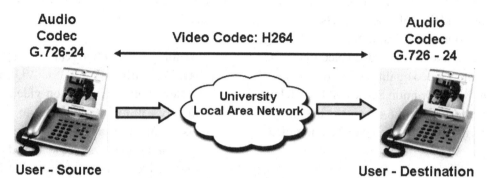

Figure 6. Evaluation process of video quality

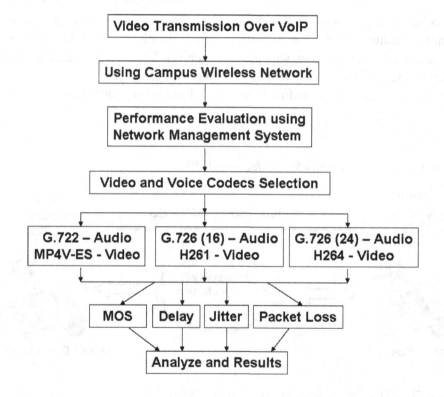

- **Audio and video codec experiment (G.726 - 16 with H.261):** Using this audio and video codec, it generates delay on video approximately 400 to 500 ms (refer to Figure 11). While, jitter and packet loss are achieved approximately 5 to 7 ms and 1% (refer to Figure 12 and Figure 13). Mean opinion score (MOS) generated by VQ manager system is rated approximately 2.50 to 2.70 (refer to Figure 14).

- **Audio and video codec experiment (G.726 - 24 with H.264):** VQ manager system generates delay on video approximately 400 to 700 ms (refer to Figure 15). Using this audio and video codec selection, jitter and packet loss are achieved approximately 5 to 7 ms and 0% (refer to Figure 16 and Figure 17). While, Mean opinion score (MOS) generated by VQ manager system is rated approximately 3.5 (refer to Figure 18).

Figure 7. Delay – Audio Codec G.722 with Video Codec MP4V-ES

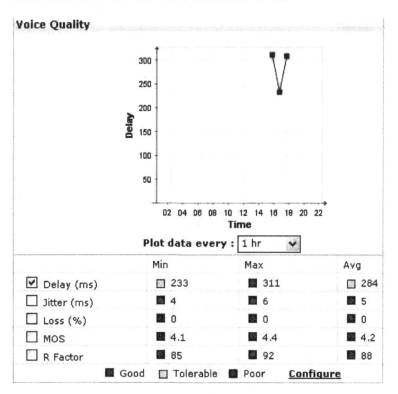

Figure 8. Jitter – Audio Codec G.722 with Video Codec MP4V-ES

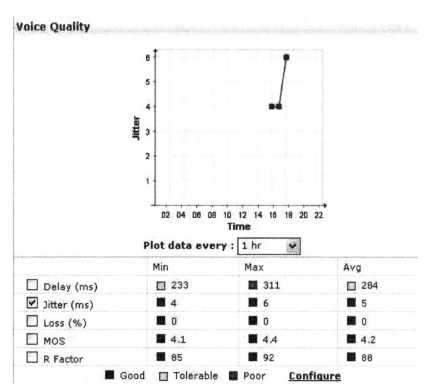

Figure 9. Packet Loss – Audio Codec G.722 with Video Codec MP4V-ES

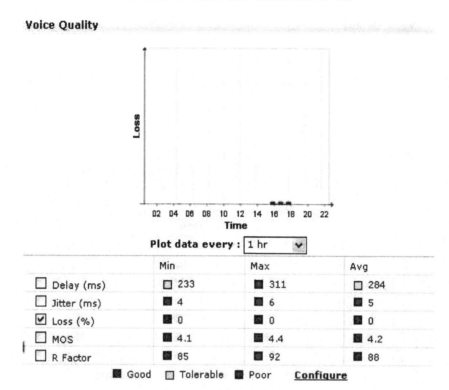

Figure 10. MOS – Audio Codec G.722 with Video Codec MP4V-ES

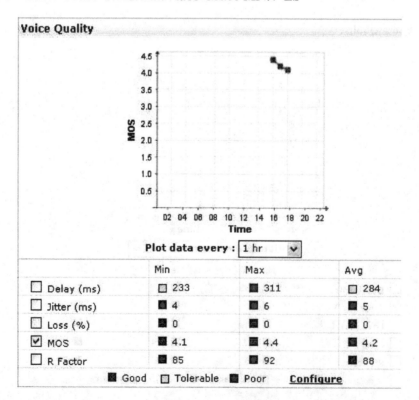

Table 1. Mean Opinion Score (MOS) ratings

	Mean Opinion Score (MOS) Ratings
Rate	5 (Perfect. Like face-to-face conversation or radio reception)
Good	4 (Fair. Imperfections can be perceived, but sound still clear. This is (supposedly) the range for cell phones)
Fair	3 (Annoying)
Poor	2 (Very annoying. Nearly impossible to communicate)
Bad	1 (Impossible to communicate)

Figure 11. Delay – Audio Codec G.726 -16 with Video Codec H.261

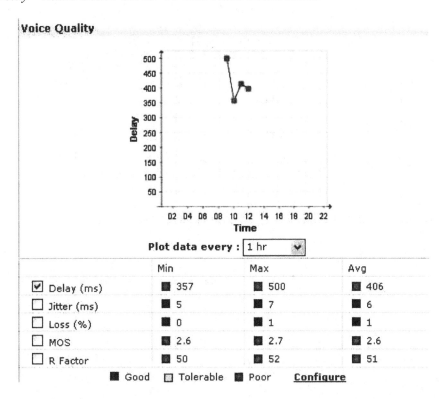

- **Overall Results:** Finally, overall results can be summarized as follows: i) Figure 19 shows the comparison of video quality through campus wireless network. The result shows that video codec MP4V-ES over VoIP transmission using G.722 has generated lower delay compare to other audio and video codec, G.726-16 with H.261 and G.726-24 with H.264. Codec G.722 with MP4V-ES able to give better video quality over VoIP; ii) Figure 20 shows a jitter occur on video transmission over VoIP.

The results also shos that video codec using G.722 with MP4V-ES has generated lower jitter compare to other audio and video codec; iii) Audio and video codec, G.726-16 with H.261 has generated higher packet loss compare to others

Figure 12. Jitter – Audio Codec G.726 -16 with Video Codec H.261

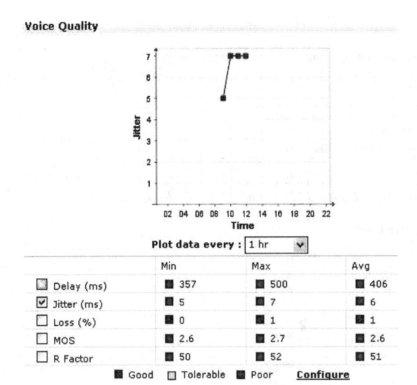

Figure 13. Packet Loss – Audio Codec G.726 -16 with Video Codec H.261

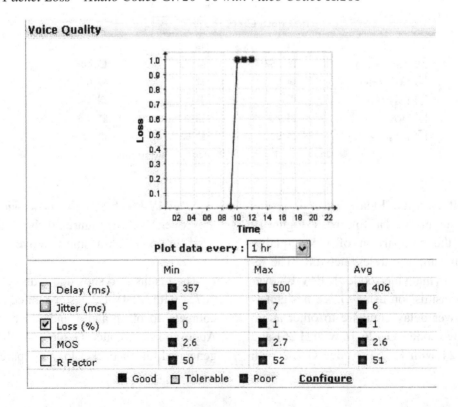

Figure 14. MOS – Audio Codec G.726 -16 with Video Codec H.261

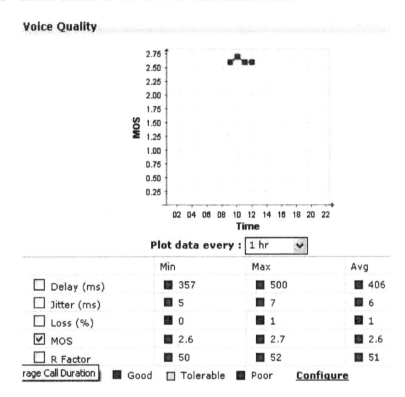

Figure 15. Delay – Audio Codec G.726 - 24 with Video Codec H.264

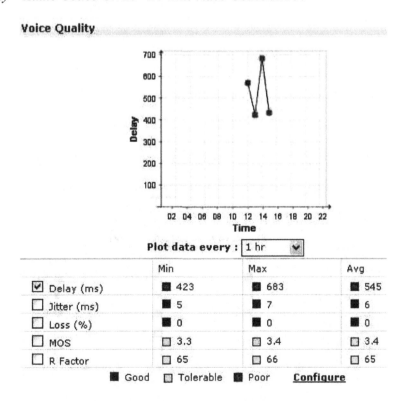

Figure 16. Jitter – Audio Codec G.726 - 24 with Video Codec H.264

Voice Quality

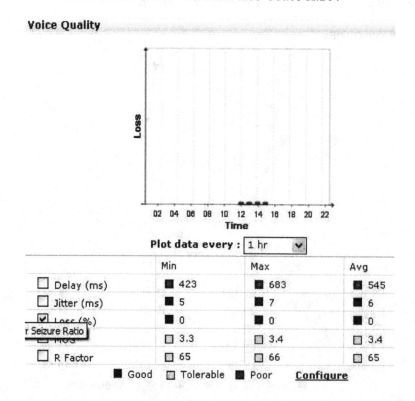

	Min	Max	Avg
☐ Delay (ms)	■ 423	■ 683	■ 545
☑ Jitter (ms)	■ 5	■ 7	■ 6
☐ Loss (%)	■ 0	■ 0	■ 0
☐ MOS	☐ 3.3	☐ 3.4	☐ 3.4
☐ R Factor	☐ 65	☐ 66	☐ 65

■ Good   ☐ Tolerable   ■ Poor   **Configure**

Figure 17. Packet loss – Audio Codec G.726 - 24 with Video Codec H.264

Voice Quality

	Min	Max	Avg
☐ Delay (ms)	■ 423	■ 683	■ 545
☐ Jitter (ms)	■ 5	■ 7	■ 6
☑ Loss (%)	■ 0	■ 0	■ 0
r Seizure Ratio			
MOS	☐ 3.3	☐ 3.4	☐ 3.4
☐ R Factor	☐ 65	☐ 66	☐ 65

■ Good   ☐ Tolerable   ■ Poor   **Configure**

Figure 18. MOS – Audio Codec G.726 - 24 with Video Codec H.264

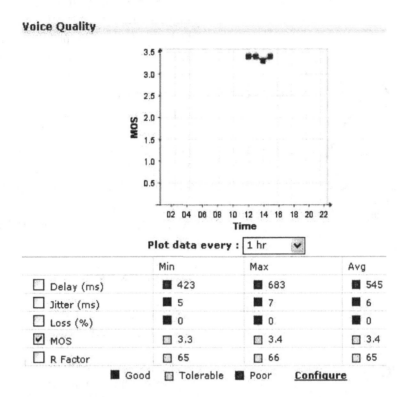

Figure 19. Audio and video codec - delay comparison

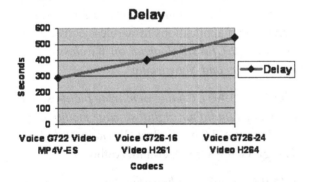

Figure 20. Audio and video codec - jitter comparison

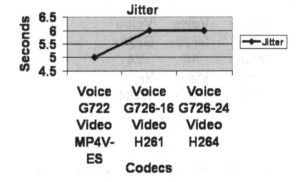

but this packet loss in minimum value (refer to Figure 21); iv) Automated MOS is used to confirm that video codec using G.722 with MP4V-ES able to provide a good video quality over VoIP. The result shows that G.722 with MP4V-ES achieves better performance compare to G.726-24 with H.264 and G.726-16 with H.261 (refer to Figure 22).

CONCLUSION AND FUTURE WORK

Based on the results, audio and video codec G.726-24 with H.264 and G.726-16 with H.261 are not able to generate higher video quality over VoIP conversation. The result shows that G.722 with MP4V-ES codec is able to achieve a good

Figure 21. Audio and video codec – packet loss comparison

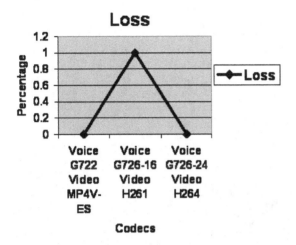

Figure 22. Audio and video codec – mos comparison

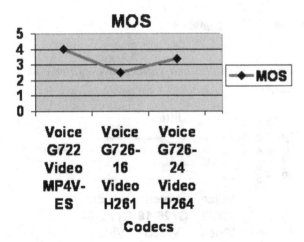

video quality performance over VoIP using campus wireless network. There are several factors that can affect and degrade transmission of audio and video quality over campus wireless network such as: a) queuing delay; b) serialization delay; c) propagation delay; d) transmission delay; and e) codec delay.

Future work for this study can extend to the implementation of VoIP over mesh wireless technology in a campus network environment. Also, we will propose an architectural solution to implement voice over IP services over IPv6 in campus environment network. IPv6 is considered to be the next-generation Internet protocol.

There are several techniques should be studied and analyzed in order to increase performance of VoIP over IPv6 in campus environment in future as follows: i) dejitter buffer; ii) Type of Service (ToS); iii) Weighted Fair Queuing (WFQ); and iv) Random Early Detection (RED).

REFERENCES

Alcatel. (2005). *Open IMS solutions for innovative applications*. Paris, France: Author.

Cole, R. G., & Rosenbluth, J. H. (2001). Voice over IP performance monitoring. *ACM SIGCOMM Computer Communication Review, 31*(2), 9–24. doi:10.1145/505666.505669

Fernandez-Duran, A., Perez Leal, R., & Alonso, J. I. (2008). Dimensioning method for conversational video applications in wireless convergent networks. *EURASIP Journal on Wireless Communications and Networking.*

ISO/IEC. (1993). *Coding of moving pictures and associated audio for digital storage media at up to about 1.5 Mbit/s – Part 2: Video (ISO/IEC 11172)*. Geneva, Switzerland: Author.

ISO/IEC. (1999). *Coding of audio-visual objects – Part 2: Visual (ISO/IEC 14496-2)*. Geneva, Switzerland: Author.

ITU-T. (1993). *Video codec for audiovisual services at px64 kbits/s*. Geneva, Switzerland: International Telecommunications Union.

ITU-T. (2000). *Video coding for low bit rate communication*. Geneva, Switzerland: International Telecommunications Union.

ITU-T and ISO/IEC. (1994). *Generic coding of moving pictures and associated audio information – Part 2: Video (ISO/IEC 13818-2)*. Geneva, Switzerland: International Telecommunications Union.

Masuda, M., & Ori, K. (2001). Delay Variation Metrics for Speech Quality Estimation of VoIP. *IEIC Technical Report, 101*(11), 101–106.

Moura, N. T., Vianna, B. A., Albuquerque, C. V. N., Rebello, V. E. F., & Boeres, C. (2007). MOS-Based Rate Adaption for VoIP Sources. In *Proceedings of the IEEE International Conference on Communication* (pp. 628-633).

Narbutt, M., & Davis, M. (2006). Gauging VoIP Call Quality from 802.11 WLAN. In *Proceedings of the 2006 International Symposium on World of Wireless, Mobile and Multimedia Networks* (pp. 315-324).

Pérez, L., & Fernández, P. C. (2006). Aplicaciones innovadorasen el entorno IMS/TISPAN. In *Proceedings of the Telecom I+D Conference*, Madrid, Spain (p. 9).

Stuedi, P., & Alonso, G. (2007). Wireless ad hoc VoIP. In *Proceedings of the 2007 Workshop on Middleware for Next-Generation Converged Networks and Applications*, Newport Beach, CA (p. 8).

Sullivan, G. J., Topiwala, P., & Luthra, A. (2004, August). The H.264/AVC Advanced Video Coding Standard: Overview and Introduction to the Fidelity Range Extensions. In *Proceedings of the SPIE Conference on Applications of Digital Image Processing XXVII Special Session on Advances in the New Emerging Standard: H.264/AVC* (pp. 1-24).

Wang, J., & Wu, Q. (2008). Porting VoIP applications to DCCP. In *Proceedings of the International Conference on Mobile Technology, Applications, And Systems*, Yilan, Taiwan (p. 8).

This work was previously published in the International Journal of Interdisciplinary Telecommunications and Networking, Volume 3, Issue 1, edited by Michael R. Bartolacci and Steven R. Powell, pp. 36-49, copyright 2011 by IGI Publishing (an imprint of IGI Global).

Chapter 4
Fuzzy QoS Based OLSR Network

G. Uma Maheswari
Vellore Institute of Technology, India

ABSTRACT

Quality-of-Service (QoS) routing protocol is developed for mobile Ad Hoc Networks. MANET is a self configuring network of mobile devices connected by wireless links. Each device in the MANET is free to move independently in any direction; therefore, it changes links to other devices frequently. The proposed QoS-based routing in the Optimized Link State Routing (OLSR) protocol relates bandwidth and delay using a fuzzy logic algorithm. The path computations are examined and the reason behind the selection of bandwidth and delay metrics is discussed. The performance of the protocol is investigated by simulation. The results in FQOLSR indicate an improvement in mobile wireless networks compared with the existing QOLSR system.

INTRODUCTION

The routing protocols for the mobile Ad hoc networks (MANETs) (IETF, n.d.), such as OLSR (Tan, 2001), AODV (Clausen & Banerjee, 2003), DSR (Perkins, Royer, & Das, 2002), are designed without explicitly considering QoS of the routes they generate. Routing is primarily concerned with connectivity. Routing protocols such as OSLR usually characterize the network with a single metric such as hop-count or delay, and use the shortest–path algorithms for path computation. The use of single metric is that it can only be used for satisfying one criteria either maximize throughput or minimize delay. Hence the use of a mixed metric that can be generated based on

multiple primitive metrics becomes attractive for the basis of possible improved routing decisions. QoS routing requires not only finding a route from a source to a destination, but a route that satisfies the end-to-end QoS requirement. The QoS requirement chosen are distinctively different routing metric and neither of these metrics is inferable from each other. These metrics should use separate rules for defining the best route (Tan, 2001). The value of a metric over any directed path $p = (i, j, k...q, r)$ can be any one of the following compositions.

- **Additive Metric:** Metric is additive if metric $(p) = met_{ij} + metjk +...+met_{qr}$. Delay, delay jitter, hop-count and cost follow the additive composition rule.

DOI: 10.4018/978-1-4666-2154-1.ch004

- **Multiplicative Metric:** Metric is multiplicative if Metric (p) = met_{ij} X met_{jk} X...X met_{qr}. The probability of successful transmission follows the multiplicative composition rule
- **Concave Metric:** Metric is concave if metric (p)=min {met_{ij}, met_{jk},..,met_{qr}}. Bandwidth follows the concave composition rule.

The use of fuzzy logic greatly simplifies the process of associating two inputs through the use of the straightforward membership function and the linguistic types of fuzzy rules make the performance fine-tuning process. The output generated is able to give a good approximation on what is to be expected.

Qos is more difficult in Ad hoc networks than in other networks, because network topology changes as the nodes move. This paper aims at specifying in FQOLSR in Mobile Ad Hoc Networks. The implementation is done in limited available resources.

The author has proposed the FQOLSR protocol, which is an enhancement of the QOLSR routing protocol prescribed in Johnson and Maltz (1996). This paper is organized as follows. The next section presents the network parameters chosen. Existing system method to relate two metrics is presented, as well as the design of fuzzy logic algorithm. The author also describes FQOLSR protocol, which is an enhancement of the QOLSR routing protocol to support multiple-metric routing criteria. Therefore, the author validates the proposal by means of performance evaluation. Finally, the author presents the conclusion.

QOS PARAMETERS

Bandwidth, delay, delay jitter, queue and loss probability are among the network parameters that are qualified to be used in the computation of a single mixed metric. It is possible that all these parameters are used at once, but it will make the computation more complex. It is unnecessary due to the redundant information present between two (or more) of the network parameters (Tan, 2001).

Delay is a critical element in QoS for many applications. Some applications, like Voice Over IP (VOIP) and video conferencing, indeed have specific delay bound where quality is considered intolerable when this limit is exceeded. By taking into account, delay becomes an important starting point to consider. Link delay is an additive metric that can be used to estimate the delay along a particular route when summed together. However, the delay along a particular route is lower when compared to others does not necessarily mean that this particular route is better or have more available bandwidth than other routes (Tan, 2001).

In order to improve the accuracy of the measured delay, one more parameter is used to generate the single mixed metric. Bandwidth available on a link, considered as an important resource required by most applications in a quest to realize better QoS. The lower the link utilization, the more likely that packets can be transmitted with low queuing delay, even if the packet arrival rate might subject to large variation (Tan, 2001). In a wide-ranging traffic characteristic suggests that it is probably a good idea to associate two correlated metrics to produce a metric that can better describe the link state. An important point to note here is that the measured delay and utilization are not synchronous to each other (Tan, 2001).

Delay peaks after link utilization. Hence link utilization also has the advantage to be used as an indicator for the upcoming delay.

EXISTING SYSTEM

OLSR protocol is a routing protocol for mobile ad hoc networks. The routing table calculation followed in the existing systems is, in order to improve quality requirements in routing information; more than one metric is considered (Badis, Munaretto, Agha, & Pujolle, 2003).

The approach treats each metric individually. Using a single metric, the best path can be easily defined. While considering multiple metrics, the best path with all parameters at their optimal values will not necessarily exist. Hence precedence is defined between bandwidth and delay. Since queuing delay is more dynamic, bandwidth is considered as more important. The strategy is to find a path with maximum bandwidth (a widest path), and when there is more than one widest path, choose the one with the shortest delay. Such a path is referred to as shortest-widest path. It is solved with shortest path algorithm such as Dijkstra routing algorithm or Bellman-Ford Algorithm. It calculates the average delay using the method RTT. The average delay (Badis et al., 2003) is calculated using the formula:

Average delay = α x Average delay + $(1-\alpha)$ x measured delay

The available bandwidth (Badis et al., 2003) is calculated using the formula:

Bw $(i,j) = (1-u)$ x Throughput (i,j)

where $\alpha=0.4$, u is the utilization, and i,j refers to nodes.

PROPOSED SYSTEM

The delay and bandwidth metrics are taken into account as QoS constraints. In this paper the two metrics are related using Fuzzy Logic algorithm to produce fuzzy metric. Fuzzy systems theory or "fuzzy logic" is a linguistic theory that models how the author reason with vague rules of thumb and commonsense (Kaehler, 1998). The basic unit of fuzzy function approximation is *"If then"* rules. A fuzzy system is a set of if- then rules that maps input to output.

Figure 1. Generalized fuzzy system

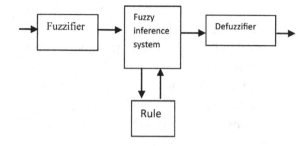

The steps involved in the Fuzzy inference system design are as follows (Figure 1).

Step 1: Fuzzy Inputs

This step will obtain inputs and determine the degree to which they belong to each of the appropriate fuzzy sets via membership functions. Fuzzification of the input amounts to either a table lookup or a function evaluation.

Step 2: Apply Fuzzy Operator

This step determines the degree to which each part of the antecedent has been satisfied for each rule. If the antecedent of a given rule has more than one part, the fuzzy operator is applied to obtain one number that represents the result of the antecedent for that rule. This number will then be applied to the output function. The input to the fuzzy operator is two or more membership values from fuzzified input variables. The output is a single truth-value. The method used may be either AND or OR operation.

Step 3: Apply Implication Method

Before applying implication proper weights are assigned to each rule. The input for the implication process is a single number given by the antecedent, and the output is a fuzzy set.

Step 4: Aggregate all outputs

Aggregation is the process by which the fuzzy sets that represent the outputs of each rule is combined into a single fuzzy set. Aggregation only occurs once for each output variable, prior to the final step, defuzzification. The input of the aggregation process is the list of truncated output functions returned by the implication process for each rule. The output of the aggregation process is one fuzzy set for each output variable.

Step 5: Defuzzify

The input for the defuzzification process is a fuzzy set and the output is a single number. The aggregate of a fuzzy set encompasses a range of output values, and so must be defuzzified in order to resolve a single output value from the set.

FUZZY INFERENCE SYSTEM

The fuzzy system with two inputs and one output. The system inputs are delay and bandwidth. The both inputs are characterized by the fuzzy membership function. The simulation of the Fuzzy inference system is to be done using NS-2 and the values are obtained.

Fuzzy inference is the process of formulating the mapping from a given input to an output using fuzzy logic. The mapping then provides a basis from which decisions can be made. The process of fuzzy inference involves membership functions (Figure 3 and Figure 4), fuzzy logic operators, and if-then rules (Figure 2). The output membership function is being a spike (Klis & Yuan, 2001) and it is known as a singleton function. The fuzzy operator used for the AND method is "min" and the operator used for the OR method is "max" (Brule, 1985).

The implication function (Horstkotte, 2000) modifies that fuzzy set to the degree specified by

Figure 2. Fuzzy rules

	Link Utilisation		
	Low	**Medium**	**High**
Low	Fuzzy Rule 1	Fuzzy Rule 2	Fuzzy Rule 3
Medium	Fuzzy Rule 4	Fuzzy Rule 5	Fuzzy Rule 6
High	Fuzzy Rule 7	Fuzzy Rule 8	Fuzzy Rule 9

(**Link Delay** labels the rows)

Figure 3. Fuzzification for link delay

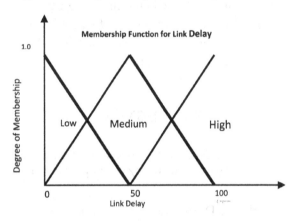

Figure 4. Fuzzification for link utilization

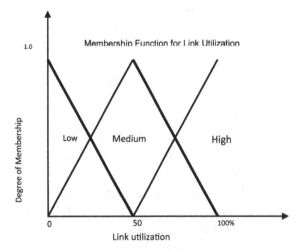

the antecedent. The implication function used in this work is "min" which simply truncates the output member function. The aggregation function used is probabilistic OR which is used to aggregate the output fuzzy sets to form a single fuzzy set.

The Defuzzification (Horstkotte, 2000) is the process of conversion of fuzzy output set into a single number. The method used for the defuzzification is, smallest of minimum. The input membership functions are delay and bandwidth. The shape of the membership function is triangular (Horstkotte, 2000). The output membership function is a fuzzy metric, and they are also designed as triangular. The rules play an important role in the fuzzy logic design (Horstkotte, 2000). The rules for fuzzy inference system in Figure 2 are made on the basis of the inputs delay and the bandwidth.

FUZZY RULE STRUCTURE

1. IF Link Delay is LOW AND Link Utilisation is LOW THEN FUZZY Metric is VERY LOW
2. IF Link Delay is LOW AND Link Utilisation is MEDIUM THEN Fuzzy Metric is LOW
3. IF Link Delay is LOW AND Link Utilisation is HIGH THEN Fuzzy Metric is MEDIUM
4. IF Link Delay is MEDIUM AND Link Utilisation is LOW THEN Fuzzy Metric is LOW
5. IF Link Delay is MEDIUM AND Link Utilisation is MEDIUM Fuzzy Metric is MEDIUM
6. IF Link Delay is MEDIUM AND Link Utilisation is HIGH THEN Fuzzy Metric is HIGH
7. IF Link Delay is HIGH AND Link Utilisation is LOW THEN Fuzzy Metric is MEDIUM
8. IF Link Delay is HIGH AND Link Utilisation is MEDIUM THEN Fuzzy Metric is HIGH
9. IF Link Delay is HIGH AND Link Utilisation is HIGH THEN Fuzzy Metric is VERY HIGH

PERFORMANCE ANALYSIS

In the simulation 50 mobile nodes move within the region of size $1000^2 m^2$. All the 50 nodes are packet-generating sources. Only 20% of the nodes are mobile and the rest are stationary. Each mobile node selects its speed and direction, which remains valid for next 60 seconds. As the mobility increases, the number of packets, delivered to the destination decreases. QOLSR with bandwidth and delay as metrics has the highest packets delivered because routes are chosen with minimal interferences. When the mobility is 500 meters/minute, 88% of packets delivered for QOLSR with delay and bandwidth (Badis et al., 2003) (considered as individual metrics). The performance is expected to improve more than 88% when the metrics are related using fuzzy algorithm.

CONCLUSION

This paper presents a Fuzzy Link state QoS routing protocol for Ad hoc networks. The incentive of using this fuzzy metric is for obtaining 'better" routing decision. Real world networks have limited resources and resource bottlenecks, and need QoS policies to ensure proper resource allocation. The proposal implements the QoS functionality to deal with limited available resources in a dynamic environment. In order to analyze the performance of the proposed QoS-based routing, the author changes the existing algorithm in (QOLSR) protocol. The evaluation performance to be estimated by the proposed QoS-based routing is expected to show improvements with the existing system. The achieved gain by our proposal can be an important improvement in such mobile wireless networks

REFERENCES

Badis, H., Munaretto, A., Agha, K. A., & Pujolle, G. (2003). QoS for Ad hoc Networking Based on Multiple Metrics: Bandwidth and Delay. In *Proceedings of the 5th IFIP-TC6 International Conference on Mobile and Wireless Communication Networks* (pp. 15-18).

Brule, J. F. (1985). *Fuzzy Systems – A Tutorial.* Retrieved from http://www.ortech-engr.com/fuzzy/tutor.txt

Clausen, T., & Banerjee, P. J. (2003). *Optimized Link State Routing Protocol.* Retrieved from http://www.ietf.org/rfc/rfc3626.txt

Horstkotte, E. (2000). *Fuzzy Logic Overview.* Retrieved fromhttp://www.austinlinks.com/Fuzzy/overview.html

IETF. (n.d.). Mobile. *Ad Hoc Networks.* Retrieved from http://www.ietf.org/html.charters/manet-charter.html.

Information Sciences Institute. (2010). *Ns Manual.* Retrieved from http://www.isi.edu/nsnam/ns/ns/documentation.html

Johnson, D., & Maltz, D. A. (1996). Dynamic Source Routing in Ad Hoc Wireless Networks. In *Mobile Computing* (pp. 153–181). Dordrecht, The Netherlands: Kluwer Academic Publishers. doi:10.1007/978-0-585-29603-6_5

Kaehler, S. D. (1998). *Fuzzy Logic – An Introduction.* Retrieved from http://www.seattlerobotics.org/encoder/mar98/fuz/fl_part1.html

Klis, G. J., & Yuan, B. O. (2001). *Fuzzy sets and Fuzzy Logic Theory and applications.* Delhi, India: Prentice Hall India.

Perkins, C., Royer, E. M., & Das, S. R. (2002). *Ad Hoc On - Demand Distance Vector routing.* Retrieved from http://www.ietf.org/rfc/rfc3561.txt

Tan, C. K. (2001). *The Use of Fuzzy Metric in QoS Based OSPF Network.* Retrieved from http://www.ee.ucl.ac.uk/lcs/previous/LCS2001/LCS054.pdf

This work was previously published in the International Journal of Interdisciplinary Telecommunications and Networking, Volume 3, Issue 1, edited by Michael R. Bartolacci and Steven R. Powell, pp. 50-55, copyright 2011 by IGI Publishing (an imprint of IGI Global).

Chapter 5
Development of a Complex Geospatial/RF Design Model in Support of Service Volume Engineering Design

Erton S. Boci
ITT Information Systems, USA

Shahram Sarkani
The George Washington University, USA

Thomas A. Mazzuchi
The George Washington University, USA

ABSTRACT

Today's National Airspace System (NAS) is managed using an aging surveillance radar system. The radar technology is not adequate to sustain the support of aviation growth and cannot be adapted to use 21st century technologies Therefore, FAA has begun to implement the Next Generation Air Transportation System (NextGen) that would transform today's aviation and ensure increased safety and capacity. The first building block of the NextGen system is the implementation of the Automatic Dependent Surveillance-Broadcast (ADS-B) system. One of the most important design aspects of the ADS-B program is the design of the terrestrial radio station infrastructure. This design determines the layout of the terrestrial radio stations throughout the United States and is optimized to meet system performance, safety and security. Designing this infrastructure to meet system requirements is at the core of Service Volume (SV) Engineering. In this paper, the authors present a complex Geospatial/RF design ER Development model-based ADS-B SV Engineering design that captures radio sites layout and configuration parameters. CORE software is selected to implement the model-based SV Engineering environment.

DOI: 10.4018/978-1-4666-2154-1.ch005

INTRODUCTION

Today's National Airspace System (NAS) is managed using an aging surveillance radar system. The radar technology is not adequate to sustain the support of the aviation growth and cannot be adapted to use the 21st century technologies Therefore, FAA has undertaken the implementation effort of the Next Generation Air Transportation System (NextGen) that would transform today's aviation and ensure increased safety and capacity in our NAS.

The first building block of the NextGen system is the implementation of the Automatic Dependent Surveillance-Broadcast (ADS-B) system. One of the most important design aspects of the ADS-B program is the design of the terrestrial radio station infrastructure. This design must determine the layout of the terrestrial radio stations throughout the US optimized to meet system performance, safety and security. Enabled by the Global Positioning System (GPS) satellite system and a nationwide radio stations terrestrial infrastructure, ADS-B will enhance surveillance capabilities and improve aviation safety and capacity in US (FAA, 2008, 2009, 2010). ADS-B services in the United States are based upon two non-interoperable data link technologies (Bruno & Dyer, 2008):

- 1090 MHz Extended Squitter (1090ES) this data link is applicable primarily to commercial aviation aircraft.
- 978 MHz Universal Access Transceiver (UAT): this data link operates at 978 MHz and is applicable primarily to general aviation aircraft.

These two data link technologies will enable the following four ADS-B services that are required within NAS.

- **ADS-B:** Surveillance of 1090ES and UAT aircraft to FAA Air Traffic Control is a service that receives position broadcasts from ADS-B equipped aircraft and distributes this information to ATC automation systems for providing separation assurance and traffic flow management.
- **ADS-R:** ADS-B Rebroadcast is a service that receives ADS-B position broadcasts, and rebroadcasts the same information to near-by aircraft that are equipped with a different ADS-B data link.
- **TIS-B:** Traffic Information Services Broadcast is a surveillance service that derives traffic information from radar/sensor sources and uplinks this traffic information to ADS-B equipped aircraft.
- **FIS-B:** Flight Information Service Broadcast is an uplink service that provides aeronautical and flight information such as textual and graphical weather reports and Notice to Airmen (NOTAM) (Gilbert & Bruno, 2009).

The SV Engineering RF design includes the layout of the terrestrial radio stations throughout the US. This layout is optimized to meet system performance, safety, and security at a minimum cost. The ADS-B RF design and optimization approach is given in (Boci, Sarkani, & Mazzuchi, 2009). The SV Engineering design effort culminates with the generation of the final SV design data package that is used for implementation and configuration of the SV based on the final radio station layout and configuration to include but not limited to radio station location, radio station's antenna system and radio channel assignments to name a few.

A MODEL-CENTRIC APPROACH TO SV DESIGN

One of the SV design team objectives was the acquisition of a software environment that would

Figure 1. Required data exchange between SV Engineering and other design teams (a document-centric approach)

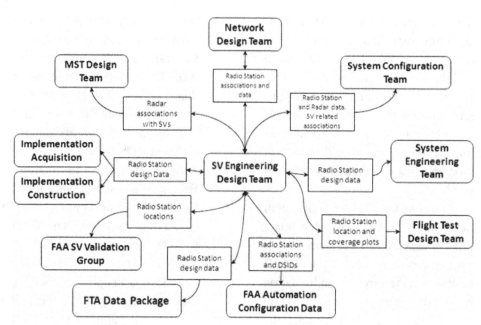

be used to capture the SV design data. There are several factors that influence the decision process for selection of such an environment. Consistent with ISO 42010 (Emery & Hilliard, 2009), the most important factors in this determination are the concerns of stakeholders. In addition, several engineering teams within the ADS-B project have a need for SV Engineering data. The potential users of the data captured from the SV Engineering design are illustrated in Figure 1.

It must be noted that the SV Engineering design team has to maintain several design data packages to accommodate the needs of different design teams. A model-centric approach (Bayer et al., 2010) was adopted to capture and maintain SV design and configuration data.

The model-centric approach aligns with INCOSE "System Engineering Vision 2020" (INCOSE, 2007). The goal is to develop a model-centric tool that would enable SV Engineering to capture, manage and distribute its data. It must be noted that the SV Engineering design team has

to maintain several design data packages to accommodate the needs of different design teams. A model-centric approach (Bayer et al., 2010) was adopted to capture and maintain SV design and configuration data (see Figure 2).

We selected CORE Software (Vitech, 2010) as the environment for capturing the SV Engineering ADS-B requirements and design. CORE software is a collaborative environment with a single database repository. It provides an entity-relationship-attribute (ER) environment and the ability to develop and maintain a database schema where the user can define custom entities, attributes and relationships (Fisher, 1998; Herzog & Torne, 2000; Kordon et al., 2007; Merida & Saha, 2005). CORE software already provides an integrated schema in support of system engineering and it utilizes a model-based systems engineering (MBSE) approach (Estefan, 2007) in support of system requirements, design, analysis, verification, and validation activities. Due to its adaptive modeling schema (Kordon et al., 2007; Fisher,

Figure 2. Geospatial / RF design ER data model facilitates the design data exchange (a model-centric approach)

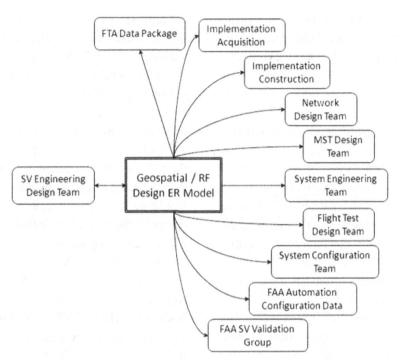

1998), CORE software provides a powerful ER data modeling environment for capturing the SV Engineering design and configuration data.

The concept of ER data modeling was first introduced by Chen (Chen, 1976). The ER model, according to Chen, adopts a more natural view of the real world. In a physical system one can identify with "ease" the entities and relationships governing the system and use ER modeling techniques to organize and capture data. Chen also provides a detailed description on the use of the English sentence structure as applied to ER model development (Chen, 1983).

...DATA MODELING refers to the process of transforming expressions in loose natural language communications to formal diagrammatic or tabular expressions; it may be considered a "vibrant, central element of information systems development and implementation work" (Teo, Chan, & Wei, 2006)

Since its invention, ER data modeling has become the most popular conceptual modeling methodology in the database community (Chaomei, Song, & Zhu, 2007). Furthermore, the development of such Geospatial / RF design ER model aligns with Spatially-aware systems engineering design modeling as described by Eveleigh et a. (Eveleigh, Mazzuchi, & Sarkani, 2007).

Geospatial / RF Design ER Development

The first step in ER modeling is identifying the entities, relationships, and attributes that will be captured. It is important to follow a methodology and/or process when designing such data model. Below we show the design methodology proposed in (Storey, 1991).

It can be observed that, ER design is an iterative process that tries to refine the selection of entities and relationships that best match the system that it represents. The model-based Geospatial / RF design ER can be partitioned into an SV requirements ER and an SV Engineering design ER. The SV requirements ER model captures entities provided in requirements such as service volumes, victim receivers etc. The SV Engineering design ER identifies entities such as radios, antennas, data source identifiers (DSID) assigned to each data stream etc. The identification of the entities and relationships can be carried out in parallel for each ER sub-model (Figure 3). The next step is the identification of the relationships that connect SV requirements entities with SV design ER entities.

In Figure 4, we present the SV Engineering data model for all entities and their relationships.

The process of developing such schema starts by identifying the entities of the ER model. The most important entities are the "Service Volume" and "Radio Station". Strain defines a SV as "an airspace volume within which a specific broadcast services application is supported and the associated performance requirements are achieved" (Strain, 2007)

In the NAS, there are three types of SV domain: En Route, Terminal and Surface. The En Route SV represents a large section of airspace stretching across several US states. Terminal SVs have the shape of a cylinder with the center typically at the terminal radar serving the SV and a radius of 60nmi. The floor for both these SVs is defined by the coverage provided by the radars serving these SVs. The ceiling for these SVs range from 24,000ft MSL to 60,000ft MSL depending on the SV domain and broadcast service provided. A large number of SV instances, under design, would need to be captured in the data model to represent the entire US air traffic control system. There is a total of 40 En Route SVs, 236 Terminal SVs and 35 Surface SVs (Broderick, 2008).

The estimated number of ADS-B radio stations required to provide compliant coverage throughout NAS is 800 (Boci, 2009). The ER model schema shown in Figure 7 will be used to capture all the instances of these radio stations and the configuration parameters.

Figure 3. Geospatial / RF design ER design methodology

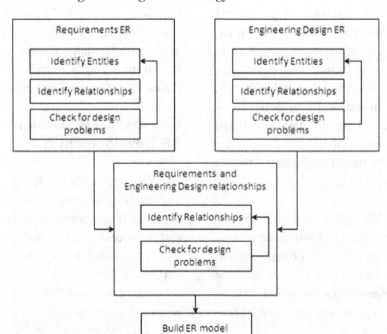

Figure 4. Geospatial / RF design ER model schema

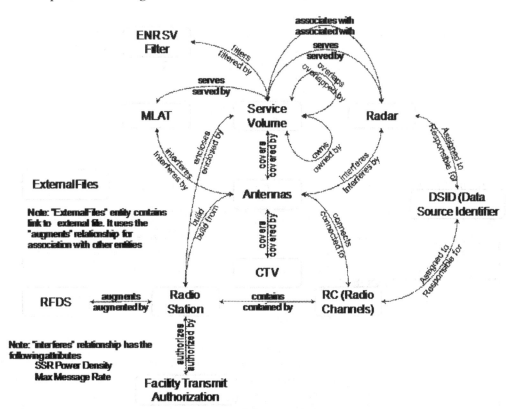

The next step was the identification of a relationship that would associate the entities. For example, the relationship named "covers" was used to capture the association between the "Antenna" entity and "Service Volume" entity. Similarly, the relationship named "build" associates the "Antenna" entity with "Radio Station" entity.

Figure 5 shows the entity-relationship diagram for the "Radio Station" entity. Figure 6 shows the entity-relationship diagram for the "Antenna" entity.

A typical ADS-B radio station infrastructure consists of 1090MHz directional antennas, 978MHz Omni antenna, and one 1030MHz di-

Figure 5. "Radio Station" entity-relationship diagram

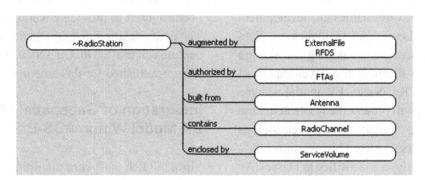

Figure 6. "Antenna" entity-relationship diagram

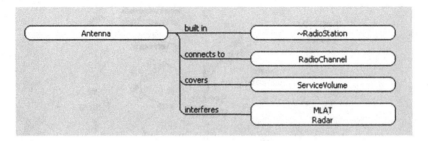

Figure 7. "1090 Ant 3 SV028-03" Instance in SV Engineering CORE data model

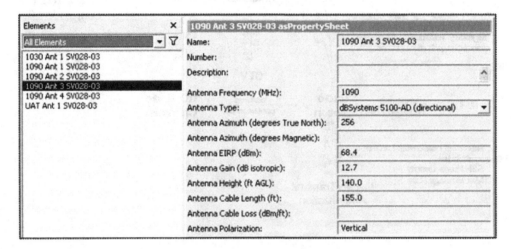

rectional antenna. Consequently, for a given instance of the "Antenna" entity an attribute will be used to distinguish among the antennas.

Implementation Using CORE Software

Vitech's CORE software was selected to implement the SV Engineering data model described above. CORE software enables the development of custom ER schemas and facilitates a collaborative environment for developing and sharing such models. To illustrate the development of our Geospatial / RF design ER model schema using CORE we will present the implementation of "Antenna" and "Service Volume" entities, as well as the "covers / covered by" relationship.

In CORE, entities are identified as classes. An instance of a class represents a specific element which we want to capture in our Geospatial / RF design ER data model. For example, the "1090 Ant 3 SV028-03" instance of class "Antenna" represents the third 1090MHz antenna installed on the SV028-03 radio station.

Figure 7 shows the "1090 Ant 3 SV028-03" instance as well as several attributes of this antenna. The following hierarchy diagram (see Figure 8) shows that this radio station's antenna is configured to provide coverage for four SVs. The entity-relation diagram depicted in Figure 9, captures all the "1090 Ant 3 SV028-03" relationships established for this antenna instance.

Integration of Geospatial / RF Design ER Model Within ADS-B System

Since CORE software facilitates incremental schema development, initially, we created only

Figure 8. Antenna "covers" hierarchy diagram for "1090 Ant 3 SV028-03"

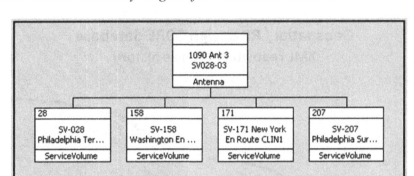

Figure 9. "1090 Ant 3 SV028-03" entity-relationship diagram

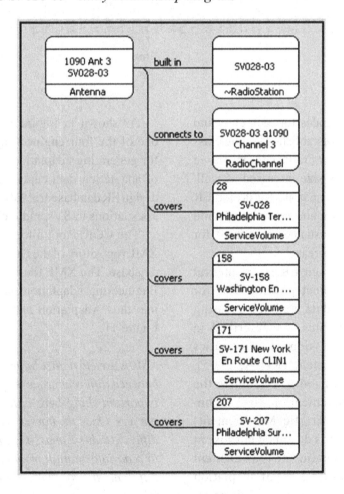

one entity, "Radio Station", in the Geospatial / RF design ER model. With time, Geospatial / RF design ER model evolved with more instances and schema enhancements. In addition, it gained recognition within the ADS-B project as the sole source of SV Engineering data. Figure 10 shows the file size of the Extended Markup Language (XML) file repository which was regularly generated as a backup for the Geospatial / RF design ER database. The repository XML file size depends on project

Figure 10. XML Repository file size history

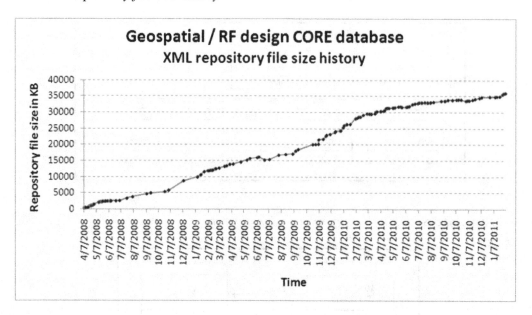

schema, number of instances for each class and number of relationships established among class instances. As the number of radio sites increased the XML file repository size increased as well. The enhancement of Geospatial / RF design ER schema with new classes and relationships and the generation of new instances accounted for the repository file size increase.

Geospatial / RF design ER is an integral part of the ADS-B project process flow for capturing adaptation data and for creating system configuration files. As described in Radio Technical Commission for Aeronautics (RTCA) DO-278 document (RTCA, 2005), "adaptation data is utilized to customize the elements of the [Communication, Navigation, Surveillance, and Air Traffic Management] CNS/ATM system for it's designed purpose at a specific location." A radio site is an element of ADS-B system as such a set of adaptation data must be used to configure it. Geospatial / RF design ER database provides a mechanism for capturing, maintaining and generating adaptation data. The inclusion of the Geospatial / RF design ER into the adaptation data generation process indicates its importance.

As shown in Figure 11, SV Engineering is one of the four engineering groups responsible for generating adaptation data. Some examples of adaptation data captured in Geospatial / RF design ER database are: SV attributes, radio station associations to SVs, radio channel attributes etc.

The CORE tool allows the generation of an XML repository of the Geospatial / RF design ER database. The XML file is used to extract the SV Engineering adaptation data which are entered into the "Adaptation Database" as indicated in Figure 11.

XML is used to realize information communication between different process operation systems. The important thing here is to develop appropriate parsers. Once the parser is developed and put on different sub-systems, it is easy for the sub-system to read information from other sub-system. For different sub-systems, they can communicate without knowing the data style and data model. They only need to act as a messenger to send information (Wu, Tao, & Qian, 2010)

We will continue the development of the Geospatial / RF design ER model and instantiation of

Figure 11. The adaptation data generation process for ADS-B system configuration

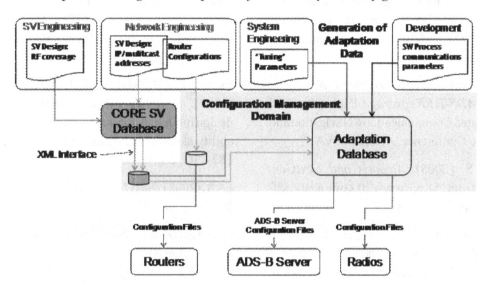

new elements and relationships as additional radio stations are designed in support of new SVs. We estimate that the final XML repository size of the Geospatial / RF design ER database to be three times its current size.

CONCLUSION

In this paper, we presented the development of a complex Geospatial / RF design ER data model in support of ADS-B SV Engineering. CORE software was used to implement this data model. This model is a strong tool that not only uniquely captures and preserves the data integrity of the ADS-B SV Engineering design, but allows an efficient use and distribution of data across other ADS-B design teams.

The successful integration of this complex Geospatial / RF design model with the ADS-B system configuration process is an important milestone achievement of the overall ADS-B SV Engineering approach. It emphasis the benefits realized by following a data-centric approach in managing the ever increasing complexity of ADS-B SV design.

Further work includes the extension of the SV Engineering ER model with new entities and relationships, as well as the integration of this model with Google Earth application in an effort to facilitate and enhance the visualization of captured data.

ACKNOWLEDGMENT

This paper is based on work leading to a dissertation submitted to The George Washington University, Washington, D.C. in partial fulfillment of the Doctor of Philosophy.

REFERENCES

Bayer, T. J., Cooney, L. A., Delp, C. L., Dutenhoffer, C. A., Gostelow, R. D., Ingham, M. D., et al. (2010). An operations concept for Integrated Model-Centric Engineering at JPL. In *Proceedings of the 2010 IEEE Aerospace Conference,* Big Sky, MT (pp. 1-14).

Boci, E. (2009). RF Coverage analysis methodology as applied to ADS-B design. In *Proceedings of the 2009 IEEE Aerospace Conference,* Big Sky, MT (pp. 17).

Boci, E., Sarkani, S., & Mazzuchi, T. A. (2009). *Optimizing ADS-B RF coverage.* Paper presented at the Integrated Communications, Navigation and Surveillance Conference, Arlington, VA.

Broderick, S. (2008). *Airports and NextGen.* Retrieved from http://www.itt.com/adsb/pdf/ITT-AirportsandNextGen.pdf

Bruno, R., & Dyer, G. (2008). *Engineering a US national Automatic Dependent Surveillance - Broadcast (ADS-B) radio frequency solution.* Paper presented at the Tyrrhenian International Workshop on Digital Communications - Enhanced Surveillance of Aircraft and Vehicles, Capri, Italy.

Chaomei, C., Song, I.-Y., & Zhu, W. (2007, June 25-27). *Trends in Conceptual Modeling: Citation Analysis of the ER Conference Papers (1979-2005).* Paper presented at the Conference on the International Society for Scientometrics and Informatrics, Madrid, Spain.

Chen, P. P.-S. (1976). The Entity-Relationship Model - Toward a Unified View of Data. *ACM Transactions on Database Systems, 1*(1), 9–36. doi:10.1145/320434.320440

Chen, P. P.-S. (1983). English sentence structure and entity-relationship diagrams. *Information Sciences, 29*(2-3), 127–149. doi:10.1016/0020-0255(83)90014-2

Emery, D., & Hilliard, R. (2009, September 14-17). *Every architecture description needs a framework: Expressing architecture frameworks using ISO/IEC 42010.* Paper presented at the IEEE/IFIP European Conference on Software Architecture (WICSA/ECSA).

Estefan, J. A. (2007). *Survey of Model-Based Systems Engineering (MBSE) Methodologies.* Retrieved from http://syseng.omg.org/MBSE_Methodology_Survey_RevA.pdf

Eveleigh, T. J., Mazzuchi, T. A., & Sarkani, S. (2007). Spatially-aware systems engineering design modeling applied to natural hazard vulnerability assessment. *Systems Engineering, 10*(3), 187–202. doi:10.1002/sys.20073

FAA. (2008). *2008 NextGen Implementation Plan.* Washington, DC: Author.

FAA. (2009). *2009 NextGen Implementation Plan.* Washington, DC: Author.

FAA. (2010). *2010 NextGen Implementation Plan.* Washington, DC: Author.

Fisher, G. H. (1998). Model-based systems engineering of automotive systems. In *Proceedings of the 1998 AIAA/IEEE/SAE Digital Avionics Systems Conference,* Bellevue, WA (pp. B15/1-B15/7).

Fisher, J. (1998). *Model-Based Proposal Development.* Retrieved from http://www.vitechcorp.com/whitepapers/files/200701031635570.fisher98.pdf

Gilbert, T., & Bruno, R. (2009). *Surveillance and Broadcast Services - An effective nationwide solution.* Paper presented at the Integrated Communications, Navigation and Surveillance Conference, Arlington, VA.

Herzog, E., & Torne, A. (2000). Support for representation of functional behaviour specifications in AP-233. In *Proceedings of the IEEE International Conference and Workshop on the Engineering of Computer Based Systems (ECBS 2000),* Edinburgh, UK (pp. 351-358).

INCOSE. (2007). *Systems Engineering Vision 2020.* San Diego, CA: International Council of Systems Engineering.

Kordon, M., Wall, S., Stone, H., Blume, W., Skipper, J., Ingham, M., et al. (2007). Model-Based Engineering Design Pilots at JPL. In *Proceedings of the 2007 IEEE Aerospace Conference,* Big Sky, MT (pp. 1-20)

Merida, S. N., & Saha, R. A. (2005). An operations based systems engineering approach for large-scale systems. In *Proceedings of the 2005 IEEE Aerospace Conference,* Big Sky, MT (pp. 4239-4250).

RTCA. (2005). *DO-278, Guidelines for Communication, Navigation, Surveillance, and Air Traffic Management (CNS/ATM) Systems Software Integrity Assurance.* Washington, DC: Author.

Storey, V. C. (1991). Relational database design based on the entity-relationship model. *Data & Knowledge Engineering, 7*(1), 47–83. doi:10.1016/0169-023X(91)90033-T

Strain, R. (2007). *Surveillance and Broadcast Services Coverage.* Retrieved from http://www.faa.gov/about/office_org/headquarters_offices/ato/service_units/enroute/surveillance_broadcast/program_office_news/ind_day/media/MITRE.pdf

Teo, H. H., Chan, H. C., & Wei, K. K. (2006). Performance effects of formal modeling language differences: a combined abstraction level and construct complexity analysis. *IEEE Transactions on Professional Communication, 49*(2), 160–175. doi:10.1109/TPC.2006.875079

Vitech. (2010). *CORE Software.* Retrieved from http://www.vitechcorp.com/products/index.html

Wu, Z., Tao, J.-S., & Qian, Y. (2010). An information model for process operation system integration based on XML and STEP standard. In *Proceedings of the 2nd International Conference on Computer Engineering and Technology (ICCET),* Chengdu, China (pp. V4-407-V4-410)

This work was previously published in the International Journal of Interdisciplinary Telecommunications and Networking, Volume 3, Issue 1, edited by Michael R. Bartolacci and Steven R. Powell, pp. 56-67, copyright 2011 by IGI Publishing (an imprint of IGI Global).

Chapter 6

Location Management in PCS Networks Using Base Areas (BAs) and 2 Level Paging (2LP) Schemes

Hesham A. Ali
Mansoura University, Egypt

Ahmed I. Saleh
Mansoura University, Egypt

Mohammed H. Ali
Mansoura University, Egypt

ABSTRACT

The main objective of PCS networks is to provide "anytime-anywhere" cellular services. Accordingly, lost calls as well as the network slow response have become the major problems that hardly degrade the network reliability. Those problems can be overcome by perfectly managing the Mobile Terminals (MTs) locations. In the existing location management (LM) scheme, Location Area (LA) is the smallest unit for registration. A MT must register itself when passing through its LA boundary to a neighboring one. Moreover, such registration takes place at the MTs' master HLR (even though currently managed by another HLR), which increases communication costs. As a result, existing LM scheme suffers from; (1) excessive location registrations by MTs located around LA boundaries (ping-pong effect) and (2) requiring the network to poll all LA cells to locate the callee MT. In this paper, a novel LM strategy is introduced by restructuring LAs into smaller areas called Base Areas (BAs), which impacts the paging cost. The proposed LM strategy uses caching to reduce unwanted updates and 2LP to reduce paging cost. Experimental results show that the proposed scheme introduces a distinct improvement in network response and tracing process.

DOI: 10.4018/978-1-4666-2154-1.ch006

INTRODUCTION

Providing a transparent cellular communication services all over the world is the most human aim over the past few years. Such services are the ones that guarantee a reliable exchange of information in any form (voice, data, video, image, etc.) with no worry about the real-time distribution of MTs. Hence, they should be independent of service time, user's location and the underlying network access arrangement. This aim is continuously promoted by the tremendous growth in wireless communications (Brown & Mohan, 1997; Fang, Chlamtac, & Lin, 2000; Fang, Chlamtac, & Fei, 2000).

A PCS network is the integration of cellular (Wireless) and conventional (wired) networks. It provides wireless communication services that enable Mobile Terminals (MTs) to communicate and exchange any form of information on the move "anytime-anywhere services". One of the key issues in the design of PCS networks is the efficient management of real-time locations for MTs. To efficiently establish any service in the PCS network, the location of the MT should be clearly identified. Consequently, when an MT moves, it should inform the system with its new location (Gibson, 1996; Wong & Leung, 2000, Zhang, 2002). There are two basic operations in location management (LM); Location Registration (Location Update) (LR) and Call Delivery (CD). Location registration is the process through which the cellular system tracks the continuously changed locations of its MTs. On the other hand, when an incoming call arrives, the system has to search for the callee MT. This process is defined as call delivery or paging (Li, Kameda, & Li, 2000; Lo, Wolff, & Bernhardt, 1992; Meier-Hellstern & Alonso, 1992). To the best of our knowledge, there are two standard schemes for PCS location management, which are; Interim Standard 41 (IS-41) for North America Digital Cellular system and Global System for Mobile (GSM) for Pan-European Digital Cellular. Both schemes use a two-tier infrastructure of Home Location Reg-

ister (HLR) and Visitor Location Register (VLR) databases. Moreover, in the existing LM schemes, only the master HLR for an MT is used for storing and updating any MT's data even though that MT moves to another Service Area (SA) served by another HLR. This increases the communication costs for accessing the master HLR for both LR and CD dramatically. Also, Location Area (LA) is considered as the smallest unit to make LR.

In this paper, we have introduced a novel strategy for LM in PCS networks by restructuring the LAs of PCS networks into smaller areas called Base Areas (BAs), which in turn minimizes the paging cost. Moreover, registration is done through the current HLR instead of the master HLR, which improves not only the network response but also the network quality of service (QoS). Also, the proposed scheme uses caching to reduce unwanted LRs (Location Updates), read/write from/to databases (accessing database), and signaling costs. While existing LM scheme locates any MT by polling all LA cells, the proposed scheme uses 2LP (2-level paging) scheme to reduce the paging cost. An analytical model is developed to study the performance of the proposed frameworks. Experimental results have shown that the proposed scheme introduces a distinct improvement in network response and tracing process as well as reducing LM costs.

The rest of this paper is organized as follows: Background and Basic Concepts and Related Works are presented first. Next we describe the existing Location Management scheme. We present the proposed framework and an analytical model is presented. The performance comparisons between the proposed location management scheme and the existing one are discussed and we conclude the paper.

Background and Basic Concepts

Figure 1 describes the PCS networks structure. As depicted in such figure, PCS network is divided into service areas (SAs), which in turn are sub-

Figure 1. The structure of the PCS network

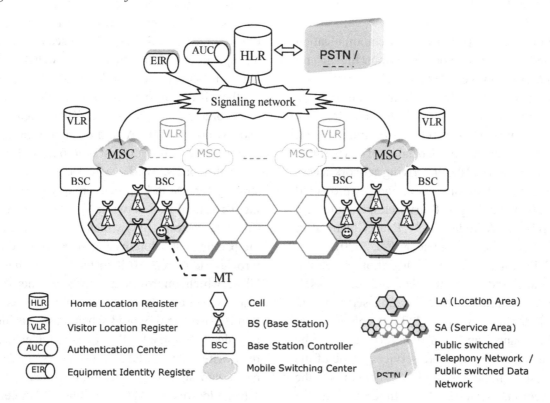

divided into cells. A cell is the communication area serviced by one Base Station (BS). Hence, the location of any MT is thus the address of the cell in which it is currently located. In each cell, there is a Base Transceiver Station (BTS) that used to communicate with MTs over pre-assigned radio frequencies.

Each group of cells is connected to a Base Station Controller (BSC). Each group of BSCs is connected to a Mobile Switching Center (MSC). MSC is a telephone exchange specially assembled for mobile applications that is responsible for setting up, routing, and supervising calls to/from MTs. The area serviced by an MSC is called a LA, which is managed by one VLR. On the other hand, each group of LAs composes a SA, which is serviced by one HLR. The HLR is a global database that maintains MT identity information including the permanent data (e.g., service subscribed, billing information, directory number,

profile information, current location, roaming-limits, VLR-address, MSC-address, and validation period) of the MTs whose primary subscription is within that HLR service area. A VLR contains temporary records for all MTs currently visit the area serviced by its MSC(s). Each VLR is associated with one or more (MSC) in the network(s) (Li, Pan, & Xiao, 2004).

A PCS network includes several SAs and thus several HLRs. The service area managed by the MTs master HLR is called the master SA for that MT, while the new SA where the MT is currently residing is called the current SA and its HLR is called the current HLR. When a MT visits a new LA, a temporary record is created in the VLR serving that LA to mirror its current location before receiving any cellular service. Also, the MTs master HLR should be updated to reflect its new location. When a MT leaves its LA, the corresponding record in the VLR is deleted. When there is a call, PCS network checks HLR to know the current VLR serving the called MT, and then the call is delivered to the current VLR.

Related Work

Due the importance of PCS networks, great efforts have been conducted for enhancing their LM schemes (Li, Pan, & Xia, 2004; Ho & Akyildiz, 1997; Jain, Lin, & Mohan, 1994; Shivakumar & Widom, 1994; Morris & Aghvami, 2008). In Li, Pan, and Xiao (2004) a dynamic HLR location management scheme has been introduced to reduce the signaling cost for both LR and CD. Such scheme provided a dynamic copy of the MT location information in the nearest HLR. So, any MT can access the location data in its nearest HLR for performing both LR and CD. In Ho and Akyildiz (1997) dynamic hierarchical database architecture for LM has been introduced. In such structure, a new tier of databases called directory registers has been added. However, this increases the systems complexity as well as posing several difficulties in the design and implementation of the system. Each directory register should determine the location information distribution strategies for its associated MTs. Moreover, it should set up the location pointers at the selected remote directory registers for each MT periodically, which dramatically increases the network bandwidth and computational overheads.

In Jain, Lin, and Mohan (1994) a per-user location caching strategy has been introduced to reduce the communication cost for CD. Such strategy reuses the cached data for the called MTs location from its last call. However, its performance depends on the probability that the cached data still valid. In Shivakumar and Widom (1995) a user profile replication scheme has been proposed. In such scheme, a user's profile is replicated at preselected locations, which in turn reduces the CD delay. The replication made by a center point, which continuously collects the mobility and calling parameters of the whole user population. However, successfully generating and distributing the replication decisions for a large user population will be computationally intensive and time-consuming. Moreover, it may incur significant amount of network bandwidth. In Morris and Aghvami (2008). a new LM strategy for cellular overlay networks has been proposed. It provides an efficient MT location registration and paging across an inter-worked network consisting of a Digital Video Broadcast (DVB) & Universal Mobile Telecommunications System (UMTS) networks. Although it achieves a four times reduction in paging cost across the inter-network compared with independent paging systems, such strategy poses several challenges and open problems as it increases the system complexity.

Existing Location Management Scheme

Currently, most LM strategies use a combination of paging and LR strategies. The traditional LM scheme used in the existing PCS networks updates MT's location when it crosses the boundary of any LA. Such process is done by comparing MT's registered LAI with current broadcast LAI, if the two IDs are different, LR takes place. On the other hand, to deliver a call to a MT, the network pages all the cells within the LA in parallel (Zhang, 2002; Ng & Chan, 2005). The Location Area Identifier (LAI) composed of three parameters as illustrated in Figure 2.

Existing Location Registration Scheme

LR is the process through which the network tracks the continuously changed locations of MTs. Figures 3 and 4 illustrate the traditional LR process when a MT passes through the boundary of its hosting LA to one of the neighboring LAs within the same SA or belongs to other SAs. Also, a detailed description included in Table 1.

Figure 2. Location area ID

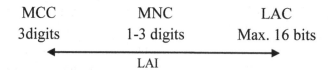

Figure 3. LR diagram for k LAs within SA

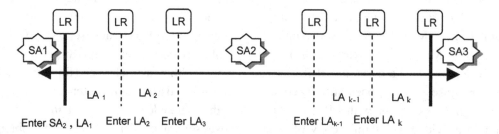

Figure 4. Location registration scheme in PCS network

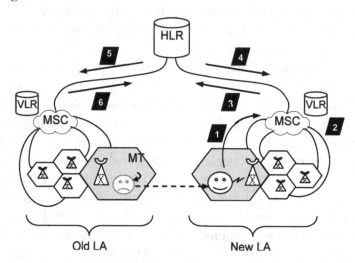

Table 1. Existing LR scheme in PCS network

Step	Description
(1)	When the *MT* detects that it has entered a new LA, it sends a *LR message* to the new MSC serving the new LA through the *BS* and *BSC*.
(2)	The *MSC* updates its associated *VLR* indicating that the MT is residing in its serving area and sends an *LR message* to the MT's *master HLR*.
(3)	The message is routed to an *STP*, which determines the MT's *master HLR* from its *MIN* (Mobile Identification Number) by a table lookup procedure called Global Title Translation (*GTT*). *LR message* is then forwarded to the master HLR.
(4)	The MT's master HLR updates its record indicating the current serving MSC of the MT then sends a *registration acknowledgment* message to the new MSC.
(5)	The master HLR sends a *registration cancellation* message to the old MSC to delete the MT's record from its associated VLR.
(6)	The old MSC deletes the record of the MT from its associated VLR and sends a *cancellation acknowledgment* message to the master HLR.

Existing Call Delivery Scheme

Basically, when a MT calls another one, the location of the called MT must be identified. Identifying the location of the called MT includes specifying its serving VLR as well as its currently visited cell. To deliver a call, steps illustrated in Figure 5 should be followed. Also, a detailed description for CD is included in Table 2.

PROPOSED FRAMEWORK

According to the traditional LM strategy, a SA is divided into zones called LAs where each LA contains a certain number of adjacent cells. When a MT crosses the LA boundary as it travels from one LA to another, it must register itself at the new LA. When an incoming call arrives for the MT, the network pages all the cells inside the LA where the MT last updated its location. The major advantage of such strategy is its simplicity in implementation. However, such traditional LM scheme suffers from two major hurdles, namely; (1) excessive LRs by MTs located around LA boundaries, which making frequent movements back and forth between two adjacent LAs (Ping-Pong effect), and (2) requiring the network to poll all LA cells to locate the callee MT, which results in excessive volume of traffic.

Figure 5. Call delivery scheme in PCS network

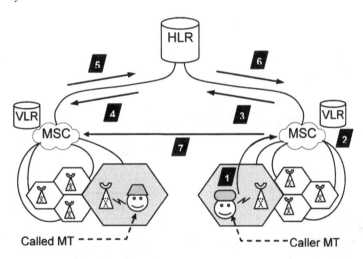

Table 2. Existing CD scheme in PCS network

Step	Description
(1)	A call is initiated by a *MT* by sending a *location request* message to its serving *BS*, which forwards it to the serving *MSC* through *BSC*.
(2)	The MSC sends a *location request* message to the *master HLR* of the *callee MT* through an STP where GTT is performed to determine the called MT's master HLR.
(3)	The *location request* is forwarded to the master HLR.
(4)	The master HLR sends a *location request* message to the MSC serving the callee MT.
(5)	The callee MT's MSC determines the *cell location* of the callee MT by polling all cells within the LA and assigns it a Temporary Location Directory Number (*TLDN*). Then, MSC sends this TLDN to the master HLR.
(6)	The master HLR forwards the *TLDN* to the calling MSC.
(7)	When the calling MSC receives *the pre-assigned TLDN*, it sets up a connection to the callee MSC using this TLDN through the SS7 network.

In this paper a novel LM strategy will be proposed by restructuring LAs of PCS networks into smaller areas composed of group of cells connected to one BSC called Base Area (BA). This makes the MTs' location more specific, which in turn minimizes the paging cost. The optimal number of a BA cells depends on the network performance, number of subscribers, and mobility rate. The proposed structure is illustrated in Figure 6.

Definition 1: Base Area (BA)

- *The Base area, denoted as (BA) is the smallest addressable area used for making LR, and paging in the proposed LM scheme.*

- *Considering a LA as a set of M adjacent cells, hence, a BA is a set of N adjacent cells$\in$ the same LA, where N$\leq$M, hence, BA$\subset$ LA.*

As illustrated in Figure 7, for the proposed LM strategy, LAI contains the following parameters; (1) *MCC,* which represents the Mobile Country Code, (2) *MNC,* identifies the Mobile Network Code, (3) *LAC,* indicates the Location Area Code, and (4) *BAC,* represents the Base Area Code (Figure 8). Lemma 1 illustrates how to calculate the exact value of BAC, which is assumed to be X bits.

Figure 6. The proposed structure of PCS network

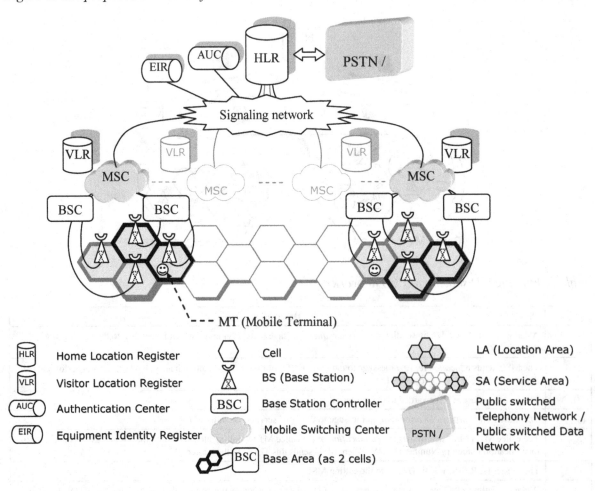

Figure 7. The new location area ID contents

MCC	MNC	LAC	BAC
3 digits	1-3 digits	max. 16 bits	X bits

LAI

Figure 8. Illustrative example for BAC

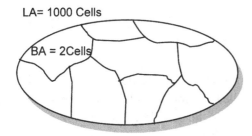

LA= 1000 Cells

BA = 2Cells

Lemma 1:

The number of bits in BAC $\geq X = \dfrac{\ln m / n}{\ln 2}$ bits where m is the number of cells in LA and n is the number of cells in each BA.

Proof:

Assume a PCS network with the following parameters:

Size (LA) = m Cells

Size (BA) = n Cells

$\therefore$ No. of BA in each LA = $\dfrac{m}{n}$

Where $\dfrac{m}{n} = 2\,X$

$\therefore \ln m/n = X * \ln 2$

$\therefore$ Size (BAC) $\geq X = \dfrac{\ln m / n}{\ln 2}$ bits

Illustrative example:

Let m = 1000 Cells and n = 2 Cells

Then BAC = $\dfrac{\ln 1000 / 2}{\ln 2} = 8.96578 = 9$ bits

In addition, in the proposed LM strategy, registration is done through the current HLR instead of the master HLR. Where, the GTT table is modified as the STP switches the LR/ location request message from the MSC to the current HLR, which improves not only the network response but also the network quality of service (QoS). The proposed LM strategy also uses caching to reduce unwanted LRs and reduce read/write from/to databases. Moreover, a new 2LP scheme has been employed to reduce the paging cost. According to 2LP, to identify the callee MT, all cells within its current BA are polled first as a level. If the callee MT still not identified, all the remaining cells of its current LA are paged simultaneously as another level. The cells in each level are paged simultaneously aiming to identify the callee MT. If it still not identified even after paging all LA cells, this indicates that the MT either left the coverage area of the network or turned off without de-registration (i.e., sudden death of the power supply).

Location Registration Scheme Using Base Area (LRBA)

According to the proposed LM strategy, HLR serving the current SA of the MT is used to perform LR, as the cost of accessing the current HLR is much smaller than the cost of accessing the master HLR. Hence, any MT can always use the location data in its nearest HLR. Also, the GTT table is modified as the STP switches the LR message from the MSC to the associated HLR, which is located in the same SA. During registration, MT records the ID of the new BA (BAC) to that VLR serving the current LA to make its location more specific. In case a MT moved from BA to another within the same LA, it overwrites its record in the associated VLR to mirror the new *BAC*. Moreover, there are 2 timers used to reduce unwanted LRs and read/write from/to databases. These timers associated with two different time delays, which are; Time-To-Cache (*TTC*) and Time-To-Delete (*TTD*).

Definition 2: Time-To-Cache; TTC

Time-To-Cache (TTC) is the period during which the new location massage sent from an MT moving inside the same LA (between BAs) is cached before overwriting the MTs' record in the current VLR.

Hence, the new location massages will reside in the VLR's cache for a period equal to TTC. If the MT's BA residence time (tB) exceeds TTC, VLR will update MT's record. On the other hand, if MT leaves the new BA before TTC expires, a new LR message will overwrite the cached one, then TTC will be reset.

Definition 3: Base Area Residence Time; tB(MTi, BAj, twall)

Is the time period during which MTi resides inside the BAj starting at the wall-clock time twall before it leaves BAj to another BA.

Definition 4: Time-To-Delete; TTD

Time-To-Delete (TTD) is the delay period before deleting the record of the moved MT from the VLR associated with the old MSC of that MT for eliminating Ping-Pong effect.

Definition 5: Location Area Residence Time; tL(MTi, LAk, twall)

Is the time period during which MTi resides inside the LAk starting at the wall-clock time twall before it leaves LAk to another LA.

During TTD, the MT's record will be virtually locked to avoid any conflict, so, it will be unseen by HLR. Also, each MT will have its *LAI history* to store the LAIs of the recently visited LAs to be used in case of ping-pong situations (assuming 3 LAIs in the history). Each record in *LAI history* uses the TTD as a timestamp. So, when an MT enters a new LA, it sends "I'm Alive Again" (*IAA*) message to the new MSC if the new LAI

is in its history and TTD period has not expired. Therefore, MSC will stop deleting the MT's record from VLR. If TTD period has expired, MT will start a new LR process. Figure 9 gives an illustration for the proposed LR. Also, Figure 10 describes the proposed framework for LR. A detailed description for the proposed LR scheme is also included in Table 3.

Call Delivery Scheme Using Base Area (CDBA)

When a MT calls another one, the location of the called MT must be identified. Identifying the location of the called MT includes specifying its serving VLR as well as its currently visited cell. Caching has been employed in the proposed LM strategy to reduce unwanted read/write from/to HLR/VLR databases. Hence, the routing address (TLDN) of a MT for its recent calls is stored in its HLR and VLR caches. If the TLDN of the callee MT is cached in the VLR of the calling MSC, then MSC contacts the callee MT directly (forgo HLR), which reduces the call setup time. Also, we have three possibilities if the TLDN is in the cache; (1) a valid cache entry, (2) an expired cache entry, (3) a valid cache entry but the MT moved to another location. However, we need to know that the cache lookups take nearly 20 microseconds and disk accesses take nearly 7 milliseconds, so look at the difference. According to the above possible results in caching, the network needs to check the cached MT's TLDN to check the MT's location validity.

Also, using 2LP scheme to reduce paging cost when an incoming call arrives. Initially all cells within BA are polled first (BA paging) as a level. If the MT isn't valid in its BA, all remaining cells within MT's LA are paged simultaneously as another level. The GTT table lookup procedure is applied to find the master HLR of the called MT, and then find its current HLR, if the called MT is in different SA. The modified GTT table lookup procedure is especially effective when the caller

Figure 9. Proposed LR diagram for N BA within LA

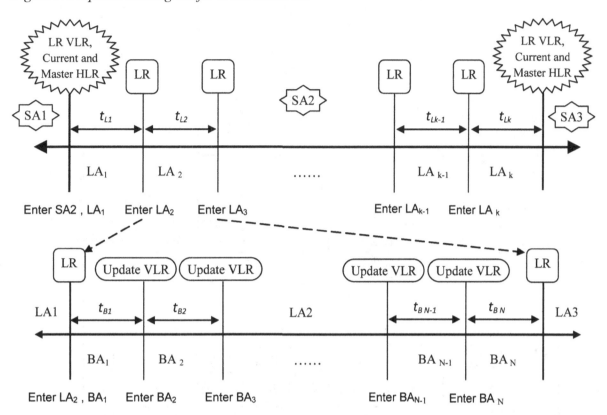

and called are in the same SA for the CD since calls are more likely to be local calls instead of remote calls. To deliver a call, steps illustrated in the proposed framework (Figure 11) should be followed. Also, a detailed description for the proposed CD scheme is included in Table 4.

Analytical Model

To study the performance of the proposed LM scheme against the existing scheme, we establish the following analytical model. Detailed description about all used parameters is illustrated in Table 5. At the first, it is required to define the mobility rate (number of movements per unit time) for MTs within each specific area. Without loss of generality, we assume that there are X of SAs and each SA has K of LAs, where each LA has d cells. Each LA is divided into N of BA where

$N <= d$, while BA has M cells, where cell $(i, j) \in$ LAK, cell $(i, j) \in$ BAm and cell (i, j) not $\in$ BAL, where m $\neq$ L and m $\in \{1,2,....., N\}$. In this analysis, we assume the database access and the signaling costs are measured by the delays required to complete the database update/query and signal transmission, respectively. These delay values can be obtained by on-line measurements, or by a table lookup process. Note that; cost for accessing HLRs and VLRs includes the communication cost (wired and wireless). We assume that the cost for a DB update is equal to the cost for a DB query. Moreover, the cost for DB update or query is much smaller than the communication cost.

To perform LR or CD, there are signaling costs over communication lines in the same SA (Intra-SA) or among two different SAs (Inter-SA). Also to calculate the paging costs for a called MT, we

Figure 10. Proposed location registration scheme in PCS network

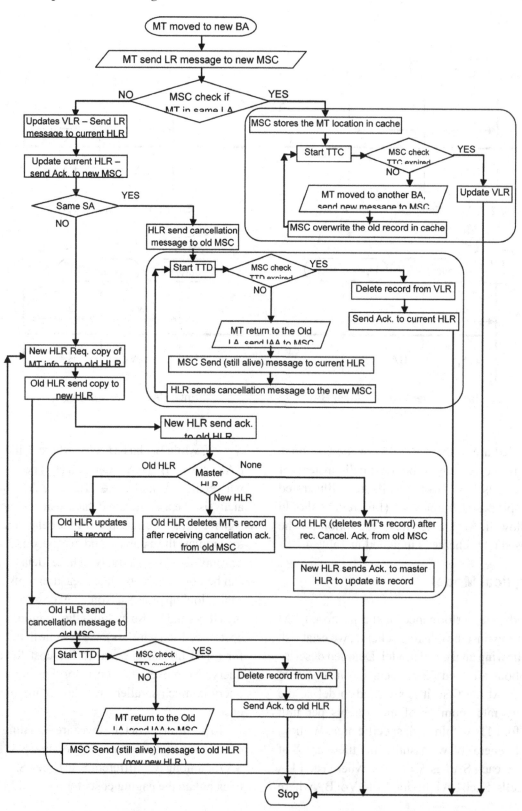

Table 3. Proposed LR scheme in PCS network

Step	Description
(1)	When a *MT* detects that it enters a new *BA*, it sends a *LR message* to the associated MSC through the *BS* and *BSC*.
(2)	If the new BA is in the same LA, MSC stores the record in its cache until *TTC* period expires then replaces the old MT's record in the associated VLR with the new one from the cache. If the MT moves to another BA within LA, it sends a new record to MSC. If the TTC period did not expire, MSC deletes the old record and stores the new one in the cache until the new TTC period expires then updates VLR.
(3)	If the new BA is in a new LA, the new MSC updates its VLR indicating that the MT is residing in a BA within its served area and in which *BA* then sends it to the STP, which forwarded it to the current HLR by the GTT table lookup procedure. The current HLR updates its record then sends a *registration acknowledgment* message to the new MSC.
(4)	If the new LA is in the same SA, the current HLR sends a *registration cancellation* message to the old MSC, which deletes the MT's record in its VLR then sends a *cancellation acknowledgment* message to the current HLR after *TTD* period expires. During this time, the record is virtually locked because it will be unseen by HLR. If the old MSC receives *IAA* message from the MT before TTD period expires, MSC stops deleting the MT's record and sends still alive message to the current HLR, which sends a *registration acknowledgment* message to the old MSC (new now) and a *registration cancellation* message to the new MSC (old now) and so on.
(5)	If the new LA is in a new SA, The new HLR sends a message to the MT's old HLR to reflect the movement of the MT into its area and requesting a copy of an MT's record. The old HLR will send the MT's record to the new HLR, also sends a *registration cancellation* message to the old MSC.
(6)	The old MSC deletes the record of the MT in its associate VLR and sends a *cancellation acknowledgment* message to the old HLR after *TTD* period has expired. During this time, the record is virtually locked because it will be unseen by HLR. If the old MSC receives *IAA* message from the MT before TTD period expires, MSC will stop deleting the MT's record and sends still alive message to the old *HLR*, which will send a *registration acknowledgment* message to the old MSC and a message to the new HLR to reflect the movement of the MT into its area and requesting a copy of an MT's record and so on.
(7)	If the new MT's HLR is the *master HLR*, it just sends the old HLR a *registration acknowledgment* message; otherwise it sends both the old HLR and the master HLR the *registration acknowledgment* message.
(8)	For the old HLR, if it is the *master HLR*, it updates its record pointing to the new HLR; otherwise it deletes the MT's record.
(9)	For the master HLR, if it is the *new HLR*, it does nothing; otherwise, it updates its record pointing to the new HLR.

need to know the incoming call arrival rates per unit time for that MT. According to the random movement behavior, the movements are independent and identically distributed. So, the probability that MTi resides within any SA = *Pr(SA, MTi)* = 1/X, and the probability that MT resides within any LA = *Pr(LA, MTi)* = 1/K. Also the probability that MT resides within any BA = *Pr(BA, MTi)* = 1/N. The proposed scheme will be compared against the existing one through the total costs for LR and CD to investigate how much we can gain or lost. The costs will be divided into signaling cost (CS) and computational cost (CLR & CCD).

For simplicity, signaling cost from MT to MSC through BS and BSC will be neglected because it approximately fixed value in most cases. The following equations can be calculated from Figure 9.

tBN (MTx, BANKX, twall) which is the residence time of MTx inside its current BA where BAN $\in$ LAK $\in$ SAX so that;

$$tBN (MTx, BANKX, twall) = \sum_{i=1}^{M} t_{si}(MT_x)$$

(1)

$$TB (MTx) = \frac{\sum_{N=1}^{n} t_{BN}(MT_x, \ BA_{NKX}, \ t_{wall})}{n}$$

(2)

tLK (MTx, LAKX, twall) which is the residence time of MTx inside its current LA where LAK $\in$ SAX so that;

Figure 11. Proposed call delivery scheme in PCS network

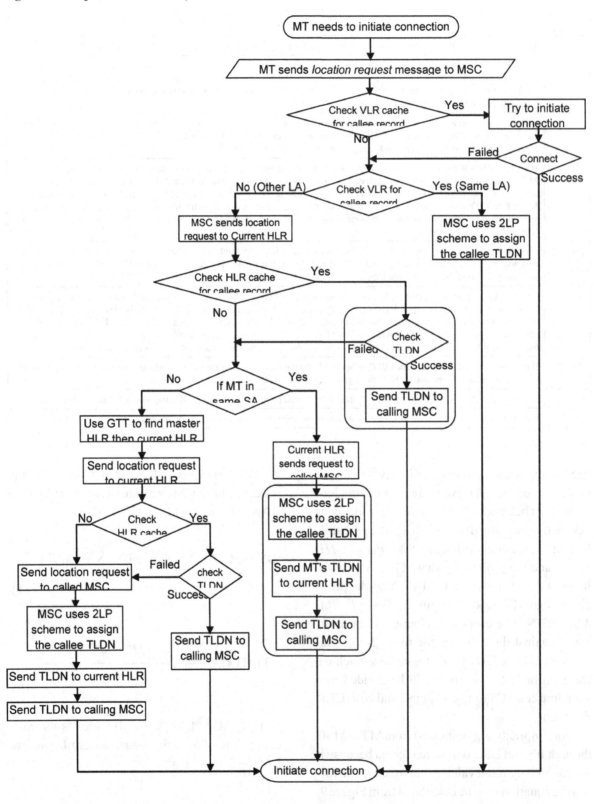

Table 4. Proposed CD Scheme in PCS Network

Step	Description
(1)	A call is initiated by a MT (*caller*) by sending *location request* to BS, which forwards the call initiation signal to the MSC through BSC.
(2)	The MSC checks its VLR cache asking about the called MT's TLDN; if there is a record try to initiate connection. While if there is no record for the callee or the connection failed, MSC checks its VLR to check if the caller within the same LA of the callee.
(3)	If a MT's record exists in VLR, MSC uses 2LP scheme to assign the TLDN of the called MT. After that MSC directly establish a connection to the callee. Otherwise the MSC sends a *location request* message to the associate current HLR.
(4)	The current HLR checks its cache. If the callee TLDN exists, checks it. If TLDN check success, the current HLR sends it to the calling MSC, which establish connection directly with the called MSC. If TLDN check failed or no record in cache, the location request is forwarded to the MSC serving the called MT, if the callee is in the same SA of the caller by checking the current HLR for MT record.
(5)	The MSC determines the cell location of the called MT by using the 2LP scheme to assigns the TLDN of the called MT. The called MSC then sends this TLDN to the current HLR. Then, the current HLR forwards the TLDN to the calling MSC, which sets up a connection to the called MSC using that TLDN.
(6)	If the callee MT is in another SA, a GTT table lookup procedure is used twice to find the master HLR of the callee MT, then to find the current HLR of the callee MT. The *location request* is forwarded to the current HLR of the callee MT which checks its cache. If the callee TLDN exists, checks it. If TLDN check success, the current HLR sends it to the calling MSC, which establishes the connection directly to the called MSC. If the TLDN check has failed or there is no record in the cache, *the location request* message is forwarded to the MSC serving the callee MT.
(7)	The MSC determines the cell location of the callee MT by using the 2LP scheme to assigns the TLDN of the called MT. The called MSC then sends this TLDN to the current HLR. Which forwards the TLDN to the calling MSC. Whose sets up a connection to the called MSC using that TLDN.

tLK (MTx, LAKX, twall) =

$$\sum_{j=1}^{N} t_{Bj}(MT_X, BA_{jKX}, t_{wall}) \qquad (3)$$

$$TL\ (MTx) = \frac{\sum_{j=1}^{k} t_{Lj}(MT_x, LA_{jX}, t_{wall})}{k} \qquad (4)$$

Also mobility rate for an MT through BAs and LAs will be:

$$\mu UB\ (MTi)\ \alpha\ \frac{1}{T_B(MT_i)}$$

where the more BA boundary crossed by an MTi the BA residence time decreased so,

$$\mu UB\ (MTi) = \frac{\tilde{A}}{T_B(MT_i)}$$

where Ã is a constant (6)

$$\mu UL\ (MTi)\ \alpha\ \frac{1}{T_L(MT_i)}$$

where the more LA boundary crossed by an MT the LA residence time decreased so,

$$\mu UL\ (MTi) = \frac{^2}{T_L(MT_i)}$$

where ² is a constant (7)
Also,

*transition rate = mobility rate * transition probability*

And, according to 2LP scheme the average number of cell paged to deliver a call to any MT is:

Cpage (MTi) = M * Pr(BA) + (d – M) * (1 – Pr(BA)) (8)

Table 5. Notation used for the analytical model

Parameters	Description
X	Number of SAs in the PCS network.
K	Number of LAs within each SA
d	Number of cells in LA.
N	Number of BAs within each LA
M	Number of cells in BA.
μUL	The LA mobility rate per unit time for MT.
μUB	The BA mobility rate per unit time for MT.
Pr(MT)	Probability of MT's movement between SAs
Pr(SA)	probability that MT resides within any SA
Pr(LA)	probability that MT resides within any LA
Pr(BA)	probability that MT resides within any BA
P	Probability that the called MT record in database cache (cache hit)
tS	Cell residence time.
tL	LA residence time.
TL	The average LA residence time.
tB	BA residence time.
TB	The average BA residence time.
n	The no. of BA boundary crossed during test period
k	The no. of LA boundary crossed during test period
Ch	Cost for updating or querying the HLR.
Cv	Cost for updating or querying the VLR.
Cg	Cost for performing Global Title Translation to determine the master or current HLR.
$\lambda L, \lambda o$	Incoming call arrival rates per unit time from the same SA and from other SAs, respectively.
λC	The total incoming call arrival rate per unit time. $\lambda C = \lambda L + \lambda o$.
Cp	Cost of paging.
δp	The unit cost for each paging in a cell.
Cpage	The average no. of cells paged.
CS1	Signaling costs for the communication links in the same SA (intra-SA).
CS2	Signaling costs for the communication links between two SAs (inter-SA).
CLR	Computational cost for LR
CCD	Computational cost for CD
CS	Signaling cost
τc	Cache access time needed to access databases cache.
TTC	Time To Cache.
TTD	Time To Delete.
η	Probability that tB > TTC
ζ	Probability that tL > TTD

Location Registration Cost

In this part the location registration cost for the existing scheme as well as the new proposed one will be analyzed in more details.

Existing Location Registration Scheme

When a MT moves between LAs inside or outside its master SA there are fixed computational costs. The total computational cost, denote as CLR can be calculated as;

$$CLR = \mu UL \, [\, 2Cv + Ch + Cg \,] \qquad (9)$$

Where; 2Cv, one to inform the new MSC to serve the moved MT and updating its associated VLR and one when sending cancellation message to the old MSC to delete the MT's record from its associated VLR. Ch when updating the master HLR for that MT and Cg for performing GTT to determine the master HLR. On the other hand, the signaling cost differs according to the MT movement pattern. The costs for existing scheme can be calculated from the following cases according to Figure 4:

Case 1: Movement inside the same LA

When a MT moves inside the same LA there is no change in LAI, so there is no LR.

$$CLR = 0 \; CS = 0$$

The computational cost for the next four cases can be calculated using Equation (9).

Case 2: Movement inside Same SA between LAs (inside its master SA)

When a MT moves between LAs inside its master SA there is signaling cost for communication link, two intra-SA signaling costs between the new MSC and the master HLR and another two between the old MSC and master HLR. Hence,

$$CS = \mu UL * \frac{K-1}{K} * \frac{1 - Pr(MT)}{X} * 4 \, Cs1$$

Case 3: Movement inside Same SA between LAs (outside its master SA)

When a MT moves between LAs outside its master SA there are signaling costs for communication link with the master HLR that is located at another SA. Two inter-SA signaling costs between the new MSC and the master HLR and another two between the old MSC and master HLR. Hence,

$$CS = \mu UL * \frac{K-1}{K} * \frac{X-1}{X} * (1 - Pr(MT)) * 4 \, Cs2$$

Case 4: Movement between two SAs (one of them is the master SA)

When a MT moves between two SAs in which one of them is its master one, two signaling costs between the new MSC and the master HLR and another two between the old MSC and master HLR. Two signaling costs are inter-SA and two are intra-SA. Hence,

$$CS = \mu UL * \frac{K-1}{K} * \frac{2}{X} * Pr(MT) * 2[Cs1 + Cs2]$$

Case 5: Movement between two SAs (no one of them is the master SA)

When a MT moves between two SAs in which none of them is its master SA, there are inter-SA signaling costs for communication link, two of them between the new MSC and the master HLR and two between the old MSC and master HLR. Hence,

$$CS = \mu UL * \frac{K-1}{K} * (1 - \frac{1}{X} + \frac{1}{X}) * Pr(MT) * 4 \, Cs2$$

So the total signaling cost can be calculated as:

$$CS = \mu UL * \frac{K-1}{K} * \frac{1-\Pr(MT)}{X} * 4\,Cs1 +$$

$$\mu UL * \frac{K-1}{K} * \frac{X-1}{X} * (1-\Pr(MT)) * 4\,Cs2 +$$

$$\mu UL * \frac{K-1}{K} * \frac{2}{X} * \Pr(MT) * 2[Cs1 + Cs2] +$$

$$\mu UL * \frac{K-1}{K} * (1 - \frac{1}{X} + \frac{1}{X}) * \Pr(MT) * 4\,Cs2$$

After modification the total signaling cost for the existing LR scheme can be calculated using Equation (10).

$$CS = \mu UL * \frac{\mathbf{K-1}}{\mathbf{K}} * \frac{\mathbf{4}}{\mathbf{X}} * (Cs1 + (X-1)*Cs2) \tag{10}$$

Proposed Location Registration Scheme

According to our proposal there are no fixed costs. The costs will be multiplied in different mobility rates for different cases to calculate the total cost for every case. The costs for the proposed scheme can be calculated from the following cases according to the proposed scheme in Figure 10:

Case 1: Movement inside the same LA between BAs

When a MT moves between two BAs inside its home LA, there is Cv cost used to update its associated VLR (overwrite MT's record), if tB(MTi) > TTC else there is no update will happen so the cost will be only the cache access cost which is very small so we neglect it. Hence,

$$CLR = \mu UB * \frac{N-1}{N} * \frac{1-\Pr(MT)}{K} * \eta * Cv$$

$$CS \approx 0$$

Case 2: Movement inside Same SA between LAs

When a MT moves between two LAs inside the same SA. In case of tL(MTi) > TTD, there are 2CV, one to inform the new MSC to serve the moved MT and updating its associated VLR and one when sending cancellation message to the old MSC to delete the MT's record from its associated VLR. Also, one Ch to update the current HLR associated to that SA and one Cg for performing *GTT* to determine the current HLR. As well as there are 4 intra-SA signaling costs for communication link, two of them between the new MSC and current HLR and another two between the old MSC and current HLR.

$$CLR = \mu UL * \frac{K-1}{K} * (1 - \Pr(MT)) * \zeta * [2Cv + Ch + Cg]$$

$$CS = \mu UL * \frac{K-1}{K} * (1 - \Pr(MT)) * \zeta * 4\,Cs1$$

On the other hand, if the MT enters LA form its history list (Ping-pong), MSC stops deleting the MT record and sends still alive message to the current HLR to update its record which will cost Ch, then sends a registration acknowledgment message to the old MSC and a registration cancellation message to the new MSC which will cost Cv as well as 3 intra-SA signaling costs.

$$CLR = \mu UL * \frac{K-1}{K} * (1 - \Pr(MT)) * (1 - \zeta) * [Cv + Ch]$$

$$CS = \mu UL * \frac{K-1}{K} * (1 - \Pr(MT)) * (1 - \zeta) * 3\,Cs1$$

Case 3: Movement between two SAs

When a MT moves between two SAs there are computational costs. In case TTD expired there are 2CV, one to inform the new MSC that MT residing in its serving area and updating its associated VLR and one when sending cancellation message to the old MSC to delete the MT record from its associated VLR. Also, 4 Ch one to update the current HLR associated to the current SA, the second for accessing the old current HLR from the new current HLR, the third for accessing the new current HLR from the old current HLR and the fourth for updating the master HLR in case neither the new current HLR nor the old current HLR is the master HLR as well as Cg for performing *GTT* to determine the current HLR. There are 4 intra-SA signaling costs for communication link, two of them between the new MSC and current HLR and two between the old MSC and the old current HLR and 3 inter-SA signaling costs for communication link between new current HLR and old current HLR.

$$CLR = \mu UL * \frac{K-1}{K} * Pr(MT) * \zeta * [2Cv + 4Ch + Cg]$$

$$CS = \mu UL * \frac{K-1}{K} * Pr(MT) * \zeta * [4\ Cs1 + 3\ Cs2]$$

On the other hand, if the MT returns to its previous SA and TTD period did not expire (Ping-pong case), MSC stops deleting the MT record and sends still alive message to the old current HLR to update its record which will repeat the previous steps but there is no need to perform GTT because current HLR known and no need to update the serving VLR.

$$CLR = \mu UL * \frac{K-1}{K} * Pr(MT) * (1-\zeta) * [Cv + 4Ch]$$

$$CS = \mu UL * \frac{K-1}{K} * Pr(MT) * (1-\zeta) * [4\ Cs1 + 3\ Cs2]$$

So the total computational cost can be calculated as:

$$CLR = \mu UB * \frac{N-1}{N} * \frac{1-Pr(MT)}{K} * \eta * Cv +$$
$$\mu UL * \frac{K-1}{K} * (1 - Pr(MT)) * \zeta * [2Cv + Ch$$
$$+ Cg] + \mu UL * \frac{K-1}{K} * (1 - Pr(MT)) * (1-\zeta)$$
$$* [\ Cv + Ch\ ] + \mu UL * \frac{K-1}{K} * Pr(MT) * \zeta *$$
$$[2Cv + 4Ch + Cg] + \mu UL * \frac{K-1}{K} * Pr(MT) *$$
$$(1 - \zeta) * [Cv + 4Ch]$$

After modification the total computational cost for the proposed LR technique can be calculated using Equation (11).

$$\mathbf{CLR = \mu UB} * \frac{\mathbf{N-1}}{\mathbf{N}} * \frac{1-\mathbf{Pr(MT)}}{\mathbf{K}} * \eta * Cv$$
$$+ \mu UL * \frac{\mathbf{K-1}}{\mathbf{K}}$$
$$* \{\ Cv * (\zeta + 1 + \mathbf{Pr(MT)}) + \zeta * Cg + Ch * (1 + 3\ \mathbf{Pr(MT)}\ \} \qquad (11)$$

And the total signaling cost can be calculated as:

$$CS = \mu UL * \frac{K-1}{K} * (1 - Pr(MT)) * \zeta * 4$$
$$Cs1 + \mu UL * \frac{K-1}{K} * (1 - Pr(MT)) * (1-\zeta) *$$
$$3\ Cs1 + \mu UL * \frac{K-1}{K} * Pr(MT) * \zeta * [4\ Cs1$$
$$+ 3\ Cs2] + \mu UL * \frac{K-1}{K} * Pr(MT) * (1-\zeta) *$$
$$[4\ Cs1 + 3\ Cs2]$$

After modification the total signaling cost for the proposed LR technique can be calculated using Equation (12).

$$CS = \mu UL * \frac{K-1}{K} * \{ Cs1 * (\zeta - Pr(MT) * \zeta + 3 + Pr(MT)) + Pr(MT) * 3\, Cs2 \} \quad (12)$$

Call Delivery Cost

In this part the CD cost for the existing scheme as well as the new proposed one will be analyzed in more details.

Existing Call Delivery Scheme

When a MT calls another one there are fixed computational costs. 2CV, one for Calling MT to querying VLR about its data and another one for called MT when requesting its location from the called MSC. Also, Ch when querying the master HLR for the serving MSC to the called MT, and Cg for performing *GTT* to determine the master HLR. The signaling cost differ according to the MT movement pattern. As well as the paging cost will be equal $Cp = \delta p * d$. The cost for the existing scheme can be calculated from the following cases according to Figure 5:

Case 1: Inside the same LA (inside the master SA of the callee)

When a MT calls another one within the same LA and within the called MT master HLR, there are 4 intra-SA signaling costs for communication link exploited. Two of them between the calling MSC and the master HLR and another two between the called MSC and master HLR. Hence,

$$CCD = \lambda L * \frac{1}{X} * \frac{1}{X} * \frac{1}{K^2} * [2Cv + Ch + Cg]$$

$$CS = \lambda L * \frac{1}{X} * \frac{1}{X} * \frac{1}{K^2} * [4\, Cs1 + Cp]$$

Case 2: inside the same LA (outside the master SA of the callee)

When a MT calls another one within the same LA but outside the called MT master HLR, there are 4 inter-SA signaling costs for communication link, two of them between the calling MSC and the master HLR and another two between the called MSC and the master HLR. Hence,

$$CCD = \lambda L * \frac{X-1}{X} * \frac{1}{X} * \frac{1}{K^2} * [2Cv + Ch + Cg]$$

$$CS = \lambda L * \frac{X-1}{X} * \frac{1}{X} * \frac{1}{K^2} * [4\, Cs2 + Cp]$$

Case 3: Inside Same SA between LAs (inside the master SA of the callee)

When a MT calls another one within different LA but inside the called MT master HLR, there are 4 intra-SA signaling costs for communication link, two of them between the calling MSC and the master HLR and another two between the called MSC and master HLR. Hence,

$$CCD = \lambda L * \frac{1}{X} * \frac{1}{X} * \frac{K-1}{K^2} * [2Cv + Ch + Cg]$$

$$CS = \lambda L * \frac{1}{X} * \frac{1}{X} * \frac{K-1}{K^2} * [4\, Cs1 + Cp]$$

Case 4: Inside Same SA between LAs (outside the master SA of the callee)

When a MT calls another one within the same LA and same SA but outside the called MT master HLR, there are 4 inter-SA signaling costs for communication link, two of them between the calling MSC and the master HLR and another two between the called MSC and master HLR. Hence,

$$CCD = \lambda L * \frac{X-1}{X} * \frac{1}{X} * \frac{K-1}{K^2} * [2Cv + Ch + Cg]$$

$$CS = \lambda L * \frac{X-1}{X} * \frac{1}{X} * \frac{K-1}{K^2} * [4\,Cs2 + Cp]$$

Case 5: Between two SAs (one of them is the master SA)

When a MT calls another one inside another SA and one of them is the called MT master HLR, there are two inter-SA signaling costs and two intra-SA signaling costs for communication link, two between the calling MSC and the master HLR and two between the called MSC and master HLR. Hence,

$$CCD = \lambda o * \left(\frac{X-1}{X} * \frac{1}{X} + \frac{1}{X} * \frac{X-1}{X} \right) * \frac{K-1}{K^2} * [2Cv + Ch + Cg]$$

$$CS = \lambda o * \left(\frac{X-1}{X} * \frac{1}{X} + \frac{1}{X} * \frac{X-1}{X} \right) * \frac{K-1}{K^2} * [2\,Cs1 + 2\,Cs2 + Cp]$$

Case 6: Between two SAs (none of them is the master SA)

When a MT calls another one inside another SA and none of them is the called MT master HLR, there are 4 inter-SA signaling cost for communication link, two of them between the calling MSC and the master HLR and two between the called MSC and master HLR. Hence,

$$CCD = \lambda o * \frac{(X-1)(X-2)}{X^2} * \frac{K-1}{K^2} * [2Cv + Ch + Cg]$$

$$CS = \lambda o * \frac{(X-1)(X-2)}{X^2} * \frac{K-1}{K^2} * [4\,Cs2 + Cp]$$

So the total computational cost can be calculated as:

$$CCD = \lambda L * \frac{1}{X^2} * \frac{1}{K^2} * [2Cv + Ch + Cg] + \lambda L * \frac{X-1}{X^2} * \frac{1}{K^2} * [2Cv + Ch + Cg] + \lambda L * \frac{1}{X^2} * \frac{K-1}{K^2} * [2Cv + Ch + Cg] + \lambda L * \frac{X-1}{X^2} * \frac{K-1}{K^2} * [2Cv + Ch + Cg] + \lambda o * \frac{2(X-1)}{X^2} * \frac{K-1}{K^2} * [2Cv + Ch + Cg] + \lambda o * \frac{(X-1)(X-2)}{X^2} * \frac{K-1}{K^2} * [2Cv + Ch + Cg]$$

After modification the total computational cost for the existing CD scheme can be calculated using Equation (13).

$$\mathbf{CCD} = \frac{1}{\mathbf{X*K}} * [2Cv + Ch + Cg] * \{ \lambda L + \lambda o * (X + \frac{1-\mathbf{X}}{\mathbf{K}} - 1) \} \qquad (13)$$

Also the total signaling cost can be calculated as:

$$CS = \lambda L * \frac{1}{X^2} * \frac{1}{K^2} * [4\,Cs1 + Cp] + \lambda L * \frac{X-1}{X^2} * \frac{1}{K^2} * [4\,Cs2 + Cp] + \lambda L * \frac{1}{X^2} * \frac{K-1}{K^2} * [4\,Cs1 + Cp] + \lambda L * \frac{X-1}{X^2} * \frac{K-1}{K^2} * [4\,Cs2 + Cp] + \lambda o * \frac{2(X-1)}{X^2} * \frac{K-1}{K^2} * [2\,Cs1 + 2\,Cs2 + Cp] + \lambda o * \frac{(X-1)(X-2)}{X^2} * \frac{K-1}{K^2} * [4\,Cs2 + Cp]$$

After modification the total signaling cost for the existing CD technique can be calculated using Equation (14).

$$CS = \lambda L * \frac{1}{X^2 * K} * \{ 4\,Cs1 + 4\,Cs2 * (X-1)$$
$$+ X * Cp \} + \lambda o * \frac{X-1}{X^2} * \frac{K-1}{K^2} * \{ 4\,Cs1 +$$
$$4\,Cs2 * (X-1) + Cp * X \} \tag{14}$$

Proposed Call Delivery Scheme

According to the proposed scheme there are no fixed costs. The costs will be multiplied in incoming call arrival rate per unit time to calculate the total cost for every case. Also paging cost will be equal $Cp = \delta p * Cpage$ (MTi) where $Cpage$(MTi) $= M * Pr(BA) + (d - M) * (1 - Pr(BA))$ which is the average no. of cells paged to deliver a call to MTi. Also for simplicity we neglect the cache access time from calculation because it is very small value with respect to other values. The costs for the proposed scheme can be calculated from the following cases according to the proposed scheme in Figure 11:

Case 1: Inside the same LA

When a MT calls another one inside the same LA, there are computational costs. 2CV, for calling MT to querying VLR about its data and for called MT when requesting its *TLDN* from the called MSC, if there is no record for the called MT in the VLR cache. There is no need to access HLR so there is no signaling cost other than paging cells to assign the TLDN to called MT. Hence,

$$CCD = \lambda L * 2Cv * \frac{1}{K^2} * \frac{1}{X} * (1-P)$$

$$CS = \lambda L * Cp * \frac{1}{K^2} * \frac{1}{X} * (1-P)$$

Case 2: Between LAs inside the same SA

When a MT calls another one inside another LA but inside the same SA, and in case of MT's record isn't in HLR cache there are 2CV, for Calling MT to querying VLR about its data and for called MT when requesting its location from the called MSC. Also there are Cg for performing GTT to determine the current HLR and Ch when querying the current HLR for the called MT's MSC. Also there are 4 intra-SA signaling costs for communication link, two of them between the calling MSC and the current HLR and another two between the called MSC and current HLR. Hence,

$$CCD = \lambda L * [\,2Cv + Ch + Cg\,] * \frac{K-1}{K^2} * \frac{1}{X} * (1-P)$$

$$CS = \lambda L * [\,4\,Cs1 + Cp\,] * \frac{K-1}{K^2} * \frac{1}{X} * (1-P)$$

On the other hand, in case of MT's record in HLR cache, there are CV for calling MT and Cg for GTT to determine the current HLR. Also there are two intra-SA signaling costs for communication link between the calling MSC and the current HLR. Hence,

$$CCD = \lambda L * [\,Cv + Cg\,] * \frac{K-1}{K^2} * \frac{1}{X} * P$$

$$CS = \lambda L * 2\,Cs1 * \frac{K-1}{K^2} * \frac{1}{X} * P$$

Case 3: Between different SAs

When a MT calls another one inside different SA and in case of MT's record isn't in HLR cache there are 2CV, for calling MT to querying VLR about its data and for called MT when requesting its location from the called MSC. As well as there are Cg for performing GTT to determine the master HLR and 2Ch, one when querying the called MT's master HLR about its current HLR and another one for querying the current HLR for the called MT's MSC. Also there are 2 intra-SA signaling costs for communication link between the called MSC and its current HLR and 4 inter-SA signaling costs for communication link between the calling MSC, master HLR and current HLR. Hence,

$$CCD = \lambda o * [\, 2Cv + 2Ch + Cg\,] * \frac{K-1}{K^2} * \frac{x-1}{X} * (1-P)$$

$$CS = \lambda o * [\, 2\,Cs1 + 4\,Cs2 + Cp\,] * \frac{K-1}{K^2} * \frac{x-1}{X} * (1-P)$$

On the other hand, in case of MT's record in HLR cache, there is Cv for calling MT and Cg for performing GTT to determine the current HLR. Also Ch when querying the master HLR about the current HLR for called MT. As well as there are 4 inter-SA signaling costs for communication link between the calling MSC, master HLR and current HLR. Hence,

$$CCD = \lambda o * [\, Cv + Ch + Cg\,] * \frac{K-1}{K^2} * \frac{x-1}{X} * P$$

$$CS = \lambda o * 4\,Cs2 * \frac{K-1}{K^2} * \frac{x-1}{X} * P$$

So the total computational cost can be calculated as:

$$CCD = \lambda L * 2Cv * \frac{1}{K^2} * \frac{1}{X} * (1-P) + \lambda L *$$
$$[2Cv + Ch + Cg\,] * \frac{K-1}{K^2} * \frac{1}{X} * (1-P) + \lambda L$$
$$*[Cv + Cg] * \frac{K-1}{K^2} * \frac{1}{X} * P + \lambda o * [\, 2Cv +$$
$$2Ch + Cg\,] * \frac{K-1}{K^2} * \frac{x-1}{X} * (1-P) + \lambda o * [$$
$$Cv + Ch + Cg\,] * \frac{K-1}{K^2} * \frac{x-1}{X} * P$$

After modification the total computational cost for the proposed CD technique can be calculated using Equation (15).

$$\mathbf{CCD} = \lambda L * \frac{1}{\mathbf{K^2}} * \frac{1}{\mathbf{X}} * \{\, Cv * (2 * K - K * P - P) + Ch * (K-1) * (1-P) + Cg * (K-1)\} + \lambda o * \frac{\mathbf{K-1}}{\mathbf{K^2}} * \frac{\mathbf{x-1}}{\mathbf{X}} * \{\, (Cv + Ch) * (2-P) + Cg\,\} \tag{15}$$

And the total signaling cost can be calculated as:

$$CS = \lambda L * Cp * \frac{1}{K^2} * \frac{1}{X} * (1-P) + \lambda L * [\, 4$$
$$Cs1 + Cp\,] * \frac{K-1}{K^2} * \frac{1}{X} * (1-P) + \lambda L * 2\,Cs1$$
$$* \frac{K-1}{K^2} * \frac{1}{X} * P + \lambda o * [\, 2\,Cs1 + 4\,Cs2 + Cp\,]$$
$$* \frac{K-1}{K^2} * \frac{x-1}{X} * (1-P) + \lambda o * 4\,Cs2 *$$
$$\frac{K-1}{K^2} * \frac{x-1}{X} * P$$

After modification the total signaling cost for the proposed CD technique can be calculated using Equation (16).

$$\mathbf{CS} = \lambda L * \frac{1}{\mathbf{K^2}} * \frac{1}{\mathbf{X}} * \{\, 2\,Cs1 * (K-1) * (2 - P) + Cp * K * (1-P)\} + \lambda o * \frac{\mathbf{K-1}}{\mathbf{K^2}} * \frac{\mathbf{x-1}}{\mathbf{X}} \{\, 2\,Cs1 * (1-P) + 4\,Cs2 + Cp * (1-P)\,\} \tag{16}$$

PERFORMANCE COMPARISONS

Through the next subsections the performance of the proposed LM scheme will be compared against the existing scheme. Initially, we establish a system model to simulate the existing PCS networks, and then a set of experiments will be illustrated to validate the work introduced in this paper.

System Model

According to the difficulty of applying the proposed LM scheme in the real life and the need to verify it in a realistic way, an application developed to simulate a PCS network using Visual Basic given. Using 3 SAs, SAX $\forall$ X $\in$ (Brown & Mohan, 1997; Fang, Chlamtac, & Fei, 2000). Each SA has 3 LAs denoted as LAij which describes the jth LA within ith SA $\forall$ i,j $\in$ (Brown & Mohan, 1997; Fang, Chlamtac, & Fei, 2000). Each LA has N BAs denoted as BAijk which describes the kth BA within jth LA within ith SA. Assuming that 100 MTs moving among SAs, LAs or BAs randomly and can call or receive calls anywhere.

Every BA has *M* cells from *d* cells in its LA. Our experiments assuming that there are 15 cells in each LA and every BA may have 1, 2, or 5 cells. Master HLR for called MT may be the current HLR for called MT or maybe not. With the possibility of changing Mobility Rate, Call Rate and Inter-SA Signaling Cost. Also assuming that $\eta = 0.5$, $\zeta = 0.5$, $P = 0.5$ and Pr(MT) = 0.05. For comparison, we define the total cost of the proposed LM scheme and the total cost of the existing one per unit time as C', C for computational cost and S', S for signaling cost, compared with mobility rate which changed from 0.1 to 10 (Figure 12).

Also the signaling costs are normalized to CS1 such that CS1 =1. As well as to study the effect of varying CS2, we consider the cases that CS2 =2, 3 and 4 CS1 according to SA location and if there is a direct connection between 2 SAs or indirect connection as described in Figure 13.

In the following section we can see the graphs which describe the computational and signaling costs with respect to the mobility rate in different cases for the proposed LR scheme over the existing one. All the results depend on the previous

Figure 12. The employed system model

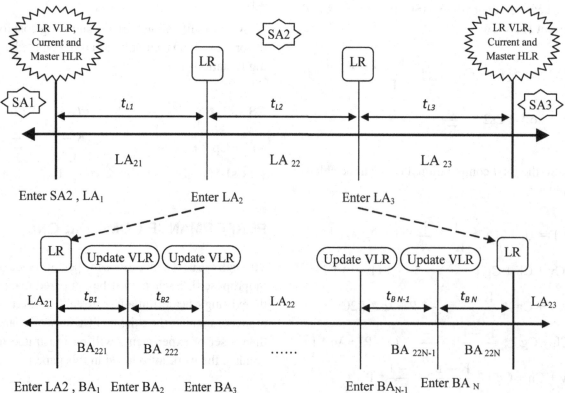

Figure 13. Direct and indirect connection between SA

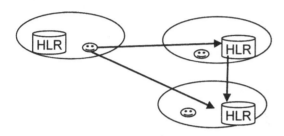

equations. Also we can observe that the computational cost for the proposed scheme higher than it for the existing scheme and grows rapidly in high mobility rate. But for signaling cost it is nearly equal in low mobility rate and grows rapidly in high rate.

In the following section we can see the graphs which describe the computational and signaling costs with respect to the mobility rate in different cases for the proposed CD scheme over the existing one. All the results depend on the previous equations. Also we can observe that the costs for the proposed scheme better than them for the existing one specially the big difference in the signaling cost.

In the following section we can see the graphs which describes the computational and signaling

costs with respect to the mobility rate in different cases for the proposed CD scheme over the existing one with changing λc . Also we can observe that the costs for the proposed scheme still better than them for the existing one especially the big difference in the signaling cost.

Also in the following section we can see the graphs which describe the total computational and total signaling costs with respect to the mobility rate in different cases for the proposed scheme over the existing one. All the results depend on the previous equations. Also we can observe that the costs for the proposed scheme better than them for the existing one in the low and medium mobility rates specially the big difference in the signaling cost.

In General, from Figures 14 through 24 we observed that LR cost for proposed scheme is higher than it for existing scheme but in CD there is saving in computational cost from 10% to 40% and saving in signaling cost from 43% to 61%. In general there is saving in total computational cost about 20% and about 52% in total signaling cost. Although the increasing of LR costs but it is not detectable by users and is not happen a lot because users stay for a period in their work, home or …. But the detectable cost for users is the CD costs.

Figure 14. Location registration cost with CS2 = 2 CS1

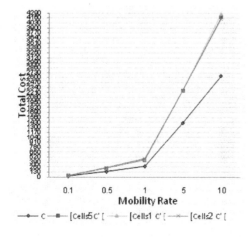

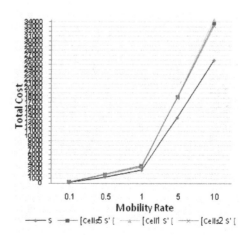

Figure 15. Location registration cost with CS2 = 3 CS1

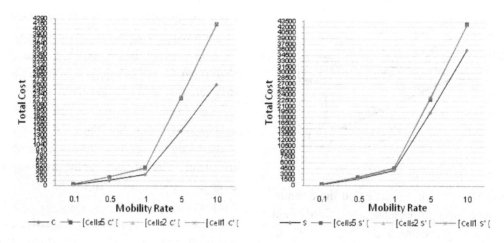

Figure 16. Location registration cost with CS2 = 4 CS1

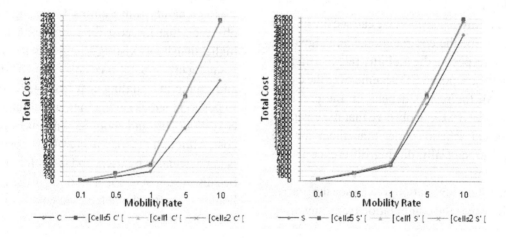

Figure 17. Call delivery cost with λc =60 and CS2 = 2 CS1

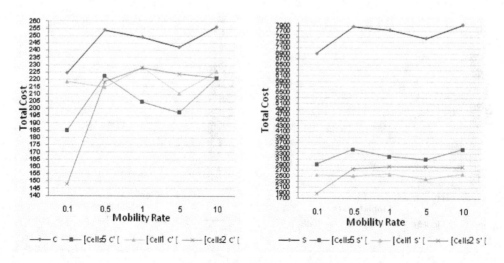

Figure 18. Call delivery cost with λc =60 and CS2 = 3 CS1

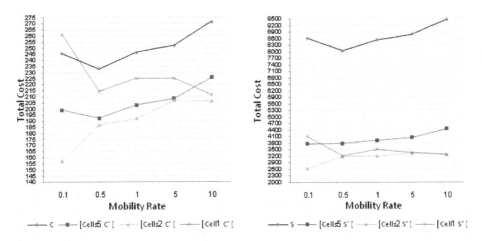

Figure 19. Call delivery cost with λc =60 and CS2 = 4 CS1

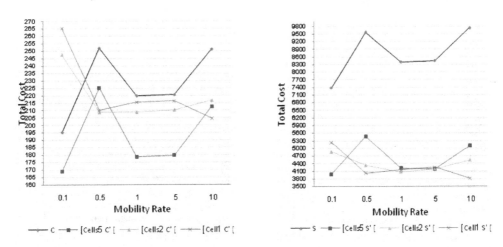

Figure 20. Call delivery cost with λc =30 and CS2 = 2 CS1

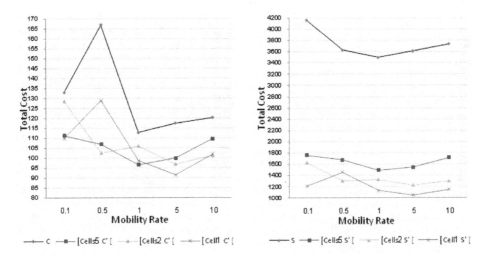

Figure 21. Call delivery cost with λc =15 and CS2 = 2 CS1

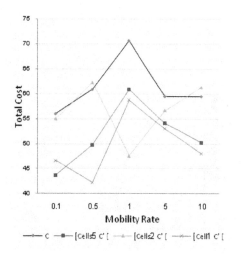 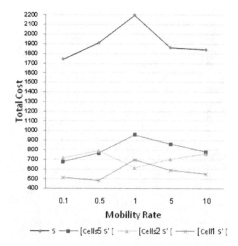

Figure 22. Total cost with λc =60 and CS2 = 2 CS1

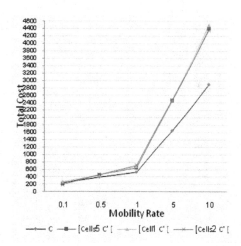 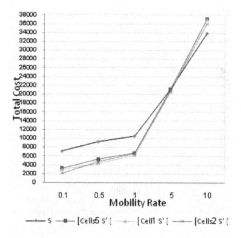

Figure 23. Total cost with λc =60 and CS2 = 3 CS1

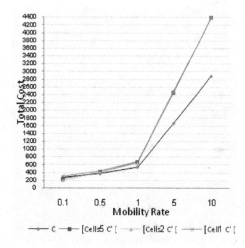 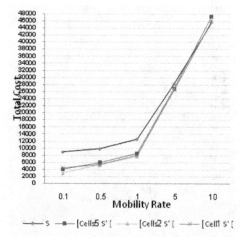

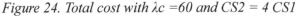

Figure 24. Total cost with λc =60 and CS2 = 4 CS1

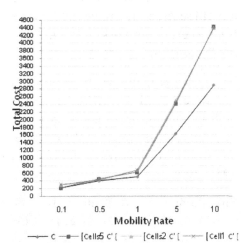
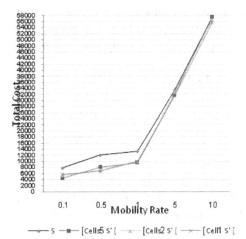

CONCLUSION

In this paper a simple and effective schema for LM in PCS networks has been presented. It provides restructuring the LAs of PCS networks into small areas called Base Areas (BAs). Moreover, it uses a 2LP scheme to accurately specify the MT's location, which will have a great impact in minimizing the paging cost. Both LR and CD incur signaling cost across the wireless channel. The more the LRs, the less the paging signaling cost, thus there is a tradeoff in terms of signaling cost. Due to the scarcity of PCS wireless bandwidth and for more scalable mobile services, it is important to reduce that signaling cost. So, using the current HLR instead of the master HLR can improve the network response by reducing the signaling cost not only in LR but also in the paging process. As well, the proposed LM strategy also employs caching to reduce unwanted LRs (updates), signaling costs, and accessing databases, which accordingly enhances the system performance. An analytical model and a simulator application are developed to study the performance of the proposed framework in a realistic way, which have shown a distinct improvement in the system

performance both in paging cost and tracing process. The performance study shows that the proposed scheme can reduce the total costs of LR and CD against the existing LM scheme significantly from about 20% to 52%.

REFERENCES

Brown, T. X., & Mohan, S. (1997). Mobility management for personal communications systems. *IEEE Transactions on Vehicular Technology, 46*(2), 269–278. doi:10.1109/25.580765

Fang, Y., Chlamtac, I., & Fei, H. (2000). Analytical results for optimal choice of location update interval for mobility database failure restoration in PCS networks. *IEEE Transactions on Parallel and Distributed Systems, 11*(6), 615–624. doi:10.1109/71.862211

Fang, Y., Chlamtac, I., & Lin, Y. (2000). Portable movement modeling for PCS networks. *IEEE Transactions on Vehicular Technology, 49*(4), 1356–1363. doi:10.1109/25.875258

Gibson, J. D. (1996). *The mobile communications handbook*. Boca Raton, FL: CRC Press.

Ho, J., & Akyildiz, I. (1997). Dynamic hierarchical database architecture for location management in PCS networks. *IEEE/ACM Transactions on Networking, 5*(5), 646–660. doi:10.1109/90.649566

Jain, R., Lin, Y., & Mohan, S. (1994). A caching strategy to reduce network impacts on PCS. *IEEE Journal on Selected Areas in Communications, 12*(8), 1434–1445. doi:10.1109/49.329333

Li, J., Kameda, H., & Li, K. (2000). Optimal dynamic mobility management for PCS networks. *IEEE/ACM Transactions on Networking, 8*(3), 319–327. doi:10.1109/90.851978

Li, J., Pan, Y., & Xiao, Y. (2004). A dynamic HLR location management scheme for PCS networks. In *Proceedings of the IEEE Annual Joint Conference INFOCOM* (Vol. 1).

Lin, Y. (1997). Reducing location update cost in a PCS network. *IEEE/ACM Transactions on Networking, 5*(1), 25–33. doi:10.1109/90.554719

Lo, C., Wolff, R., & Bernhardt, R. (1992). Expected network database transaction volume to support personal communications services. In *Proceedings of the 1st International Conference on Universal Personal Communications* (pp. 1-6).

Ma, W., & Fang, Y. (2002). Two-level pointer forwarding strategy for location management in PCS networks. *IEEE Transactions on Mobile Computing, 1*(1), 32–45. doi:10.1109/TMC.2002.1011057

Meier-Hellstern, K., & Alonso, E. (1992). The use of SS7 and GSM to support high density personal communications. In *Proceedings of the International Conference on Discovering a New World of Communications* (p. 1698).

Morris, D., & Aghvami, A. H. (2008). A novel location management scheme for cellular overlay networks. *IEEE Transactions on Broadcasting, 52*(1).

Ng, C. K., & Chan, H. W. (2005). Enhanced distance-based location management of mobile communication systems using a cell coordinates approach. *IEEE Transactions on Mobile Computing, 4*(1). doi:10.1109/TMC.2005.12

Shivakumar, N., & Widom, J. (1995). User profile replication for faster location lookup in mobile environments. In *Proceedings of the 1st Annual International Conference on Mobile Computing and Networking* (pp. 161-169).

Wong, W., & Leung, V. (2000). Location management for next generation personal communication networks. *IEEE Network*, 8-14.

Xiao, Y. (2003). A dynamic anchor-cell assisted paging with an optimal timer for PCS networks. *IEEE Communications Letters, 7*(6).

Zhang, J. (2002). Location management in cellular networks. In Stojmenovic, I. (Ed.), *Handbook of wireless networks and mobile computing* (pp. 27–49). New York, NY: John Wiley & Sons. doi:10.1002/0471224561.ch2

Zhu, Y., & Leung, C. M. (2008). Optimization of distance-based location management for PCS networks. *IEEE Transactions on Wireless Communications, 7*(9).

This work was previously published in the International Journal of Interdisciplinary Telecommunications and Networking, Volume 3, Issue 2, edited by Michael R. Bartolacci and Steven R. Powell, pp. 1-30, copyright 2011 by IGI Publishing (an imprint of IGI Global).

Chapter 7
Application of Extrinsic Information Transfer Charts to Anticipate Turbo Code Behavior

Izabella Lokshina
SUNY Oneonta, USA

ABSTRACT

This paper examines turbo codes that are currently introduced in many international standards, including the UMTS standard for third generation personal communications and the ETSI DVB-T standard for Terrestrial Digital Video Broadcasting. The convergence properties of the iterative decoding process associated with a given turbo-coding scheme are estimated using the analysis technique based on so-called extrinsic information transfer (EXIT) chart. This approach provides a possibility to anticipate the bit error rate (BER) of a turbo code system using only the EXIT chart. It is shown that EXIT charts are powerful tools to analyze and optimize the convergence behavior of iterative systems utilizing the turbo principle. The idea is to consider the associated SISO stages as information processors that map input a priori LLR's onto output extrinsic LLR's, the information content being obviously assumed to increase from input to output, and introduce them to the design of turbo systems without the reliance on extensive simulation. Compared with the other methods for generating EXIT functions, the suggested approach provides insight into the iterative behavior of linear turbo systems with substantial reduction in numerical complexity.

INTRODUCTION

Turbo codes represent a great advancement in the coding theory. Their excellent performance, especially at low and medium signal-to-noise ratios, has attracted an enormous interest for applications.

Currently, even if many research issues are still open, the success of turbo codes is growing, and their introduction in many international standards is in progress; moreover the UMTS standard for third generation personal communications and the European Telecommunications Standards Institute (ETSI) DVB-T standard for Terrestrial Digital

DOI: 10.4018/978-1-4666-2154-1.ch007

Video Broadcasting are among them (Brink, 2001; Iliev *et al.*, 2009; Lokshina, 2009).

In turbo processes the channel decoder and the demodulator at the receiver exchange extrinsic information. Such turbo processes are frequently analyzed using extrinsic information transfer (EXIT) charts (Brink, 2000; Divsalar *et al.*, 2000; Iliev & Radev, 2007; Lokshina, 2009).

This approach provides a possibility to anticipate the bit error rate (BER) of a turbo code system using only the EXIT chart. EXIT charts are powerful instruments to analyze and optimize the convergence behavior of iterative systems applying the turbo principle, i.e., systems exchanging and refining extrinsic information, without dependence on extensive simulations. Compared with the other known methods for generating EXIT functions, the proposed analytical method provides insight into the iterative behavior of linear turbo systems with substantial reduction in numerical complexity.

THE PROCESS OF EXTRINSIC INFORMATION TRANSFER

The process of extrinsic information transfer can be described with analysis of iterative turbo decoder. The block diagram of an iterative turbo decoder is shown in Figure 1, where each APP decoder corresponds to a constituent encoder and

generates the corresponding extrinsic information $\Lambda_{e,r}^{(m)}$ for $m=1,2$, using the corresponding received sequences.

The interleavers are identical to the turbo encoder's interleavers, and they are used to reorder the sequences so that the sequences at each decoder are properly aligned. The algorithm is iterated several times through the two decoders; each time the constituent decoder uses the currently calculated a posteriori probability as input.

The a posteriori probabilities produced by the first decoder are shown as (1),

$$L\left(u_r\right) = \Lambda_{e,r}^{(2)} + \Lambda_{e,r}^{(1)} + \Lambda_s \qquad (1)$$

where:

$$\Lambda_s = \log \frac{P\left(y_r^{(0)} \middle| u_r = 1\right)}{P\left(y_r^{(0)} \middle| u_r = 0\right)}$$

is the a posteriori probability of the systematic bits, which are conditionally independently distributed. Direct use of (1) would lead to an accumulation of "old" extrinsic information by calculating:

$$L_r^{(1)'} = \Lambda_{e,r}^{(1)'} + \underbrace{\Lambda_{e,r}^{(2)} + \Lambda_{e,r}^{(1)} + \Lambda_s}_{L_r^{(1)}}$$

Figure 1. Block diagram of turbo decoder with two constituent decoders

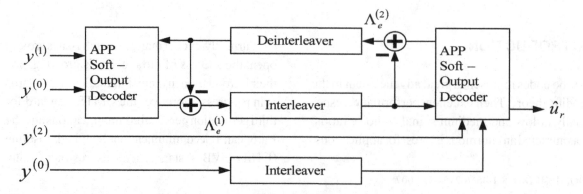

As a result, the decoders are constrained to exchange extrinsic information only, which is achieved by subtracting the input values to the APP decoders from the output values as indicated in Figure 1.

The extrinsic information is a reliability measure of each constituent decoder's estimate of the transmitted information symbols based on the corresponding received constituent parity sequence only. Since each constituent decoder uses the received systematic sequence directly, the extrinsic information allows the decoders to share information without biasing.

The effectiveness of this technique can be seen in Figure 2, which shows the performance of the original (37, 21, 65536) turbo code as a function of the decoder iterations. It is remarkable that the performance of the code with iterative decoding continues to improve with increasing number iterations, but frequently, after 6 iterations a sufficient convergence has already been reached.

CONSTITUENT DECODER AS EXTRINSIC LLR TRANSFORMER

The distance spectrum and the minimum distance of turbo codes can explain the error floor and contribute to the threshold behavior of these codes. The behavior of turbo codes at the onset of the turbo cliff is better understood by a statistical analysis, called the extrinsic information transfer (EXIT) analysis (Brink, 2001; Richardson, 2000; Iliev, 2007).

The EXIT analysis is related to similar methods that are based on the transfer of variances (Clevorn *et al.*, 2006; Iliev *et al.*, 2009). The basic philosophy of these methods is to view the constituent decoders of the turbo decoder, as a statistical processor that transforms an input value - that is, the extrinsic LLR of the information symbols - into an output value, the recomputed extrinsic LLR, as is illustrated in Figure 1.

Figure 2. Performance of the (37, 21, 65536) Turbo Code as Function of Decoder Iterations

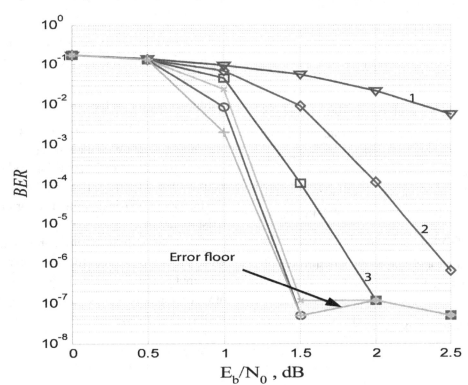

Figure 3. Constituent Decoder Considered as Extrinsic LLR Transformer

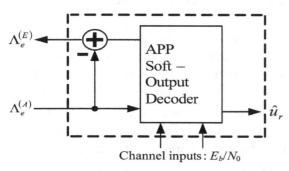

Channel inputs: E_b/N_0

- The upper limit indicates high reliability that is, if $\Lambda_e^{(m)}$ has variance σ_λ^2 then $\sigma_\lambda^2 \to 0$;
- The lower bound indicates low reliability, and $\sigma_\lambda^2 \to \infty$;
- The measure is monotonically increasing.

In order to capture the input–output behavior of the constituent decoder, simulated extrinsic values $\Lambda_e^{(A)}$ are generated at its input, according to the Gaussian distribution (2) with independent realizations. This independence models the effect of the interleaver, which, ideally, destroys any dependence between successive LLR values.

In the case of Gaussian model input LLR values, the mutual information measure, in bits, can be calculated as:

$$I_A = I(u; \Lambda_e^{(A)}) =$$
$$= \frac{1}{\sqrt{2\pi}\sigma_\lambda} \int_{-\infty}^{\infty} \exp\left(-\frac{(\lambda - \mu_\lambda)^2}{2\sigma_\lambda^2}\right)$$
$$(1 - \log_2[1 + \exp(-\lambda)])d\lambda$$

The constituent decoder is after seen as a nonlinear LLR transformer, as is shown in Figure 3. The input LLR is denoted by $\Lambda_e^{(A)}$ since it takes on the function of a priori probability, and the output extrinsic LLR is denoted by $\Lambda_e^{(E)}$.

Evidently, we expect the constituent decoder to improve the extrinsic LLR in the course of the iterations, so that $\Lambda_e^{(E)}$ is better than $\Lambda_e^{(A)}$, in some sense, since otherwise the iterations will lead nowhere. Measurements of $\Lambda_e^{(m)}$ show that it has an approximately Gaussian distribution, presented as (2),

$$\Lambda_e^{(m)} = \mu_\lambda u + n_\lambda; \; n_\lambda \approx N\left(0, \sigma_\lambda^2\right) \qquad (2)$$

where $u \in \{-1, 1\}$ is the information bit whose LLR is expressed by $\Lambda_e^{(m)}$, μ_λ is the mean, and n_λ is a zero-mean independent Gaussian random variable with variance σ_λ^2.

One of the main questions is how to measure the reliability of $\Lambda_e^{(m)}$. The mutual information $I\left(u, \Lambda_e^{(m)}\right)$ between $\Lambda_e^{(m)}$ and u has proven to be the most accurate and convenient measure, which has the following advantages:

- The measure is bounded $0 \le I\left(u, \Lambda_e^{(m)}\right) \le 1$;

The mutual information of the output extrinsic LLR and the information bit u is more complex to evaluate since $\Lambda_e^{(E)}$ is not exactly Gaussian. Additionally, the values of $\Lambda_e^{(E)}$ corresponding to $\Lambda_e^{(A)}$ must be found when simulating the constituent decoder. Other than in very simple cases, such as the rate $R=1/2$ repetition code, there is neither closed form nor even analytical formulas for the output extrinsic mutual information that exist to date.

Then, the output extrinsic information I_E is an empirical function of the constituent decoder, the input I_A and the channel signal-to-noise ratio at which the decoder operates, formally provided by the EXIT function:

$$I_E = T\left(I_A, E_b/N_0\right) \qquad (3)$$

To evaluate this function the following steps were executed. At the first step a Gaussian model input LLR with mutual extrinsic information value I_A was generated and sent into the decoder and the output extrinsic LLRs $\Lambda_e^{(E)}$ were achieved. At the second step the numerical estimation of mutual information between the information symbols u and the obtained extrinsic output LLRs was completed.

$\Lambda_e^{(E)}$ yields I_E given as:

$$I_E =$$

$$\frac{1}{2}\sum_{u\pm 1}\int_{-\infty}^{\infty} p_E\left(\xi|u\right)\log_2\left|\frac{2p_E\left(\xi|u\right)}{\begin{array}{c}p_E\left(\xi|U=-1\right)\\+p_E\left(\xi|U=1\right)\end{array}}\right|d\xi$$

(4)

where $p_E\left(\xi|u\right)$ is the empirical distribution of $\Lambda_e^{(E)}$, as measured at the output of the APP decoder.

NUMERICAL RESULTS

The recommended method of assessing the output extrinsic information is demonstrated in Figure 4. The EXIT function $T(I_A, E_b/N_0)$ is considered for a memory 4, rate $R=1/2$ constituent convolutional code with $h_0(D)=D^4+D^3+1$ and $h_1(D)=D^4+D^3+D^2+D+1$. The numbers on the trajectories are correspondent to the E_b/N_0 calculated in dB.

In a full turbo decoder this extrinsic information is exchanged between the decoders, and the output extrinsic LLRs become input extrinsic LLRs for the next decoder. In suggested approach these iterations are captured by a sequence of applications of constituent decoder EXIT functions, presented as (5),

$$0 \xrightarrow{T_1} I_E^{(1)} = I_A^{(2)} \xrightarrow{T_2} I_E^{(2)} = I_A^{(2)} \xrightarrow{T_1} I_E^{(1)} = I_A^{(2)} \xrightarrow{T_2} ..,$$

(5)

where T_1 is the EXIT function of decoder 1 and T_2 is that of decoder 2. The behavior of this exchange can be visualized in the EXIT chart, where the

Figure 4. EXIT Function of a Rate 1/2, 16-state Convolutional Code

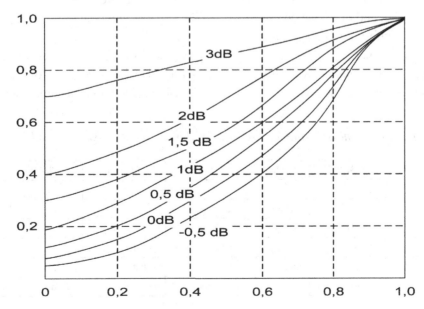

Figure 5. EXIT Chart Combining EXIT Functions of Two 16-State Decoders Involved in Turbo Decoder for $E_b/N_0 = 0.8dB$

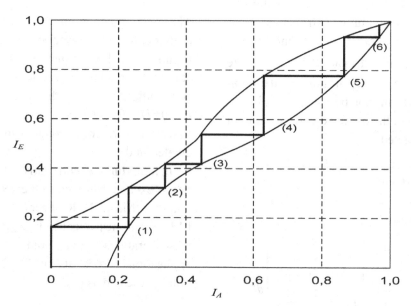

EXIT functions of both constituent decoders are plotted that is shown in Figure 5.

Since the iterations start with zero LLRs - that is, with zero mutual information at point (0, 0) in the chart - the iterations progress by bouncing from one curve to the other as is illustrated in Figure 5. The first iteration through decoder 1 takes an input value of $I_A=0$ and produces an output value of $I_E=0.18$. This value then acts as the input to decoder 2, and after one completed iteration the mutual information has a value of 0.22, at point (1) in the plot.

Analogously, the iterations proceed through points (2), (3), and so on, until they reach $I_A=I_E=1$

after about 6 iterations. The trajectory plotted in Figure 5 is a measured characteristic of a single decoding cycle of the turbo decoder, and the accuracy with respect to the prediction of the EXIT chart is impressive, confirming the robustness of this method. The growing discrepancies between the measured characteristic and the EXIT chart starting at iteration 7 are the outcomes based on the finite-size interleaver, in this case $N=60,000$. The superior interleaver selected produces more accurate EXIT chart, which later is used as a prediction tool. The obtained results are shown in Table 1, where the pinch-off values are given for all combinations of constituent codes. The

Table 1. Pinch-off signal-to-noise ratios for various combinations of constituent codes

v	C_1/C_2	(3, 2)	(7, 5)	(13, 15)	(23, 37)	(67, 45)	(147, 117)
1	(3, 2)	>2 dB					
2	(7, 5)	1.49 dB	0.69 dB				
3	(13, 15)	1.14 dB	0.62 dB	0.62 dB			
4	(23, 37)	1.08 dB	0.65 dB	0.64 dB	0.68 dB		
5	(67, 45)	0.86 dB	0.62 dB	0.66 dB	0.70 dB	0.77 dB	
6	(147, 117)	0.84 dB	0.63 dB	0.67 dB	0.72 dB	0.81 dB	0.84 dB

most of processed iterations occur at very high bit error rates, that is, $P_b > 10^{-2}$, in fact $I_E \approx 0.9$ corresponding to $P_b = 10^{-2}$.

The analysis shows that the EXIT chart is an effective instrument to evaluate constituent decoders, their cooperative statistical behavior such as onset of the turbo cliff, number of iterations to convergence, etc, but on the other side delivers no information about the error floor performance of a code and its ultimate operational performance. As an example of using the EXIT chart to predict start of the turbo cliff behavior, the signal-to-noise ratio E_b/N_0 for the codes in Figure 5 was reduced until the open channel was closed at $E_b/N_0 = 0.53$ dB. This demonstrates the pinch-off signal-to-noise ratio of the original turbo code combinations for encoders.

The sharpness of this pinch-off signal-to-noise ratio depends on the size of the interleaver, as seen in Table 1, since finite-size interleavers always leave some residual correlation, in particular for larger iteration numbers, which violates the assumption of independence used to generate the EXIT functions of the constituent decoders.

CONCLUSION

One of the most important parameters to evaluate the channel coding methods is the bit error rate (BER) vs. E_b/N_0, where E_b denotes the energy per binary information symbol, and N_0 denotes the noise power spectral density, that is the probability for bit error at a given ratio E_b/N_0. During transmission the digital data (e.g., the encoded signal) is affected by different factors, including interference, fading, and channel noise, that distort the signal.

If the receiver is to reach a given BER (that is a residual bit error rate as the BER after enhancement, error detection and recovery is considered), for example, 10^{-5}, then it is necessary either to increase the transmitter power, or to use proper methods for channel coding. As of all practical

error correction methods known to date, turbo codes and low-density parity-check codes (LDPCs) come closest to approaching the Shannon limit (≈ 0.5dB), the theoretical limit of maximum information transfer rate over a noisy channel.

This is possible through iterative decoding process, which allows correction of larger number of error bits during the transmission in the channel, but also increases decoding time. After 18 iterations, a BER of 10^{-5} is reached at an E_b/N_0 ratio at only 0.5dB from the theoretical limit predicted by the Shannon capacity.

A new analytical approach to analysis and optimization of the convergence behavior of iterative systems utilizing the turbo principle, i.e., systems exchanging and refining extrinsic information with EXIT charts, was proposed. An implementation of the suggested approach for determining the pinch-off signal-to-noise ratios for the turbo codes, using the constituent codes and their EXIT functions was demonstrated.

We considered results presented in Table 1 as pinch-off values for all possible combinations of constituent codes. As can be seen in the numerical examples, the use of larger codes, e.g. v>4 did not produce stronger turbo codes since the pinch-off was not improved. A combination of two strong codes provided a worse pinch-off value when to compare with one of the best combinations that had a pinch-off signal-to-noise ratio of 0.53 dB, as shown in Table 1. That has demonstrated the likelihood for a superior choice of constituent codes with v=4, (37, 21).

REFERENCES

Brink, S. (2000). Design of serially concatenated codes based on iterative decoding convergence. In *Proceedings of the Second International Symposium on Turbo Codes and Related Topics*, Brest, France (pp. 319-322).

Brink, S. (2001). Convergence behavior of iteratively decoded parallel concatenated codes. *IEEE Transactions on Communications, 49*(10), 1727–1737. doi:10.1109/26.957394

Clevorn, T., Godtmann, S., & Vary, P. (2006). BER prediction using EXIT charts for BICM with iterative decoding. *IEEE Communications Letters, 10*(1), 49–51. doi:10.1109/LCOMM.2006.1576566

Divsalar, D., Dolinar, S., & Pollara, F. (2000). Serial turbo trellis coded modulation with rate-1 inner code. In *Proceedings of the International Symposium on Information Theory*, San Francisco, CA (pp. 194-200).

Iliev, T. (2007). Analysis and design of combined interleaver for turbo codes. In *Proceedings of the Romanian Technical Sciences Academy*, Bacau, Romania (Vol. pp. 148-153).

Iliev, T., Lokshina, I., & Radev, D. (2009). Use of extrinsic information transfer chart to predict behavior of turbo codes. In *Proceedings of the IEEE Wireless Telecommunications Symposium*, Prague, Czech Republic (pp. 1-4).

Iliev, T., & Radev, D. (2007). Turbo codes in the CCSDS standard for wireless data. In *Proceedings of the Papers ICEST*, Ohrid, Macedonia (pp. 31-35).

Lokshina, I. (2009). Analysis of Reed-Solomon codes and their application to digital video broadcasting systems. In *Proceedings of NAEC*, Riva del Garda, Italy (pp. 31-36).

Richardson, T. (2000). The geometry of turbo decoding dynamics. *IEEE Transactions on Information Theory, 46*, 9–23. doi:10.1109/18.817505

This work was previously published in the International Journal of Interdisciplinary Telecommunications and Networking, Volume 3, Issue 2, edited by Michael R. Bartolacci and Steven R. Powell, pp. 31-37, copyright 2011 by IGI Publishing (an imprint of IGI Global).

Chapter 8
A Femtocellular–Cabled Solution for Broadband Wireless Access:
A Qualitative and Comparative Analysis

Dario Di Zenobio
Fondazione Ugo Bordoni, Italy

Lorenzo Pulcini
Fondazione Ugo Bordoni, Italy

Massimo Celidonio
Fondazione Ugo Bordoni, Italy

Arianna Rufini
Fondazione Ugo Bordoni, Italy

ABSTRACT

Broadband Wireless Access is a strategic opportunity for mobile operators which aim to provide connectivity in digital divide areas, in order to accelerate speed of deployment and save in installation costs. This paper presents an innovative approach to access the end user, relying on infrastructural integration of femtocellular technology with existing cabled network. Usually, the adoption of Femtocell Access Points, operating in the licensed cellular bands typically designed to be used in SOHO, improves the radio coverage and the building penetration of the existing mobile networks, based on macrocells. In the proposed solution, the peculiar functionality of femtocells is further improved using a MATV/SMATV cabled infrastructure which facilitates the signal connection inside the building. The potentiality of the solution is even more evident, taking into account the growing interest towards the possible deployment of new mobile technologies, like LTE in both the last portion of the UHF band V and the GSM frequency band, resulting from the re-farming process.

INTRODUCTION

During the last decades people have experienced a relevant diffusion of wireless devices. However, although about 70% of calls occur indoors, cellular phones continue to face issues such as poor signal strength and service quality when calls are made into buildings.

In order to improve signal quality in critical areas, mobile operators are always looking for new architectural solutions and the deployment of femtocells is a possible answer to this key issue.

Technically, femtocells are low power wireless access point installed in Small Office-Home office (SOHO) environment to provide voice and broadband services increasing throughput for mobile data services.

DOI: 10.4018/978-1-4666-2154-1.ch008

They are similar in size to a router and offer excellent indoor signal coverage (2G/3G/4G), thereby reducing traffic load on the external macrocell. In particular the femtocell approach leads to an increased data capacity of the overall system, determined by offloading traffic load from the mobile network towards the fixed one (xDSL connections, etc.), as well as a reduction of radiated power, infrastructure deployment and maintenance costs.

Strategic positioning of femtocell systems inside buildings, combined with their ability to deliver high-quality voice and data services, and to bridge mobile terminals with the rest of the home or office networks, permits to build compelling business cases for both landline system replacement and fixed/mobile services convergence. In fact, through the stipulation of a specific contract with the mobile operator it is possible to make the mobile terminal working as a domestic/office handset.

On one side the key benefits of adopting femtocells are numerous for both operators and users. On the other side operators must still face with several challenges in order to be able to deploy a large number of Femtocell Access Points (FAPs) operating jointly to the existing macrocells. In this context, for example, handovers (femto to femto), handouts (femto to macro), handins (macro to femto), electromagnetic interference mitigation and management, device miniaturization and general costs optimisation, still remain the major problems that strongly prevent a large deployment of multiple-tier cellular networks. In addition, mobile service providers do not have a plan to deploy these network devices, so they are doing the job of getting it in the customer's hands. On the contrary, the customers bristle at the idea to pay an extra cost for a provider's inability to guarantee service to them.

To pave the way to an extensive femtocells deployment different solutions are under study. Among them, the one proposed in the present paper, based on the reuse of cabled infrastruc-tures already existing in the customer premises, requires low financial impact and minimum time commitment. More in detail, the cabled-wireless solution makes use of an "enhanced femtocell", as described in the following, integrated into a cabled distribution network (e.g. condominium SMATV/MATV infrastructure). This system might result advantageous in buildings located in extremely critical areas, where the customer, at his apartment, suffers a poor radio coverage from macrocell base station or the signal is definitely not present, while other customers in the same condominium, but at different floors, are still able to receive and transmit from/to the same base station.

The structure of the paper is as follows. At first, a brief introduction relating to the femtocellular technology is provided. After that a description of the mentioned solution based on a specific architectural configuration (Femto/Cable), with a detailed characterization concerning the propagation models adopted for different applicative scenarios is given. Successively, an extensive comparison, based on simulation results concerning signal power reception in different operative scenarios between the enhanced femtocell solution and the traditional macrocell one, is reported. In conclusion, some final comments aiming at analysing the system features and performance are reported.

LTE FEMTOCELL ARCHITECTURES: A BRIEF OVERVIEW

As previously mentioned, a femtocell is a small cellular base station, typically designed to be implemented in the customer premises (home or small business) to connect mobile users to the service provider's network through a broadband fixed connection (such as DSL or cable). Current designs typically support from 2 to 4 active mobile phones in a residential setting, and from 8 to 16 active mobile phones in enterprise settings. Femtocells operate on licensed spectrum and provide excellent user experience through better coverage for voice and very high data throughputs.

The advantages of adopting a femtocellular technology are numerous for a mobile operator: from the ability to improve both coverage and system capacity, especially indoors, reducing both capital expenditures and operating expenses, to the opportunity to provide new services to the customers. On the other hand, also the consumers could benefit of improved coverage, potential better quality of service and longer battery life in the mobile terminal. In addition more attractive tariffs (e.g. discounted calls from home) may also be offered.

Typically femtocells are connected both to the Internet, by making use of a broadband connection such as Digital Subscriber Line (DSL), cable or optical fiber (which will be typically shared with other network devices in the home), and to the cellular operator's network as well (Figure 1).

Concerning the air interface, FAPs provide radio coverage using specific wireless technologies. In this paper we will consider femtocells (as well as macrocells) using LTE as wireless technology.

Since indoor traffic will be transmitted over the Internet Protocol backhaul, this femtocellular configuration will help the operator to manage the expected exponential growth of traffic reducing the overall network cost.

As LTE is based on a flat all-IP architecture, the interfaces are the same for femtocells as for macrocells. LTE femtocells (Home eNodeBs) require no new interfaces to be defined and no changes are required to core network elements.

There are multiple possible architectures for connecting LTE femtocells to the core network. A new optional element, called Home eNodeB Gateway (HeNB GW), is defined to provide aggregation of multiple Home eNodeBs traffic data to the core network, as shown in Figure 2. The HeNB GW aggregates S1 interfaces (S1-MME and S1-U), potentially improving the scalability of the core network with regards to femtocells.

An alternative approach to achieve a better signal coverage is based on the usage of Relay Stations (RSs) which extend the macro/micro cellular networks in indoor areas. Relay transmission can be seen as a kind of collaborative communication, in which a RS helps to forward user data from neighbouring user equipment (UE) to a macro base station (macro BS), extending its signal and service coverage and enhancing the

Figure 1. Typical femtocellular access network

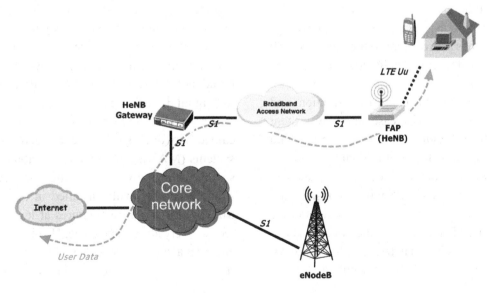

Figure 2. LTE Femtocell Architecture

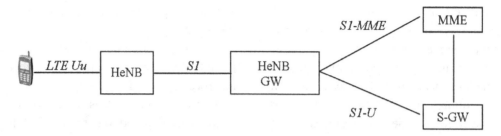

overall throughput performance of the cellular communication system. Performances of relay transmissions are greatly affected by the collaborative strategy, which includes the selection of relay types and relay partners (i.e., to decide when, how, and with whom to collaborate). Two types of RSs (Type-I and Type-II) have been defined in 3GPP LTE-Advanced standards.

Specifically, a Type-I RS can help a remote UE unit to access to a BS which is located far away from it. So a Type-I RS needs to transmit the common reference signal and the control information for the BS, and its main objective is to extend signal and service coverage. Type-I RSs mainly perform IP packet forwarding in the network layer (layer 3) and can make some contributions to the overall system capacity by enabling communication services and data transmissions for remote UE units.

On the other hand, a Type-II RS can help a local UE unit, which is located within the coverage of a BS and has a direct communication link with it, to improve its service quality and link capacity. So a Type-II RS does not transmit the common reference signal or the control information, and its main objective is to increase the overall system capacity by achieving multipath diversity and transmission gains for local UE units.

In the peculiar approach described in the following, only the Type-I RS will be considered.

Although on this paper main attention is focused on LTE technology, the concept is applicable to all mobile standards, including GSM, CDMA2000, TD-SCDMA, and WiMAX.

THE PROPOSED FEMTO/ CABLE SOLUTION

In 2012 should be concluded the transition process from analog to digital television system: the switch-off process. In many countries it will make possible the release of the last portion of the UHF band V, that ranging from 790 to 862 MHz (or from channel 61 to channel 69), where are currently broadcasted TV signals both in analog and digital formats, to provide new generation mobile services. In addition, in December 2013, it will be concluded the refarming process of the entire GSM band, to allow the introduction of 3G, and eventually 4G, services in the 900 MHz (880-915 MHz, 925-960 MHz) and 1800 MHz (1710-1785 MHz, 1805-1880 MHz) frequency bands with coverage and cost-saving benefits.

In this context, femtocells can find a large application in all circumstances customer experiences a poor indoor macrocell signal.

The peculiar solution, proposed in this paper and named "Femto/Cable", allows to extend broadband mobile services inside buildings found in "digital divide" areas.

Essentially, the idea is to exploit the existing cabled networks, such as centralized television systems (MATV, SMATV), to distribute 3G/4G signals within individual apartments/offices. This solution can be easily installed in the existing MATV/SMATV infrastructures at low cost and requires very low maintenance. Thanks to the fixed end-user access network (e.g. MATV), and to the frequency bands considered for the 3G/4G technologies, no serious problems of signal attenuation are expected. Moreover, the potential presence of deterministic signal degradation elements, intrinsic in the cabled infrastructure, could be eliminated through an appropriate design and installation. To this aim, specific guidelines could be found in the detailed description of the *Cabled-Wireless END USER ACCESS (C-W EUA)*, that is a Fondazione Ugo Bordoni patented solution (2002).

In the mentioned operative circumstances (absence of broadband fixed connection), and even in very poor experienced radio mobile signal, a good level of performance could be achieved through the use of an "enhanced femtocell". It can transmit towards the end-users an higher power than the traditional one (e.g. 26 dBm), thanks to the cabled way, and can profit of a facilitated TX/RX way to/from the macrocell base station, as located at the roof of the end-user building, in quasi-Line Of Sight (LOS) conditions.

The Femto/Cable solution (the architectural configuration is shown in Figure 3) is particularly efficient in terms of indoor coverage area extension, because avoids the propagation attenuation due to walls penetration.

The proposed system consists of a special interface, named Condominium Set Top Box (C-STB), usually located at the roof of the end user building, transmitting the TV UHF channels (broadcasted on the 470-790 MHz frequency band) and the 3G/4G signals in the same cabled infrastructure: the SMATV plant. The radio mobile

Figure 3. Enhanced Femtocell architectural configuration

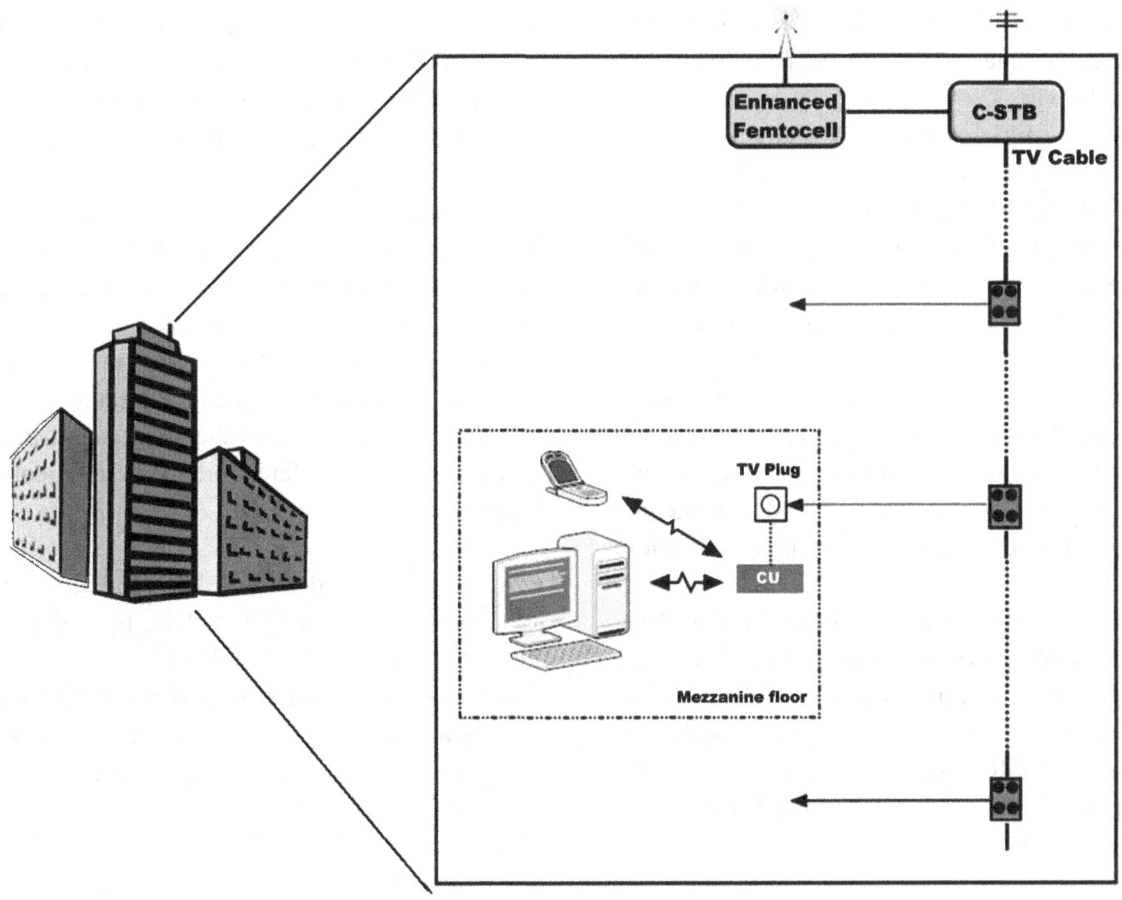

Figure 4. DVB-T and radio mobile services spectrum allocation

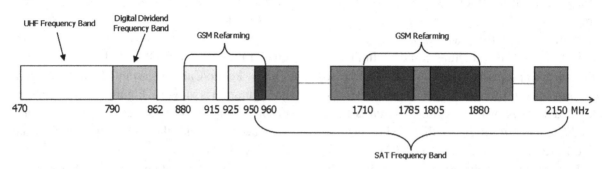

signals are relayed and amplified by a specific equipment operating in the 790-862 MHz frequency band (addressed to new mobile generation) or, alternatively, in the frequency bands involved in GSM re-farming process as well as the ones actually used for 3G mobile services.

As it is evident in Figure 4, when used in SMATV systems, this solution is applicable only for 3G/4G signals transmitted in the digital dividend spectrum, because in the other application cases, interferences in the cable with SAT signals exist.

The resulting mixed (TV+3G/4G) signal is transmitted to the Customer Unit (CU) through the existing MATV, SMATV plant. At end-user side the CU, plugged to the coax network, will separate the TV signal from the 3G/4G one, and will interface the TV set as well as a passive 3G/4G antenna.

Obviously, the system performances can be enhanced with the use of appropriate filters and power amplifiers, but in this paper they are not mentioned. Details could be found in a patented solution developed in Fondazione Ugo Bordoni (2002).

On the basis of the previous considerations, the advantages of the Femto/Cable system are evident when the macrocell base station is located quite far from the user building, so that the received 3G/4G signal is strongly attenuated and its power level is below the sensibility threshold of UE (S_{rx}).

EXPERIMENTAL SCENARIO

The analysis that we intend to carry out in this paper focuses on two case studies:

a. The first one considering the radio signal received, directly from the macrocell, through a terminal located close to the external facade of the building, at different floors;

b. The second one related to a terminal, in the same position, receiving the radio signal from the proposed femto/cable solution, at a distance of 5m from the CU.

The macro base station, in the simulated radio link, is supposed to be located at different distances from end-user building, at the roof-top level and in Line Of Sight (LOS) conditions.

In the analysis, specific path loss models, selected within a range accepted in literature, have been considered to calculate the signal attenuation as it propagates in different urban environments. Precisely:

• Outdoor to indoor path loss modelling for macrocell: Okumura – Hata, for the case study a).

• Indoor to indoor path loss modelling for femtocell: ITU P.1238, for the case study b), strictly limited to the propagation from the CU to the UE.

In detail, the expression of Okumura – Hata path loss model is provided in (1)

$$L_U = 69,55 + 26,16 * \log_{10}(f)$$
$$-13,82 * \log_{10}(h_B) - C_H + (44,9$$
$$-6,55 * \log_{10}(h_B)) * \log_{10}(d)$$

(1)

where:

- L_U: Signal attenuation in urban areas (dB);
- f: Frequency (MHz);
- h_B: Fixed station antenna height (m);
- C_H: Antenna height correction factor (dB);
- d: Distance between the fixed station and mobile terminal (km).

For medium-sized building must be considered the following expression of the correction factor C_H:

$$C_H = (1,1 * \log(f) - 0,7) * h_M - (1,56 * \log(f) - 0,8)$$

where h_M is the mobile terminal antenna height (m).

ITU P.1238 indoor propagation model (2) provides a path loss between two indoor terminals assuming an aggregate loss, which takes into account obstacles like furniture, internal walls and doors, and it is characterized by a power loss exponent factor N depending on the type of building.

$$L_{total} = 20 * \log_{10}(f) + N * \log_{10}(d)$$
$$+ L_f(n) - 28$$

(2)

where:

- L_{total}: Signal attenuation in indoor areas (dB);
- N: Distance power loss coefficient;
- f: Frequency (MHz);
- d: Distance (m) between the fixed station and mobile terminal (where $d > 1$ m);

- L_f: Floor penetration loss factor (dB), in the present analysis it should be considered not influent;
- n: Number of floors between fixed station and mobile terminal ($n \geq 1$).

In particular, in case of 800-900 MHz frequency band, the ITU P.1238 model fixes the power loss coefficient N equal to 33. In the present case study, L_{total} can be simplified to the indoor attenuation (L_{indoor}), in absence of floor to floor attenuation.

As far as the MATV signal distribution is concerned, the RG58 coax parameters have been adopted. The nominal attenuation of this cable is equal to 0.48 dB per meter. Furthermore, an additional signal attenuation of 3 dB per floor has been considered to take into account the presence of switches/splitters. Alternatively, in case of SMATV signal distribution, CCF SAT coaxial cables have been considered with an attenuation of 0.18 dB per meter.

Expressions (3) and (4) are used to calculate the received signal power at CU (P_{rx1}) and UE station (P_{rx2}) respectively, transmitted by enhanced femtocell located at the roof of the building through the cabled infrastructure and the CU.

$$P_{rx1} = EIRP\_f + G_{ua} - L_{c\_ua} - L_{ca} - L_s$$

(3)

$$P_{rx2} = P_{rx1} - L_{indoor} + G_{ua} - L_{c\_ua} + G_{ue}$$
$$-L_{c\_u}$$

(4)

In expression (3) and (4) the following parameters are considered:

- $EIRP\_f$ is the effective transmitted power by the enhanced femtocell;
- G_{ua} is the CU antenna gain;
- G_{ue} is the UE antenna gain;
- L_{c_ua} is the antenna connector loss;

- L_{ca} and L_s are the cables and splitter attenuation, respectively;
- L_{indoor} is the path loss in the domestic environment;
- L_{c_u} summarizes the UE connector and body losses.

ANALYSIS OF RESULTS

The simulation results are reported in terms of coverage areas and in terms of comparison between indoor received power at the user terminal, when the signal is received directly from the macrocell, and when it is received from the proposed femto/cable solution. Related data rates available at UE are also calculated. The tests have been carried out with a simulation tool operating in the above mentioned frequency band, considering a building of 5 floors. In the cabled infrastructure, at every floor, a switch-box routes the signal into apartments where, before reaching the CU, it passes through five splitters. For each of these devices insertion loss and attenuation have been considered.

As previously underlined, the performances have been evaluated using the above mentioned propagation loss expressions applied to the specific communication environment and to the particular frequency band involved.

Figure 5 shows the obtained results in terms of received power, with the macrocell positioned at 300 and 500 meters of distance from the end-user building. The figure details the trend of signal power received into the apartments, in correspondence of different floors of the building.

In this case the results show that, for each floor of the building, the signal power received by the end-user directly from macrocell is quite similar to the one received from femto/cable system.

On the contrary, when the macrocell is located at 3 and 5 km away from the building, as evidentiated in Figure 6, the femto/cable solution offers better results in respect to the macrocell direct link.

Based on these results and aiming to give a rough idea of the throughput values achievable through the use of the proposed system, an LTE signal, operating in the 900 MHz frequency band, with a bandwidth of 5 MHz, that propagates in a AWGN (Additive White Gaussian Noise) channel, has been considered.

Figure 5. Different P_{rx} values from near macrocell and from femto-cable at f=900 MHz

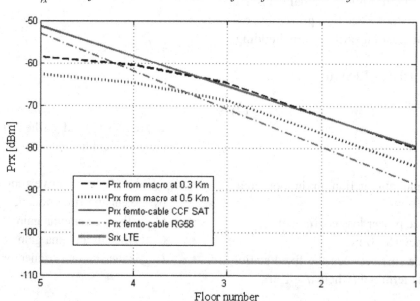

Figure 6. Different P$_{rx}$ values from macrocell and from femto-cable at f=900 MHz

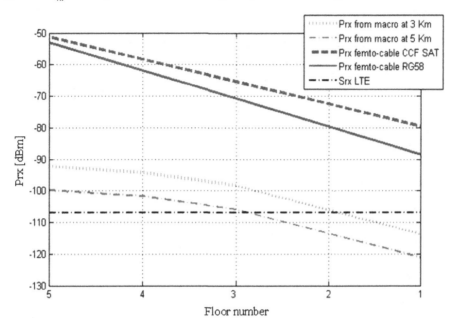

In these hypotheses, it is possible to estimate the value of the SNR (Signal to Noise Ratio) at UE receiver taking into account the White Gaussian noise calculated through the following formula:

$$N = -173.8 + F + 10 * \log_{10}(B) \qquad (5)$$

In Table 1 are reported the numerical results of the SNR values for the mentioned scenarios (assuming a noise factor $F = 9dB$) and, in Figure 7, is shown the trend of the SNR, for both RG58 and CCF SAT coaxial cables, and for different building floors.

By using a link level simulator, the throughput of the proposed system as a function of the SNR has been estimated.

In Figure 8, the resulting LTE throughputs calculated for different CQI (Channel Quality Indicator) are reported, jointly to the different type of adopted modulation scheme (QPSK, 16QAM and 64QAM), therewith including different coding rates.

By interpolating the above calculated SNR results with LTE data throughputs summarized in Figure 8, it has been possible to determine the throughput for the proposed femto/cable solution calculated at 900 MHz frequency band, considering CCF SAT coaxial cables.

These results have been compared with throughput values calculated at 2 GHz frequency band (Figure 9) to highlight the good performances the proposed solution can provide, especially

Table 1. SNR at f = 900 MHz

Floor number	5	4	3	2	1
SNR [dB] from macro at 3 Km	5.55	3.61	-0.51	-8.16	-15.81
SNR [dB] from macro at 5 Km	-1.93	-3.87	-8.01	-15.65	-23.30
SNR [dB] femto-cable (with CCF SAT)	46.57	39.49	32.41	25.34	18.26

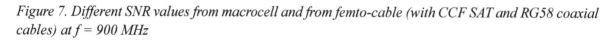

Figure 7. Different SNR values from macrocell and from femto-cable (with CCF SAT and RG58 coaxial cables) at f = 900 MHz

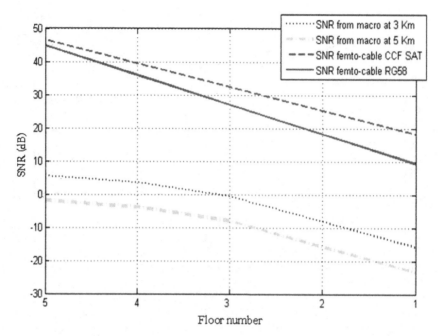

Figure 8. LTE Throughput vs SNR

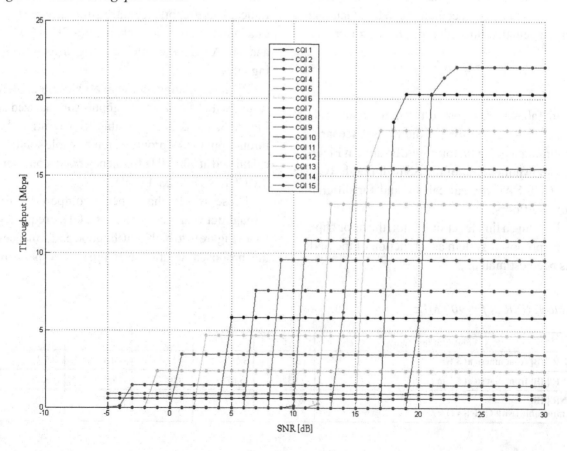

Figure 9. Femto-Cable LTE Throughput vs. UE floor number

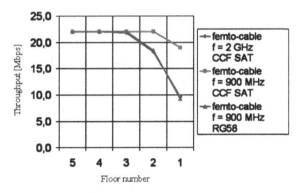

taking into account that, in absence of the femto/cable system, the LTE signal would be very low or missing.

CONCLUSION

The simulation results for the proposed innovative broadband wireless access solution, based on a femtocell Relay Station integrated with existing cabled MATV/SMATV networks, highlight that it could facilitate the provision of the expected IMT-Advanced services to users, resident in "digital divide" areas, in a cost-efficient way.

The proposed approach requires the installation of a less number of macro base stations, leading to cost savings and lower time commitments to roll-out 3G/4G services. In addition, the usage of a single enhanced femtocell per condominium simplifies the configuration and interference problems of the equipment, optimizes the roll-out planning, and can fix the investment plan of the operators.

Jointly to the mentioned advantages, the improvement of QoS performances, achievable through the use of the femto/cable solution, in sites where the radio signal is absent or very low and no alternative access techniques are offered by the fixed network, justify the effectiveness of the proposed approach.

REFERENCES

3rd Generation Partnership Project. (2008). *TS 36.401 V8.4.0: Evolved universal terrestrial radio access network (E-UTRAN); architecture description (Release 8)*. Retrieved from http://www.quintillion.co.jp/3GPP/Specs/36401-840.pdf

3rd Generation Partnership Project. (2008). *UTRA-UTRAN long term evolution (LTE) and 3GPP system architecture evolution (SAE)*. Retrieved from ftp://ftp.3gpp.org/Inbox/2008_web_files/LTA_Paper.pdf

3rd Generation Partnership Project. (2010). *TS25.104, v.9.4.0: Base station (BS) radio transmission and reception (FDD)*. Retrieved from http://www.3gpp.org/ftp/specs/html-info/TSG-WG--R4.htm

3rd Generation Partnership Project. (2010). *TS 32.571, v. 9.0.0: Technical specification group services and system aspects- telecommunication management- home node B (HNB) and home eNode B (HeNB) management- type 2 interface concepts and requirements*. Retrieved from http://www.etsi.org/deliver/etsi_ts/132500_132599/132571/09.00.00_60/ts_132571v090000p.pdf

3rd Generation Partnership Project. (2010). *TS 32.581, v. 10.0.0: Telecommunications management; home node B (HNB) operations, administration, maintenance and provisioning (OAM&P); concepts and requirements for Type 1 interface HNB to HNB management system (HMS)*. Retrieved from http://www.3gpp.org/ftp/specs/html-info/32581.htm

3rd Generation Partnership Project. (2010). *TS 32.591, v. 10.0.0: Telecommunications management; home enode b (HeNB) operations, administration, maintenance and provisioning (OAM&P); concepts and requirements for Type 1 interface HeNB to HeNB management system (HMS)*. Retrieved from http://www.3gpp.org/ftp/tsg_sa/WG5_TM/TSGS5_76/_.../32591-a10.doc

3rd Generation Partnership Project. (2010). *TS 36.300, v. 10.2.0: Evolved universal terrestrial radio access (E-UTRA) and evolved universal terrestrial radio access network (E-UTRAN); overall description; Stage 2.* Retrieved from http://www.3gpp.org/ftp/specs/html-info/36300.htm

Celidonio, M., Di Zenobio, D., & Pulcini, L. (2010, May). A broadband integrated radio LAN. In *Proceedings of the 17th IEEE Workshop on Local and Metropolitan Area Networks.*

Celidonio, M., Di Zenobio, D., Pulcini, L., & Rufini, A. (2011, June). Femtocell technology combined to a condominium cabled infrastructure. In *Proceedings of the IEEE International Symposium on Broadband Multimedia Systems and Broadcasting*, Nuremberg, Germany.

Chandrasekhar, V., & Andrews, J. (2008). Femtocells networks: A survey. *IEEE Communications Magazine, 46*(9), 59–67. doi:10.1109/MCOM.2008.4623708

Chen, J., Rauber, P., Singh, D., Sundarraman, C., Tinnakornsrisuphap, P., & Yavuz, M. (2010, Jan) *Femtocells- architecture & network aspects.* Retrieved from http://www.qualcomm.com/common/documents/white_papers/Femto_Overview_Rev_C.pdf

Femto Forum. (2009). *Femtocells - natural solution for offload.* Retrieved from http://www.femtoforum.org

Femto Forum. (2009). *Interference management and performance analysis of UMTS/HSPA + femtocells.* Retrieved from http://www.femtoforum.org

Femto Forum. (2009). *Interference management in OFDMA femtocells.* Retrieved from http://www.femtoforum.org

Femto Forum. (2009). *The best that LTE can be: Why LTE needs femtocells.* Retrieved from http://www.femtoforum.org

Fondazione Ugo Bordoni. (2002). *Italian patent IT 14114241: Cabled- Wireless END USER ACCESS (C-W EUA).* Brussels, Belgium: European Patent Office.

Gerhardt, W., & Medcalf, R. (2010, March). *Femtocells: Implementing a better business model to increase SP profitability.* Retrieved from http://www.cisco.com/en/US/solutions/ns341/ns973/ns941/femtocell_point_of_view.pdf

International Telecommunications Union. (2001). *Propagation data and prediction models for the planning of indoor radiocommunication systems and radio local area networks in the frequency range 900 MHz to 100 GHz* (Tech. Rep. No. ITU-R P.1238). Geneva, Switzerland: International Telecommunications Union.

Lopez-Perez, D., Valcarce, A., de la Roche, G., & Zhang, J. (2009, June). OFDMA femtocells: A roadmap on interference avoidance. *IEEE Communications Magazine, 47*(9), 41–48. doi:10.1109/MCOM.2009.5277454

Sundaresan, K., & Rangarajan, S. (2009). Efficient resource management in OFDMA femto cells. In *Proceedings of the Tenth ACM International Symposium on Mobile Ad Hoc Networking and Computing* (pp. 33-42).

This work was previously published in the International Journal of Interdisciplinary Telecommunications and Networking, Volume 3, Issue 2, edited by Michael R. Bartolacci and Steven R. Powell, pp. 38-50, copyright 2011 by IGI Publishing (an imprint of IGI Global).

Chapter 9

Data Transmission Oriented on the Object, Communication Media, Application, and State of Communication Systems

Mike Sabelkin
École de Technologie Supérieure, Canada

François Gagnon
École de Technologie Supérieure, Canada

ABSTRACT

The proposed communication system architecture is called TOMAS, which stands for data Transmission oriented on the Object, communication Media, Application, and state of communication Systems. TOMAS could be considered a Cross-Layer Interface (CLI) proposal, since it refers to multiple layers of the Open Systems Interconnection Basic Reference Model (OSI). Given particular scenarios of image transmission over a wireless LOS channel, the wireless TOMAS system demonstrates superior performance compared to a JPEG2000+OFDM system in restored image quality parameters over a wide range of wireless channel parameters. A wireless TOMAS system provides progressive lossless image transmission under influence of moderate fading without any kind of channel coding and estimation. The TOMAS system employs a patent pending fast analysis/synthesis algorithm, which does not use any multiplications, and it uses three times less real additions than the one of JPEG2000+OFDM.

INTRODUCTION

Development of new techniques for effective data transmission is very important now-a-days. Our proposed system architecture is called TOMAS which is data Transmission oriented on the Object, communication Media, Application, and state of communication Systems (Figure 1). Efficient data transmission could be provided if communication systems involved in data transmission would take into consideration and monitor constantly the following four aspects:

- **Object:** Its type, size, nature, etc. It can be one/two dimensional (1D/2D) signals, a three dimensional (3D) mesh, symbol data

DOI: 10.4018/978-1-4666-2154-1.ch009

Figure 1. Data transmission oriented on the object, communication media, application, and state of communication systems

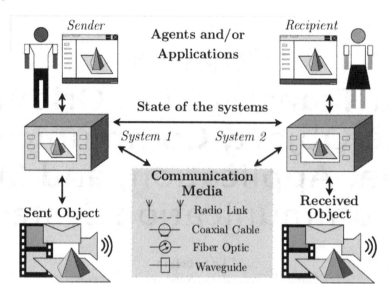

alone, or any combination of them. For example, audio and video signals or an optical image of surface and its 3D map. Object properties impose certain requirements on processing techniques. For example, the image compression standard (International Organization for Standardization, 1994) does not support very large images, therefore it cannot be used directly for processing such data object;

- **Communication Media:** In case of a complex communication system, a data object might be transmitted over different communication media such as a wireless, coaxial, fiber optic link, or a waveguide (IEEE Standards Association, 1993). Hence, the total performance of one object transmission will depend on the performance of the weakest link. For example, LOS, fair channel conditions may not require from the system to employ a 256 subcarrier OFDM in order to provide required QoS;

- **Agent and/or Application:** A human user or some application imposes certain requirements on quality and rapidity of object transmission. For example, if the

image Region Of Interest (ROI) access is required, JPEG2000 source coding might be preferable (International Organization for Standardization, 2000);

- **State of Communication Systems:** Time-varying characteristics of all systems involved in data exchange such as charge of batteries and status of all hardware, firmware and software components have to be monitored constantly. For example, it would not be useful to send a picture on some wireless terminal with a broken screen, or to employ some complex coding technique if the recipient side experiences a problem with decoder.

WIRELESS TOMAS

Among all types of communication media mentioned above, we concentrate our attention on a radio link or wireless communications. Figure 3 proposes an architecture for *wireless* data Transmission oriented on the Object, communication Media, Application, and state of communication Systems (*Wireless TOMAS*).

Figure 3. Wireless TOMAS system structure

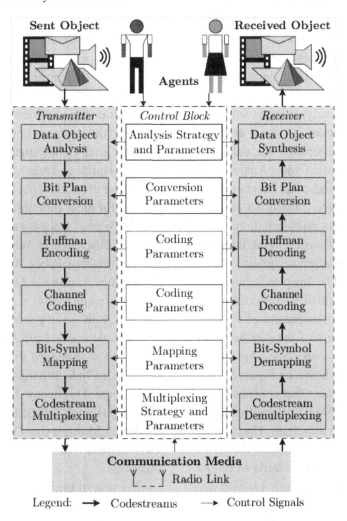

The Physical layer of Wireless TOMAS is represented by a transmitter and receiver, while the Control Block represents Media Access Control (MAC) layer in terms of the Open Systems Interconnection Basic Reference Model (OSI) (International Organization for Standardization, 1994).

At the *Data Object Analysis* stage features of the object are extracted. An appropriate technique might be used for this procedure. For example, Mel-Frequency Cepstral Coefficients (MFCC) serve well as the features of a voice signal (O'Shaughnessy, 1999). Or a two dimensional image might be represented by a set of its Dis-

crete Cosine Transform (DCT) coefficients. The right Data Object Analysis tool should be able to decompose the data object into data segments of unequal importance. The right *Data Object Synthesis* tool should be able to restore the data object from data segments with minimal loss (Pratt, 1978). An appropriate quality criterion of the restored data object should be used as well as an Error Sensitivity Descriptor (ESD) might be introduced. ESD reflects importance of the data segment. First data segment is considered to be more sensitive than the second one if corruption of this segment causes more damage to the restored data object than corruption of the second segment. The output of the Data Object Analysis is the set of data segments ranked in descending order according to their importance. The Control Block obtains the list that contains the order of the data segments from the Transmitter and provides it to the Receiver.

The Data Object Analysis often employs floating-point transformations such as DCT or wavelet transform. Therefore the data segments at the output of Data Object Analysis are represented by the floating-point numbers. The *Bit-plan Conversion* stage transforms the data segments into fixed-point representation. Truncation or rounding of floating-point numbers might cause the degradation of quality of the restored data object. The bit-plan conversion stage is the second stage of decomposition of the data object into data segments of unequal importance. The bit-plans of the data segment is formed by grouping corresponding bits of the coefficients as it is shown on Figure 2. The bit-plan of the data segment that consists of the Most Significant Bits (MSB) of coefficients is considered to be the most important. The bit-plan of the data segment that consists of the Least Significant Bits (LSB) is considered to be the least important. The bits of each bit-plan are grouped into words. The word length can differ from one bit-plan to another as well as from one data segment to another.

Figure 2. Bit-plan conversion

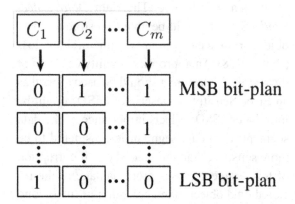

The *Huffman Encoding* serves to reduce the redundancy of the bit-plan data. First of all the histogram of the whole bit-plan data is created. From these results a Huffman code generated, and finally the bit-plan data is encoded. The Control Block obtains the histogram from the Transmitter and provides it to the Receiver. The output of the Huffman Encoder is represented by multiple binary code streams.

The Huffman code is very sensitive to the noise; therefore the *Channel Coding* should be applied for most communication environment.

The *Symbol Mapping* stage serves to improve spectral efficiency by mapping a group of bits into a complex symbol. Any appropriate type of Quadrature Amplitude Mapping or Modulation could be employed at this stage. For example, BPSK, QPSK, or in general form, M-QAM.

Multiple code streams of complex symbols should be multiplexed in order to be sent over serial radio link. Promising techniques such as Orthogonal Frequency Division Multiplexing (OFDM) (Engels, 2002; Burr, 2001) and Wavelet Packet Multiplexing (WPM) (Lindsey, 1996, 1997; Lindsey & Dill, 1995) might be used at this stage.

The task of the *Control Block* is to look for a compromise between application's demands on object transmission and communication media/communication system abilities. In order to fulfill that task, the Control Block could either assign appropriate parameters to PHY blocks or reconfigure the connections between blocks from the PHY layer toolbox. The last operation might take place only in case of Software Defined Radio (SDR) system implementation.

The technique called Unequal Error Protection (UEP) has been known for a while. Recent works (Wang et al., 2006; Soyjaudah & Fowdur, 2006; Ramzan & Izquierdo, 2006; Boeglen & Chatellier, 2006; Thomos et al., 2005) demonstrate some potential of such a technique. The common observation is that researches are concentrated on improving some parameters while others left unmonitored by authors. For example, they modify their algorithms in order to improve quality of a transmitted image. Often it leads to degrading of spectral efficiency of the system.

IMAGE TRANSMISSION USING WIRELESS TOMAS

Data Transmission Scenario

The following scenario is used to validate the proposed TOMAS technique:

- **Object:** Is a gray scale image (8 bit per pixel) derived from the image Lenna (University of Southern California, 1995). Point of origin on the source image is (250, 250). Size is 128x128 pixels. See Figure 4;
- **Communication Media:** Is a band-pass wireless LOS channel (Kjeldsen et al., 2003) with parameters presented in Table 1. Additive White Gaussian Noise (AWGN) is present in the channel. Symbol Rate is 10 Megasymbols per second;
- **Application:** Requires progressive lossless transmission. In other words, to transmit the image progressively and without losses in restored image quality;
- **State of Communication Systems:** Communication systems do not employ either channel coding or channel estimation.

Figure 4. Test Image Lena128

Communication Media

Ones of the most important factors that have influence on wireless communication system performance are multipath fading and Doppler spread. *Multipath fading* is caused by superimposition of the multiple signal reflections and scattering. *Doppler spread* is caused by relative movement between receiver and transceiver antennae.

When there are no obstacles between the transmitter and receiver antennae the communication scenario is called the Line-of-Sight (LOS) transmission, otherwise it is called the Non Line-of-Sight (NLOS) transmission.

For the wireless channels with a dominant stationary (non-fading) signal component, or a LOS communication scenario, the multipath fading can be modeled by the Ricean distribution. The Ricean distribution is represented by following expression (Rappaport, 2002):

$$f_R(r) = \begin{cases} \dfrac{r}{\sigma^2} e^{-\frac{r^2+A^2}{2\sigma^2}} I_0(\dfrac{Ar}{\sigma^2}), A \geq 0, r \geq 0, \\ 0, r < 0 \end{cases}$$

(1)

where A is the peak amplitude of the dominant LOS signal and $Io(.)$ is the modified Bessel function and in terms of Taylor's series it can be represented as follows:

$$I_0(x) = 1 + \frac{x^2}{2^2} + \frac{x^4}{2^2 \cdot 4^2} + \frac{x^6}{2^2 + 4^2 + 6^2} + \dots$$

(2)

The Ricean distribution is often characterized by a parameter K, which defines the ratio between the LOS signal power A and the variance of the multipath reflection signals sigma^2. It is known as the *Ricean factor*, and it is calculated according to:

$$K = \frac{A^2}{2\sigma^2}, \quad \text{or } K = 10 \cdot \log_{10}\left(\frac{A^2}{2\sigma^2}\right), dB$$

(3)

Decreasing the dominant component intensity leads to the transformation of Ricean distribution into Rayleigh distribution:

$$f_R(r) = \begin{cases} \dfrac{r}{\sigma^2} e^{-\frac{r^2}{2\sigma^2}}, 0 \le r \le \infty, \\ 0, r < 0 \end{cases} \quad (4)$$

where sigma^2 is the time-average power of the received signal before envelope detection.

Performance Parameters of Communication System

In order to evaluate the performance of a communication system, the following parameters are used:

Spectral Efficiency = Total Object Bits/Transmitted Symbols, (5)

Algorithm Complexity = Total Processing Operations/Total Object Bits, (6)
where

Total Object Bits = 128*128*8 = 131,082. (7)

for the particular data object described above.

In our case, the *Spectral Efficiency* (5) is measured in *bits-per-symbol*. It depends on the number of transmitted symbols. The goal of the communication system is to represent the Data Object by a minimal number of symbols. Let us note that, in case of fixed symbol mapping parameters, any kind of channel coding employed by the system will decrease the spectral efficiency.

The complexity of the communication system is measured by the *Algorithm Complexity* parameter (6). It reflects how many real additions and multiplications are required in order to process one bit of the transmitted data object.

Quality Parameters of Restored Image

Historically, researchers working in the domain of image processing use *Mean Square Error* (MSE) and *Peak Signal-to-Noise Ratio* (PSNR) as measures of restored image quality. MSE represents the normalized squared difference between the pixel values of the original image X and the restored one Y:

$$MSE = \frac{1}{N \cdot M} \sum_{i=1}^{N} \sum_{j=1}^{M} (X_{i,j} - Y_{i,j})^2 \quad (8)$$

where NxM is image size.

PSNR of an 8 bit grayscale image is calculated according to:

$$PSNR, dB = 20 \log_{10} \frac{255}{\sqrt{MSE}} \quad (9)$$

Despite obvious simplicity in calculation and interpretation, the integral image quality parameters as MSE and PSNR are not reliable image quality measures in some cases. Often, the value of PSNR is relatively high (up to 20 dB), even for the case when the restored image is totally destroyed. In case of lossless transmission MSE equals to zero, and that leads the value of PSNR to infinity. Therefore the *Visual Information Fidelity* measure (VIF) (Sheikh & Bovik, 2006) has also been measured. It is a natural measure of restored image quality. VIF lies in the interval [0,1], if VIF = 0 indicates that all sent image data has been lost in the distortion channel. If VIF = 1, the image is not distorted at all.

Structure of the Wireless TOMAS for Image Transmission

The communication system under investigation operates with two-dimensional images. Now-a-days the most popular technique to deal with such data is a Discrete Cosine Transform (DCT). It is employed by several standards: JPEG (International Organization for Standardization, 1994) for still image compression and MPEG-1,2,4 for video compression. These standards are well developed and may be employed at the Data Object Analysis stage of the wireless TOMAS algorithm. However data compression technique based on block DCT has a drawback that makes it unsuitable for our scenario. It does not support progressive transmission which is important for wireless communication scenario. This type of data transmission operates in following manner: a low resolution, low quality thumbnail of the original image is sent at the first place, and refinement data is sent later. *Progressive transmission* allows a recipient to have the whole image even if transmission is cut. The quality of restored images will depend on transmission duration before it is cut. Longer transmission provides better quality. Ideally, progressive transmission should be lossless if no interruption occurs. Progressive transmission provides also a flexible tool for the spectral efficiency manipulation. Arbitrary spectral efficiency could be obtained simply by stopping data transmission after a certain number of symbols have been transmitted (see Equation 5). Techniques that decompose the data object into segments suitable for the progressive transmission typically use wavelets and wavelet packet decompositions.

Figure 5 shows the structure of the communication system that satisfy the required scenario. Two-dimensional Wavelet Packet decomposition based on Haar wavelet has been employed for the Data Object Analysis. Wavelet packet decomposition is chosen over wavelet decomposition, employed in JPEG 2000 algorithm (International Organization for Standardization, 2000), because it provides far finer data segmentation. Losing large data segments while transmitting data in harsh channels leads to more damage, than losing

a small segment. However this choice leads to an increase of algorithm complexity. The Depth of the Image Decomposition is equal three. This choice is motivated by the compromise between data segment size and algorithm complexity.

The most vulnerable part of the wireless TOMAS system to channel noise is the Huffman encoding. Long Huffman codes are more vulnerable than the short ones. Since our experimental data transmission system does not employ channel coding, the transmitted data will be affected seriously by harsh channel conditions. Therefore the word length = 2 bits is chosen as a bit-plan conversion parameter. It means that in every bit-plan two bits are grouped into one symbol. A two-bit symbol may take $2^2 = 4$ different values. The histogram of such a source has only 4 bins. That makes a Huffman map very short and the Huffman code less vulnerable to noise. However, short Huffman code reduces spectral efficiency of the system.

In order to convert the codestream bits into complex symbols, Quadrature Amplitude Mapping is employed. QAM order = 2 provides high robustness to the channel noise, but reduces spectral efficiency of the system.

Two alternatives were considered for the Codestream Multiplexing. There are *Orthogonal Frequency Division Multiplexing* (OFDM) and *Wavelet Packet Multiplexing* (WPM). Harmonics employed by OFDM are orthogonal in frequency domain only. Wavelet Packets are orthogonal to each other both in time and frequency domain. Papers (Yang et al., 1997; Newlin, 1998; Suzuki et al., 1999) show that WPM mitigates multipath channel interference better than OFDM does. Relatively high complexity of the wavelet packet transformation algorithms used in mentioned above papers might be the reason WPM did not gain such popularity as OFDM did. We developed proprietary fast wavelet packet algorithm which allowed us to employ WPM for the wireless TOMAS at the Codestream Multiplexing stage. Results

Figure 5. Structure of the Wireless TOMAS for Image Transmission

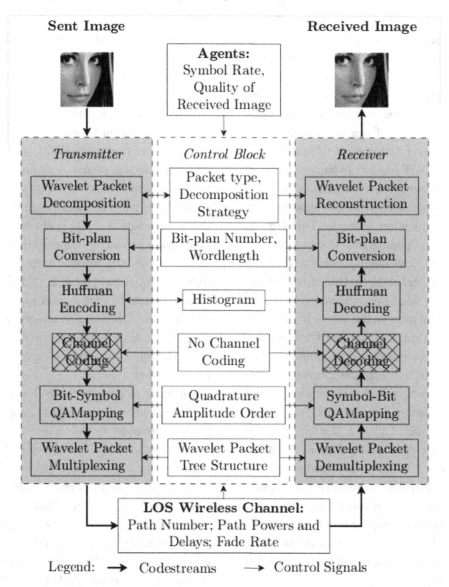

from Wang et al. (2006) show that increasing the order of wavelet does not bring significant improvement in performance, but dramatically increase algorithm complexity. Hence, to constrain complexity, the Haar wavelet is once again chosen as the basic wavelet for WPM. In order to preserve orthogonality between wavelet packets, one frame WPM should consist of the following number of wavelet packets called subbands in literature:

$$Subband\ Number = 2 \cdot 2^{ceil(\log_2 \frac{T\max}{T_s})}, \qquad (10)$$

where tau_max is maximum path delay and Ts is symbol duration.

In case of particular data transmission scenario as shown in Table 1, WPM operates with the following number of subbands:

$$Subband\ Number = 2 \cdot 2^{ceil(\log_2 \frac{5 \cdot 1 \cdot 10^{-6}}{10^{-7}})} = 128 \qquad (11)$$

Table 1. Wireless channel parameters

Path	Type	Delay, μs	Fade Rate, Hz	Loss, dB
Direct	Ricean	0		0
Second	Rayleigh	0.9	2, 37, 225	-3
Third	Rayleigh	5.1	2, 37, 225	-9

Structure of the Benchmark Communication System JPEG2000 +WirelessMAN OFDM

The performance of data transmission using wireless TOMAS has been compared to the results of the benchmark communication system. The benchmark system uses JPEG2000 algorithm for the Data Object Analysis and WirelessMAN OFDM (IEEE Standards Assocation, 2004) for the codestream multiplexing as shown on Figure 6. JPEG2000 employs wavelet decomposition based on the wavelet called in the standard *jpeg9.7*. Decomposition depth is 4. Word length equals to 2 as a bit-plan conversion parameter and QAM order equal to 2 as a bit-symbol mapping parameter. The WirelessMAN OFDM parameters are presented in Table 2.

Performance Comparison of Wireless TOMAS and JPEG2000+ WirelessMAN OFDM Communication Systems

Performances of the two communication systems are evaluated and compared for the cases of low (fade rate = 2 Hz), moderate (fade rate = 37 Hz) and high (fade rate = 225 Hz) relative mobility between the transmitter and receiver antennae. The influence of AWGN was estimated for the following values of SNR: 6, 8, 10, 17, 25, 35, 40 and 45 dB. Figures 7, 9 and 11 present performances of the two systems and the difference between them using PSNR criterion. Let us recall that lossless transmission case leads PSNR towards infinity. We assumed that PSNR = 80 dB corresponds to lossless case. Figures 8, 10, and 12 present performances of the two systems and the difference between them using VIF criterion.

Wireless TOMAS provides lossless image transmission in low and moderate fading rates. For spectral efficiencies of 2 bit/symbol and 10 bit/symbol, the two systems have the similar performance in low and moderate fading environment according to PSNR criteria. However, according to VIF criteria, the wireless TOMAS performs better than its competitor. Low spectral efficiency of 0.8 bit/symbol results demonstrate the advantage of wavelet packet decomposition employed by the wireless TOMAS over wavelet decomposition employed by JPEG2000. As it was mentioned above, wavelet packet decomposition provides finer data segmentation, than wavelet decomposition. In fading channel environment larger data segment is transmitted longer, hence it is more affected by fading than the short data segment. In case of fast fading rates (fading rate = 225 Hz), the wireless TOMAS also outperforms JPEG2000+OFDM by providing up to 7 dB more in PSNR and up to 35 percent in VIF, as shown on Figures 11c and 12c.

The complexities of algorithms are calculated theoretically for both systems. The results are presented in Table 3. TOMAS system employs a patent pending fast analysis/synthesis algorithm, which does not use any multiplications, and it requires 3 times less additions than JPEG2000+OFDM system does.

Figure 6. Structure of the Benchmark Communication System JPEG2000+ WirelessMAN OFDM

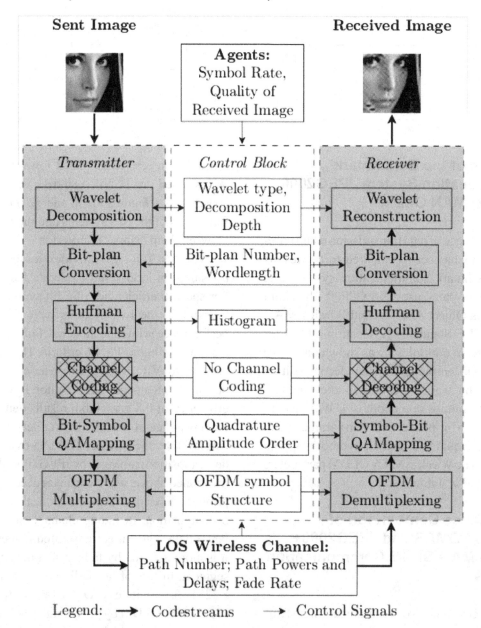

Table 2. WirelessMAN OFDM PHY Parameters

FFT size	Nfft= 256
Guard Number	Ng= 55
Pilot Number	P= 8
Cyclic Prefix Ratio	CP= 1/2

Table 3. Algorithm complexity

Operations per bit	JPEG 2000+OFDM	TOMAS
Real Multiplications	12.84	0
Real Additions	11.39	3.81

Figure 7. Comparison of system performances using PSNR criteria. Fading rate = 2 Hz

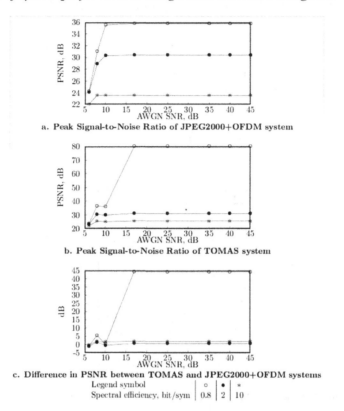

a. Peak Signal-to-Noise Ratio of JPEG2000+OFDM system

b. Peak Signal-to-Noise Ratio of TOMAS system

c. Difference in PSNR between TOMAS and JPEG2000+OFDM systems

Legend symbol	∘	●	*
Spectral efficiency, bit/sym	0.8	2	10

Figure 8. Comparison of system performances using VIF criteria. Fading rate = 2 Hz

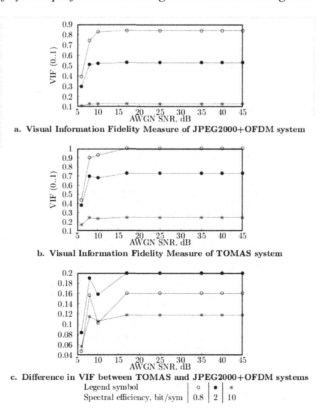

a. Visual Information Fidelity Measure of JPEG2000+OFDM system

b. Visual Information Fidelity Measure of TOMAS system

c. Difference in VIF between TOMAS and JPEG2000+OFDM systems

Legend symbol	∘	●	*
Spectral efficiency, bit/sym	0.8	2	10

Figure 9. Comparison of system performances using PSNR criteria. Fading rate = 37 Hz

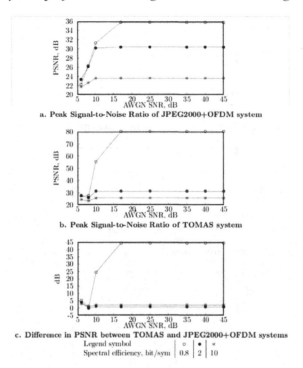

a. Peak Signal-to-Noise Ratio of JPEG2000+OFDM system

b. Peak Signal-to-Noise Ratio of TOMAS system

c. Difference in PSNR between TOMAS and JPEG2000+OFDM systems

Legend symbol	∘	•	*
Spectral efficiency, bit/sym	0.8	2	10

Figure 10. Comparison of system performances using VIF criteria. Fading rate = 37 Hz

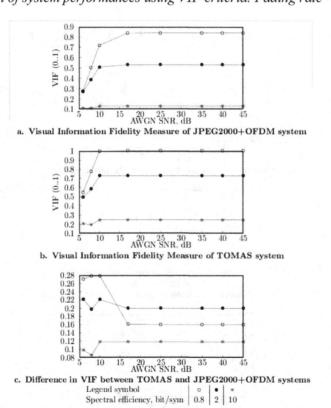

a. Visual Information Fidelity Measure of JPEG2000+OFDM system

b. Visual Information Fidelity Measure of TOMAS system

c. Difference in VIF between TOMAS and JPEG2000+OFDM systems

Legend symbol	∘	•	*
Spectral efficiency, bit/sym	0.8	2	10

Figure 11. Comparison of system performances using PSNR criteria. Fading rate = 225 Hz

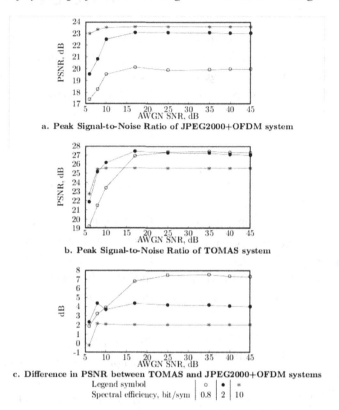

a. Peak Signal-to-Noise Ratio of JPEG2000+OFDM system

b. Peak Signal-to-Noise Ratio of TOMAS system

c. Difference in PSNR between TOMAS and JPEG2000+OFDM systems

Legend symbol	○	●	*
Spectral efficiency, bit/sym	0.8	2	10

Figure 12. Comparison of system performances using VIF criteria. Fading rate = 225 Hz

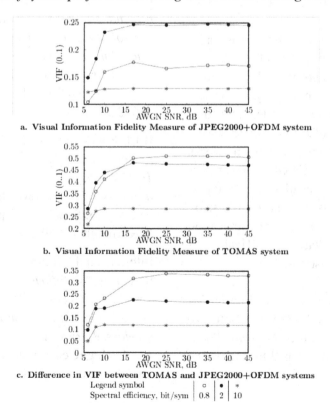

a. Visual Information Fidelity Measure of JPEG2000+OFDM system

b. Visual Information Fidelity Measure of TOMAS system

c. Difference in VIF between TOMAS and JPEG2000+OFDM systems

Legend symbol	○	●	*
Spectral efficiency, bit/sym	0.8	2	10

CONCLUSION

Results of simulation in Matlab (R) environment validate the idea of Data Transmission Oriented on the Object, Communication Media, Agent, State and Health of Communication Systems. Our research made the following contributions to knowledge:

- Given particular scenario of image transmission over a wireless LOS channel, proposed system provides superior performances compared to JPEG2000+OFDM systems in restored image quality parameters over a wide range of wireless channel parameters. JPEG2000+OFDM system is not able to provide lossless transmission at all.

- Despite common opinion that channel coding is mandatory in wireless communications, we demonstrated that without any channel coding, the proposed system provides lossless progressive image transmission under influence of moderate fading. Channel coding might mitigate effects of severe fading.

- Despite decrease of interest towards wavelets, they have high potential in telecommunication. The algorithm, derived from wavelets, does not employ any multiplications, and it uses three times less real additions than the one of JPEG2000+OFDM.

- Complex telecommunication networks that operate with multimedia content over mixture of communication media types, could profit from use of Data Transmission Oriented on the Object, Communication Media, Agent, State and Health of Communication Systems.

Future work will be concentrated on implementing of the wireless TOMAS in Software Defined Radio platform called *Software Radio Dongle*(TM) which is developed by DTMSDR Technologies INC.

ACKNOWLEDGMENT

The authors would like to thank Industrial Research Chair of Natural Sciences and Engineering Research Council of Canada (NSERC) and Ultra Electronics Tactical Communication Systems (TCS); and DTMSDR Technologies INC for support of research.

REFERENCES

Boeglen, H., & Chatellier, C. (2006). On the robustness of a joint source-channel coding scheme for image transmission over non frequency selective Rayleigh fading channels. *In Proceedings of the 2nd Conference on Information and Communication Technologies* (pp. 2320-2324).

Burr, A. (2001). *Modulation and coding for wireless communications. Upper Saddle River, NJ: Pearson/Prentice Hall. Engels, M. (2002). Wireless OFDM systems: How to make them work?* Boston, MA: Kluwer Academic.

IEEE Standards Association. (1993). *American national standard, Canadian standard graphic symbols for electrical and electronics diagrams (including reference designation letters).* Retrieved from http://ieeexplore.ieee.org/xpl/freeabs_all.jsp?isnumber=21239&arnumber=985670&count=1

IEEE Standards Association. (2004). *Standard for local and metropolitan area networks. part 16: Air interface for fixed broadband wireless access systems.* Retrieved from http://ieeexplore.ieee.org/xpl/freeabs_all.jsp?tp=&isnumber=33683&arnumber=1603394

International Organization for Standardization. (1994). *Information technology - open systems interconnection - basic reference model - conventions for the definition of OSI services.* Retrieved from http://www.iso.org/iso/iso_catalogue/catalogue_tc/catalogue_detail.htm?csnumber=18824

International Organization for Standardization. (1994). *JPEG image coding system.* Retrieved from http://www.iso.org/iso/home.html

International Organization for Standardization. (2000). *JPEG 2000 image coding system.* Retrieved from http://www.iso.org/iso/home.html

Kjeldsen, E., Dill, J., & Lindsey, A. (2003). Exploiting the synergies of circular simplex turbo block coding and wavelet packet modulation. In. *Proceedings of the Military Communications Conference, 2*, 1202–1207.

Lindsey, A. (1996). Improved spread-spectrum communication with a wavelet packet based transceiver. In *Proceedings of the IEEE-SP International Symposium on Time-Frequency and Time-Scale Analysis* (pp. 417-420).

Lindsey, A. (1997). Wavelet packet modulation for orthogonally multiplexed communication. *IEEE Transactions on Signal Processing, 45*(5), 1336–1339. doi:10.1109/78.575704

Lindsey, A., & Dill, J. (1995). Wavelet packet modulation: A generalized method for orthogonally multiplexed communications. In *Proceedings of the Twenty-Seventh Southeastern Symposium on System Theory* (pp. 392-396).

Newlin, H. (1998). Developments in the use of wavelets in communication systems. In. *Proceedings of the Military Communications Conference, 1*, 343–349.

O'Shaughnessy, D. (1999). *Speech communications: Human and machine* (2nd ed.). New York, NY: John Wiley & Sons.

Pratt, W. K. (1978). *Digital image processing.* New York, NY: John Wiley & Sons.

Ramzan, N., & Izquierdo, E. (2006). Scalable video transmission using double binary turbo code. In *Proceedings of the IEEE International Conference on Image Processing* (pp. 1309-1312).

Rappaport, T. S. (2002). *Wireless communications: Principles and practice.* Upper Saddle River, NJ: Prentice Hall.

Sheikh, H., & Bovik, A. (2006). Image information and visual quality. *IEEE Transactions on Image Processing, 15*(2), 430–444. doi:10.1109/TIP.2005.859378

Soyjaudah, K., & Fowdur, T. (2006). An integrated unequal error protection scheme for the transmission of compressed images with ARQ. In *Proceedings of the International Conference on Networking, International Conference on Systems and International Conference on Mobile Communications and Learning Technologies* (p. 103).

Suzuki, N., Fujimoto, M., Shibata, T., Itoh, N., & Nishikawa, K. (1999). Maximum likelihood decoding for wavelet packet modulation. In *Proceedings of the 50th IEEE Vehicular Technology Conference* (Vol. 5, pp. 2895-2898).

Thomos, N., Boulgouris, N., & Strintzis, M. (2005). Wireless image transmission using turbo codes and optimal unequal error protection. *IEEE Transactions on Image Processing, 14*(11), 1890–1901. doi:10.1109/TIP.2005.854482

University of Southern California. (1995). *Image database: Lenna test image.* Retrieved from http://sipi.usc.edu/database/database.php

Wang, S., Dai, J., Hou, C., & Liu, X. (2006). Progressive image transmission over wavelet packet based OFDM. In *Proceedings of the Canadian Conference on Electrical and Computer Engineering* (pp. 950-953).

Yang, W., Bi, G., & Yum, T.-S. (1997). A multirate wireless transmission system using wavelet packet modulation. In *Proceedings of the 47th IEEE Vehicular Technology Conference* (Vol. 1, pp. 368-372).

This work was previously published in the International Journal of Interdisciplinary Telecommunications and Networking, Volume 3, Issue 2, edited by Michael R. Bartolacci and Steven R. Powell, pp. 51-65, copyright 2011 by IGI Publishing (an imprint of IGI Global).

Chapter 10
Machine–to–Machine Communications and Security Solution in Cellular Systems

Mahdy Saedy
University of Texas at San Antonio, USA

Vahideh Mojtahed
Purdue University, USA

ABSTRACT

This paper introduces an efficient machine-to-machine (M2M) communication model based on 4G cellular systems. M2M terminals are capable of establishing Ad Hoc clusters wherever they are close enough. It is also possible to extend the cellular coverage for M2M terminals through multi-hop Ad Hoc connections. The M2M terminal structure is proposed accordingly to meet the mass production and security requirements. The security becomes more critical in Ad Hoc mode as new nodes attach to the cluster. A simplified protocol stack is considered here, while key components are introduced to provide secure communications between M2M and the network and also amongst M2M terminals.

1. INTRODUCTION

Conventional telecommunication networks enable Human-to-Human (H2H) communications characterized with fairly high demand for bandwidth to accommodate for the voice and data applications. The total low cost of ownership, fast network rollout and numerous capabilities of wireless data networks can bring the machine-to-machine communication known as M2M to a wider audience and exceedingly broad and diverse

application scenarios. M2M is a special type of communication where the majority of end user terminals are entities that communicate with each other and interact with the environment without human intervention in terms of collecting data and taking required actions based on received commands from the network. Large number of communicating terminals, small and infrequent traffic has already generated numerous market scenarios leading to developing the supervisory control and data acquisition (SCADA) systems in the early 1990s, based on technology in which a

DOI: 10.4018/978-1-4666-2154-1.ch010

centralized server reaches out and polls field equipment regularly. SCADA is based on proprietary technologies, so its costs never dropped enough to make widespread deployment practical. Unlike SCADA, M2M system works with standard technologies such as TCP/IP, IEEE wireless/wired LANs, and cellular technologies. Using standards allows easier device interoperation in M2M systems and provides scalable network expansion with less expensive and quicker implementation while the total cost of ownership is lower due to mass-production of M2M terminals. With the advent of new and smart applications, the need for higher data rate and ubiquitous coverage as well as the ability for M2M terminals to interconnect and expand regardless of network limitations, where needed, is well understood. In other words, today's M2M communications needs to be defined on modern wireless platforms like 4G with Ad Hoc capability. 3GPP LTE-Advanced standards can deliver the required data rate and QoS for this purpose while we also consider Ad Hoc scenario to enable the M2M terminals to interact with each other where they succeed to form a local Ad Hoc cluster within the coverage of 4G system. Enhanced QoS, mobility and resource management along with high bit rate for fixed and mobile users in 4G system, opens new horizons to M2M communications.

Security is one of the most critical issues in networks which are well established for H2H communications in 3GPP standards. For M2M terminals to be able to form Ad Hoc cluster within the coverage of 4G cellular networks, it is extremely important to revisit the security procedures since machines can interact with other machines in Ad Hoc mode as well as normal access to 4G network resources.

2. CONVENTIONAL M2M SOLUTION

Figure 1 shows a typical wireless M2M implementation based on 2G system (GSM) via protocols such as TCP/IP. The system sends the information to a back-end server, which processes the data and sends it via the Internet to a monitoring center that monitors and controls the machines.

In current networks, the M2M terminal acts similar to a normal mobile phone, which communicates with the base station and then with the M2M server via the upper layer (i.e. NAS in 3GPP). The difference might be that the machine is triggered by a specific event other than human being (Chen & Yang, 2009). The actual data to be sent may be of very small volume due to successive handshakes with the network. These transmissions and extra overheads cause power waste which makes the current system not economic for M2M communications (Cristaldi, Faifer, Grande, & Ottoboni, 2005).

Figure 1. Conventional M2M solution

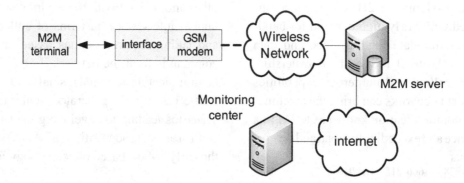

3. SECURITY IN M2M

Considering the large number of M2M terminals deployed in highly distributed networks in both cellular and Ad Hoc manner, the security Issue becomes critical since network operators do not want the hackers to break into the services. On the other hand, because taking the Required action is one of the roles of M2M terminals, any mistake in handling the imposed risks will lead to uncontrolled propagation of risk and damage (Cristaldi, Faifer, Grande, & Ottoboni, 2005). When the terminal is not authenticated by the core network, the connection request may be fraud. A possible solution might be the application layer encryption e.g. certain synchronized encryption schemes. An example is to use symmetric architecture and message authentication code (MAC), where the machine and the peer share the key –stored in the SIM- to encrypt or decrypt the message.

A. Adaptation of Level of Security

In order for the overall risk to remain manageable, in both Ad Hoc and cellular scenarios, there needs to be a balance between security provisions on the user side and those in the network: it may be possible to adapt security on the user side for M2M communication to a certain extent, but this would then have to be compensated for by access restrictions on the M2M user enforced in the network. Some of these access restrictions could be realised by dynamically configurable packet filters. The security issue becomes more critical when M2M terminals tend to communicate with each other in Ad Hoc mode since the information will be routed and exchanged through M2M terminals. The level of security then needs to be adapted based on the mode of operation as well which implies the importance of a policy maker entity in the network. The cost of additional software on M2M operator side needs to be considered as well as security level adaptation to account for both cellular and Ad Hoc modes since introducing extra protocol headers and manipulating the security on top of the 4G application layer needs additional software.

B. Security for Unattended M2M Devices

In contrast to traditional ME, which is carefully held and protected by a person, the M2M terminals will be placed in more or less accessible locations, and may be tampered with by unauthorised persons. Furthermore, theft or fraudulent modification of an M2M terminal may not be detected and reported as quickly as personal handheld ME. Fraud targets could both be the M2M user (e.g. fraudster suppresses payment messages) or the PLMN operator (fraudster uses M2M device or it's UICC for theft of service). The related work can be found in 3GPP TR 33.905 "Recommendations for Trusted Open Platforms" for M2M. One major challenge is to secure the UICC (Universal Integrated Circuit Card) in such a way that it is not trivial to tamper with or steal. On the other hand making the UICC completely theft proof challenges the flexibility for the M2M administrator/end-user to change subscription if that is desired. The work-item on "Network Improvements for Machine type Communications" is agreed in 3GPP but needs to be revisited for Ad Hoc mode of operation (Chen & Yang, 2009).

4. SECURITY MODELS

The spatial distribution of M2M terminal and different threat profile of each serving area from one side and the amount of handshakes between host cellular network and M2M terminals from the other side, brings the need for a security model into picture. Other issues like latency and database maintenance will be of importance as well.

C. Centralized Security Model

In centralized model, the security enforcement task is done in one or more nodes that are physically located as close to each other to minimize the eavesdropping and unwanted interceptions. This method is very much trusted but introduces huge amount of signaling traffic to the network in terms of protocol handshaking.

D. Distributed Security Model

Parts of the enforcement tasks can be delegated to trusted nodes dispersed across the network. This evolved security model is a useful, practical, and scalable approach for M2M security especially when nodes create distributed clusters in Ad Hoc mode, which is a critical factor for the overall success of the M2M market.

5. SECURITY THREATS AND VULNERABILITIES FOR M2M

M2M terminals are deployed in a wide area and in very large quantities. These terminals are mostly mobile devices communicating wirelessly. Because there is a lot less human attendance in M2M communications and these terminals are intended to both send information to the network and receive some information to execute commands, they require remote management and flexibility in terms of subscription management. These requirements introduce a number of security vulnerabilities.

E. Physical Attacks

Physical attack includes the insertion of valid authentication tokens into a manipulated device, inserting and/or booting with fraudulent or modified software (reflashing), and environmental/side-channel attacks, both before and after in-field deployment.

F. Compromise of Credentials

Comprising brute force attacks on tokens and authentication algorithms, physical intrusion, or side-channel attacks, as well as malicious cloning of authentication tokens residing on the machine communication identity module (MCIM).

G. Configuration Attacks

Such as fraudulent software update/configuration changes; misconfiguration by the owner, subscriber, or user; and misconfiguration or compromise of the access control lists.

H. Protocol Attacks

Directed against the device, which include man-in-the-middle attacks upon first network access, denial-of-service (DoS) attacks, compromising a device by exploiting weaknesses of active network services, and attacks on over-the-air management (OAM) and its traffic.

1. Attacks on the Core Network

The main threats to the mobile network operator include impersonation of devices; traffic tunneling between impersonated devices; misconfiguration of the firewall in the modem, router, or gateways; DoS attacks against the core network; also changing the device's authorized physical location in an unauthorized fashion or attacks on the network, using a rogue device.

2. User Data and Identity Privacy Attacks

includes eavesdropping user's or device's data sent over the access network; masquerading as another user/subscriber's device; revealing user's network ID or other confidential data to unauthorized parties.

Figure 2. M2M terminal block diagram

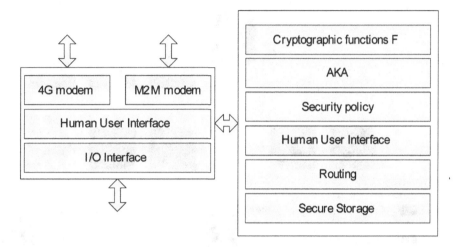

6. M2M TERMINAL FUNCTIONAL STRUCTURE

As mentioned before, M2M terminals in the proposed model are capable of communicating in both cellular and Ad Hoc mode. If the terminal is located in the coverage area of 4G network but far from other terminals, it will use 4G network resources otherwise it is more recommended to communicate with nearby terminals. There will be a case where the terminal is in range of both other terminals and 4G coverage. Figure 2 represents the functional block diagram for proposed M2M terminal. The AKA along with Cryptographic block is responsible for AAA functionality with symmetric and asymmetric encryption and decryption, hash value calculation and verification and digital signature. Because the proposed M2M will be communicating with other terminals in Ad Hoc mode, the Security policy and Routing modules control the routing procedures and apply the network security policy. All the keys, credentials, authentication and routing information are stored in a secure storage. The secure storage is also used to store the Quality of Service, resource management and remote network management information from

4G network. Human and external interface are also for local subscription and tuning purposes.

7. CELLULAR-BASED M2M

Better coverage, efficient use of network resources, scalable expansion and lower network deployment cost are the main drives for using cellular infrastructure in M2M communications. Using cellular networks for connecting machines to other machines and to the core network requires more complicated protocol than what regular H2H communications for voice/data needs. Regarding the fact that with smarter applications machines will also need to take up about the same bandwidth as in H2H, modern generations of cellular communications can be of better help in terms of high speed access and more classified quality of service. Figure 3 shows M2M communications based on 3G cellular network.

M2M terminals coexist with UEs and the back-end server communicates with the terminals through the MGW, RNC and eNodeB. Figure 4 shows the 4G network structure which is based on All-IP transport layer and fixed/mobile Soft Switch traffic handling.

Figure 3. M2M based on 3G and beyond with Ad Hoc clusters of machines

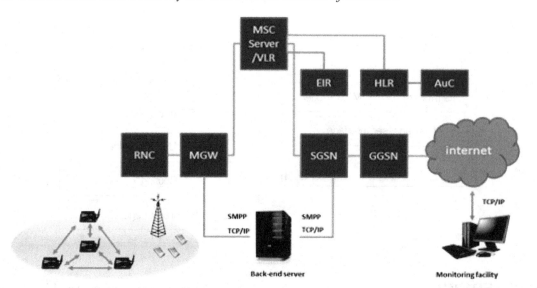

Figure 4. Migration to M2M based on 4G network structure

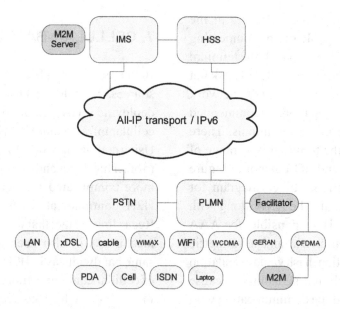

Figure 4 shows how M2M nodes are integrated into 4G network.

8. M2M IN 3GPP

Machines can either communicate with each directly if they are close or through cellular network if they are far apart. In the case where machines tend to communicate with each other, they form an Ad Hoc network wherever possible. Therefore, each machine is equipped with a modem that is capable of both communicating with other machines and the cellular network. The complexity lies in the fact that most of the protocols need to set up connections before communication. In order to transmit a small piece of data, the machine searches the network and sends uplink access preambles. When it receives the acknowledgement, it tries to establish RRC connection. When RRC connection is allowed, it needs to connect to core network (CN). After this, it needs to connect to the M2M server, involving TCP handshaking and application layer initializations before data transmission. Hence, in such a solution, the overhead is much more than the data. Another obvious drawback is power consuming. If we keep the current protocol stack, it is difficult to simplify the machine terminal, because the protocol defines the behaviours. Figure 5 shows the protocol stack peer connections between M2M terminal, eNodeB and the M2M server. This protocol handshake is for the case when M2M terminal uses the 4G network resources but if other terminals are close in a way that terminals can for an Ad Hoc network then the protocol handshakes are different since only M2M terminals are involved. In both cases the protocol stack can be simplified considering the fact that information exchanged between M2M terminal and server is transparent to 4G network.

9. NEW PROTOCOL STACK

The new protocol stack is shown in Figure 7. In the new protocol stack, the TCP/IP is not needed. The data from application layer is encrypted and sent to PDCP (Packet Data Convergence Protocol) directly. The application data may have been compressed to keep the SDU (Session Data Unit) smallest. The data is forwarded to the facilitator and then to the server. The facilitator support full protocol of application layer, e.g. FTP. It is assumed that the facilitator knows the address of the M2M server. A new connection cause can be used to indicate the eNodeB that the connection request is from a machine (Chen & Yang, 2009).

Since TCP/IP is removed, PDCP is useless (because there is no IP header) secondly; the acknowledgement of RLC is also not needed, because lower layer acknowledgement is sufficient. Such an evolution is suitable for extreme small amount of data. The data is transmitted to eNodeB by a special access preamble, or predefined preambles followed by the first access preamble. The terminal does not need to establish

Figure 5. M2M terminal connection to M2M server through 4G system

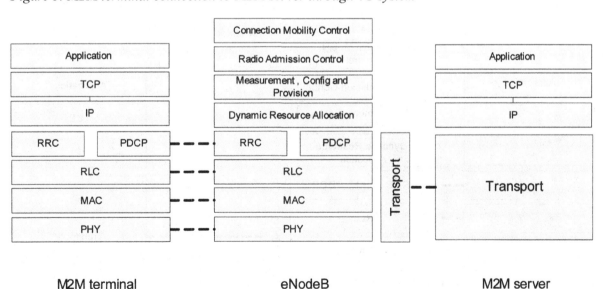

the RRC connection, and even does not need to transmit the data via the uplink channel Scheduled by the eNodeB, as the data is sent on PRACH directly. Consequently, MAC is totally removed. Considering the capacity of the access channel, only very small amount of Data can be transmitted by such a method. If Ad Hoc connection is the case and the network has some limitations in terms of resource elements, the further modifications need to be considered to dynamically allocate resources to M2M terminals (Figure 6) without service interruption for other users.

10. SECURITY SOLUTION FOR M2M

Since M2M Communication introduces new applications and on the other hand the internet is widely used to transfer the payloads all across the network among the M2M terminals and M2M servers, it is critically important to secure the information through encryption and other security policies adopted in network layer. New software patches can then be introduced in application layer and IMS/HSS interfaces towards AAA entity will be modified accordingly to accommodate the extra header volume. Although the new design gives more system level flexibility, the processing power needs to be considered to prevent the system latency from violating the QoS requirements. Considering the fact that M2M terminals happen to communicate with each other in Ad Hoc mode and that hackers and fraudsters can access the network resources though aforementioned vulnerabilities, there needs to be a continuous handshake between the facilitator and IMS/HSS in 4G part to not only verify the validity of all requests but also update the location information.

Figure 6. New protocol stack using M2M facilitator

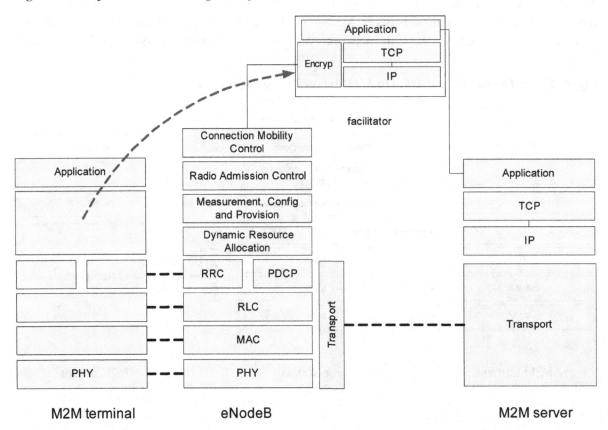

Figure 7. Security solution integration

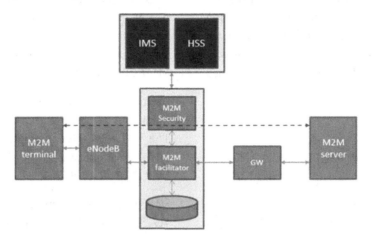

11. HIGH LEVEL PROTOCOL FOR M2M COMMUNICATIONS

As new node i attaches the network or in another way the connection fails and node i attempts to establish a new connection to maintain its communication and operation, there is a sequence of high-level handshakes proposed to get the node i properly and securely attached to the network (Table 1).

CONCLUSION

M2M communications is a new area where a lot of new applications can be introduced. Due to the fact that machines are getting smarter and more complicated every day and wireless technology is now able to provide high quality services, the security considerations seem to be of critical part in such applications.

One of the main features of next generations of wireless systems is that they provide ubiquitous coverage enabling the users/machines to also form Ad Hoc clusters within the coverage area and on the edge of coverage region to expand the serving area. This requires further network modifications and optimization to not only account for new threats but also optimize the protocols and interconnections to best integrate M2M communications into the wireless networks. One of the most important further investigations is the resource management in cases where the number of M2M terminals is greater than ordinary

Table 1. High-level M2M communication protocol based on ad hoc and cellular network structure

High-level M2M communication protocol based on Ad Hoc and cellular network structure
1. node i searches the network for both 4G RAS signal and M2M neighborhood signal.
2. send the uplink access preambles (depending on available signal from either cellular od Ad hoc network)
3. if there is idle resource in 4G or available channel from M2M nodes, the access is granted to node i.
4. AKA is initiated
5. Facilitator interrogates node i, it's data base and HSS to make sure the machine is not a potential threat.
6. Security policies and QoS consideration are done.
7. Acknowledgement is sent to node i
8. RRC connection is established through either eNodeB or PHY layer of M2M terminal.
9. Connection is registered and logged in Facilitator data base and a copy is sent to HSS.
10. Control bearer packets are broadcasted to nearby MGWs and finally to SGSN and GGSN
11. Node i connects to the Core Network
12. TCP/IP handshakes and Application layer initializations are done.
13. Connecting to M2M server

cellular users and M2M terminals do not receive decent QoS. This requires hardware modification in M2M terminal and concrete protocol to allocate resources to M2M terminals in an optimum way so RRM algorithm will also be of high importance in future works.

REFERENCES

Castelliccia, C., Hartenstein, H., Paar, C., & Westhoff, D. (Eds.). (2005). *Security in ad hoc and sensor networks*. New York, NY: Springer.

Chaouchi, H. (2009). *Wireless and mobile network security*. New York, NY: John Wiley & Sons. doi:10.1002/9780470611883

Chen, Y., & Yang, Y. (2009). Cellular based machine to machine communication with un-peer2peer protocol stack. In *Proceedings of the 70th IEEE Conference on Vehicular Technology* (pp. 1-5).

Cristaldi, L., Faifer, M., Grande, F., & Ottoboni, R. (2005, February 8-10). An improved M2M platform for multi-sensors agent application. In *Proceedings of the Sensors for Industry Conference* (pp. 79-83).

Curran, I., & Pluta, S. (2008, April 22-23). Overview of machine to machine and telematics. In *Proceedings of the IEEE 6th Institution of Engineering and Technology Water Event* (pp. 1-33).

Inhyok Cha Shah, Y., Schmidt, A. U., Leicher, A., & Meyerstein, M. V. (2009). Trust in M2M communication. *IEEE Vehicular Technology Magazine*, *4*(3), 69–75. doi:10.1109/MVT.2009.933478

Larmo, A., Lindstrom, M., Meyer, M., Pelletier, G., Torsner, J., & Weimann, H. (2009). The LTE link-layer design. *IEEE Communications Magazine*, *47*(4), 52–59. doi:10.1109/MCOM.2009.4907407

3rd Generation Partnership Project. (2009). *4G SAE specification*. Retrieved from http://www.3gpp.org

3rd Generation Partnership Project. (2009). *LTE-advanced technical specification*. Retrieved from http://www.3gpp.org/LTE-Advanced

Zheng, Y., He, D., Xu, L., & Tang, X. (2005). Security scheme for 4G wireless systems. In *Proceedings of the International Conference on Communications, Circuits and Systems* (Vol. 1, pp. 397-401).

Zheng, Y., He, D., Yu, W., & Tang, X. (2005). Trusted computing-based security architecture for 4G mobile networks. In *Proceedings of the Sixth International Conference on Parallel and Distributed Computing, Applications and Technologies* (p. 251).

This work was previously published in the International Journal of Interdisciplinary Telecommunications and Networking, Volume 3, Issue 2, edited by Michael R. Bartolacci and Steven R. Powell, pp. 66-75, copyright 2011 by IGI Publishing (an imprint of IGI Global).

Chapter 11
Sensing Technologies for Societal Well-Being:
A Needs Analysis

Elizabeth Avery Gomez
New Jersey Institute of Technology, USA

ABSTRACT

Sensing technologies by design are calibrated for accuracy against an expected measurement scale. Sensor calibration and signal processing criteria are one type of sensor data, while the sensor readings are another. Ensuring data accuracy and precision from sensors is an essential, ongoing challenge, but these issues haven't stopped the potential for pervasive application use. Technological advances afford an opportunity for sensor data integration as a vehicle for societal well-being and the focus of ongoing research. A lean and flexible architecture is needed to acquire sensor data for societal well-being. As such, this research places emphasis on the acquisition of environmental sensor data through lean application programming protocols (APIs) through services such as SMS, where scant literature is presented. The contribution of this research is to advance the research that integrates sensor data with pervasive applications.

INTRODUCTION

Sensing technologies by design are calibrated for accuracy against an expected measurement scale. Sensor calibration and signal processing criteria are one type of sensor data while the sensor readings are another. Ensuring data accuracy and precision from sensors is essential and an ongoing challenge but haven't stopped the potential for application use. The recent advent of sensor networks as enablers for completely new classes of applications has captured the imagination of scientists and engineers from different domains (in-place). To-date sensing technology for pervasive applications (microwave ovens, dusk-dawn timers) is typically transparent to the user and has a single purpose. Technological advances afford an opportunity for sensor data integration and the motivation of our research. We focus on readings (data) from distinct types of sensor as a vehicle for societal well-being.

DOI: 10.4018/978-1-4666-2154-1.ch011

Leveraging the "stand alone" design of environmental sensors, which use little battery power we focus on mobile device-based services and applications. Our integration focus begins with environmental sensors and where the data recipient is a human user. The data service that initiates our research is SMS (Short Message Service) which continues as "the most popular data service over cellular networks" and one of the most successful wireless data services in recent years (Gomez & Bartolacci, 2011; Gomez, 2010; Zerfos, Meng, & Wong, 2006; Gomez & Bartolacci, 2006). To date, the deployment of WSNs, especially in an environmental context, for pervasive system applications is limited and lacking in resilience. The Japanese 2011 crises (earthquake, tsunami, nuclear), Haiti 2010 earthquake and Pakistan 2010 floods demonstrate the resilience of SMS as a data service when transporting sensor readings (electronic and human).

Normal operating conditions, herein societal well-being, will be the baseline for our research. This research takes a bottom-up approach using lean and agile mobile technologies to increase probability during times of crisis when technology often fails. We begin with passive sensors which are typically battery powered allowing them to prevail in times of crisis. The vulnerable point then shifts to the integration between the sensor and wireless sensor network (WSN), which causes a failure to occur before reaching its destination (i.e. human user). Recognizing there is a vulnerable point, we highlight the temporary restoration of services via mobile WSNs that are rapidly deployed by crisis response teams unlike web-base applications.

The purpose of this paper is to present the critical role sensing technologies play for societal well-being and the value of sensor data for individual users. We argue that "passive" sensing needs to be extended for ubiquitous use. We also identify the challenges in achieving this goal. We highlight the complexities when taking the same instance and extending use of sensors for crisis management (lifesaving). Sensor data for its intended purpose (data relevant to the user) and then the data associated with the sensor both play an important role for individual usage behavior. This paper begins with a review of the public health and sensor technologies literature; we then introduce the need for sensor data integration, and present our analysis approach. Conclusions and next steps will be discussed in the final section.

REVIEW OF LITERATURE

Public health, as it is known in the United States, centers on preventing disease, prolonging life, and promoting physical and mental health through organized community efforts (Gomez, 2008; IOM, 2003). The public health sector aims to prepare and protect the lives of an individual, family or group against a health-related event. These efforts span governmental, nongovernmental, and private sectors. Protecting lives against health-related events in a crisis depends on environmental factors. The increase in recent natural disasters, such as Alabama (2011 tornado), Japan (2011 earthquake, tsunami, nuclear), Pakistan (2010 floods), and Haiti (2010 earthquake) highlight the role environmental factors play when protecting lives. These same events have witnessed rapid response through the use of mobile technologies with primary emphasis on SMS data services.

We base our literature review on societal well-being in the United States beginning with the events of September 11, 2001 because of the crisis management focus that evolved. For example, President Bush's executive order on October 8, 2001 included public health and our communities within the realm of Homeland Security. His directive states that "The Office shall work with executive departments and agencies, State and local governments, and private entities to ensure the adequacy of the national strategy for detecting, preparing for, preventing, protecting against, responding to, and recovering from terrorist threats or attacks within the United States and shall periodically review and coordinate revisions to that strategy as necessary" (Bush, 2001, 2002). Moreover, ensuring health preparedness for a terrorist attack includes current vaccinations, increasing vaccine and pharmaceutical supplies, and hospital capacity.

Moving to President Obama's State of the Union address on February 1, 2011, a decade later, the focus shifts to "initiatives that expand "wireless coverage to 98% of Americans" and create a "nationwide interoperable wireless network for public safety" (Obama, 2011) which support the technological infrastructure societal well-being. "Extending access to high-speed wireless not only provides a valuable service to Americans living in those areas—access to medical tests, online courses, and applications that have not yet been invented—but also catalyzes economic growth by enabling consumers and businesses living in those areas to participate in the 21st century economy."

Public Health and Surveillance

Aside from natural disasters, the rise of health-related crises of our nation continues and includes: daily health and well-being crises (i.e. obesity, diabetes, mental health, and injury), disease outbreak (i.e. West Nile and SARS); and bioterrorist induced events (i.e. anthrax and smallpox). Early and reliable detection of health crises and detection for the prevention of injury is essential, not only for best possible response and treatment, but also for economic reasons (Cooper, 2004; Halperin, 1992). Prediction plays a pivotal role for early detection and requires multiple sources of data. For instance public health surveillance focuses on the systematic collection, analysis, and interpretation of health data (Halperin, 1992). Health data is essential for planning, implementation and the evaluation of public health practices, and depends on timely and accurate dissemination of this information to identified recipients.

Surveillance of weak and noisy signals provides indicators for early detection (Cooper, 2004; Halperin, 1992). Effective use of environmental sensor readings can improve the accuracy and timeliness of individual health situations and for local communities. The importance of surveillance was presented by the Institute of Medicine (IOM), The Future of Public Health (1988), where a widening gap in training was identified (Halperin, 1992; IOM, 1988). For example, surveillance data includes passive and active sensors, such as increased over the counter drug purchases and increased emergency room visits for a single illness (food poisoning). The above addresses areas specific to surveillance which do not target the individual, who must be self-reliant on a daily basis and for the first 72 hours following a crisis. As such, we propose the need for multiple sources of real-time sensor data on a personalized basis, much like a recommender system. Individual surveillance can also complement initiatives such as personalized electronic health records.

Societal Well-Being and Geographic Boundaries

Looking at public health surveillance as our guide, we focus on the individual who needs their own "surveillance" tools. The tools an individual user needs should align with an individual's unique needs. Our focus is on the day-to-day activities for societal well-being and the role of information technologies, namely sensors play for routine individual use. We highlight the opportunities for environmental impact (global warming) and those that increase readiness for crisis response. Environmental impacts/opportunities with sensors are best represented by "surveillance" and "readiness" can best be achieved by "education" by training and practice. These health-related disasters have highlighted the need for lean, agile and timely data to save lives and return to normalcy.

Recognizing the multiple boundaries and efforts of local communities, we posit that individuals have their own personalized needs that go beyond predefined boundaries. The landscape of the public health sector has multiple boundaries, impacting the control of a health crisis. Geographic boundaries are found at the local, state, federal and global levels, whereas mission oriented, religious, cultural, and illness related boundaries also exist within each of the geographic boundaries or could span several geographic boundaries. The interaction of public health practitioners within a specified boundary, such as the community boundary, not only spans horizontally, but extends vertically depending on the nature of the public health initiative (Figure 1).

Societal well-being depends on both "passive" and "active" sensors to inform citizens and communities. Passive environmental sensors at the local level can provide important data to citizens in communities. Active sensors on the other hand for routine use may be either information technology or human invoked.

The usage behavior of individuals based on sensor data can have a pivotal effect on energy usage and global warming. Informed and engaged citizens and communities will play key roles in homeland security (McDonald, 2002). Without full-engagement from individuals and communities, homeland security is unlikely to be achieved (McDonald, 2002). McDonald (2002) further mentions the importance of access to population data and data mining to identify patterns of risk. Risk communication, as discussed by McDonald (2002), is essential for citizen understanding and we look to our communities of interest for guidance

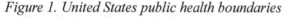

Figure 1. United States public health boundaries

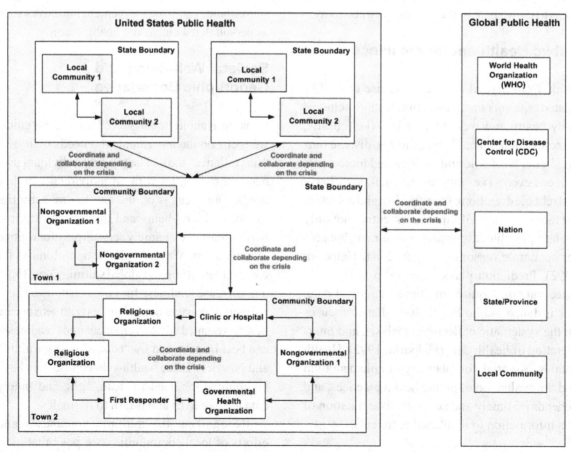

on parameters of importance. McDonald (2002) posits that the key to long-term homeland security is to design systems to engage citizens and their communities globally in a manner that is sensitive to cultural diversity and addresses causal links to health and human prosperity.

To date, academic research related to information technology is documented at the disease level within the public health domain or towards the architectural infrastructure; not the usage of the technology. The increased use of mobile technologies and social media applications over a short period of time (decade) has left little time to study and accurately document usage behavior. Sensing technologies can also reach dispersed populations. Gatenbein et al. (2004) discuss the issue of population dispersion, rural healthcare service delivery and how to overcome the geographic distance and spatial location in their paper "Establishing a Rural Telehealth Project: the Wyoming Network for Telehealth." Findings suggested the need for consortia to increase access to healthcare, create a sense of community and greater opportunities for professional and public health education in communities. Included were results from a small survey (85 respondents out of 464) conducted that revealed a high level of interest in technology. Sensing technologies for instances such as the aforementioned can assist with healthcare needs.

PUBLIC HEALTH COMMUNITY NEEDS ANALYSIS

Focusing on the crisis management domain as our backdrop, we discuss the integration of sensor data for societal well-being as seated within the public health literature. This research places emphasis on the acquisition of environmental sensor data through lean application programming protocols (APIs), where scant literature is presented. The use of APIs is a lightweight, interoperable vehicle for exchanging data through services such as SMS.

Integrating sensor data for pervasive use has societal benefits that can extend to local public health communities. Mobile device usage and wireless services are vehicles to improve effective communication and community preparedness depends on the right task-technology fit for practitioners who respond to health-related emergencies. The contribution of this research is to advance the public health domain literature with reference to information systems technology and to identify communication patterns that need further investigation. We present a business process model based on an extensive literature review of the local public health community landscape.

Sensing technologies play an integral role in data collection for measurable objectives. For instance, the missions of Healthy People 2020 includes the need 1) for "measurable objectives and goals that are applicable at the national, State, and local levels"; 2) to "engage multiple sectors to take actions to strengthen policies and improve practices that are driven by the best available evidence and knowledge"; and 3) "identify critical research, evaluation, and data collection needs."

Using Healthy People 2020 as our example, we focus on one special information technology challenge of the overarching initiative to develop "innovative approaches to help communities track their progress and develop agendas for health improvement." We focus on environmental monitoring with sensor networks to detect a "sentinel event" which can be defined as an unexpected occurrence involving death, serious physical injury, or serious psychological injury, or the risk thereof, thus signaling the need for immediate investigation and response, (Gomez & Bartolacci, 2011; JCAHO, 2002). Observations, or the disturbing lack thereof, collected from a sensor network can forewarn of a sentinel event. For example, a sensor that reports air quality for critically ill asthmatics creates a sentinel event when it detects high levels of pollutants or irritants in the air being breathed. Robustness in design and functionality would be required for such a system to work under adverse conditions. On a global scale, biosurveillence is set within the context of cooperating sensor networks (human and WSN) that deal with categories of risk requiring immediate investigation and response upon the detection of sentinel events (Kass-Hout & Zhuang, 2010). The utilization of sensor networks as "sentinels" can provide a range of information within a mobile emergency response network and adds new dimensions to the notion of crisis response.

Our analysis is structured using the Unified Modeling Language (UML) (Gomez, 2010; Fowler, 1997). UML is a standard notation (Larman, 1998) that can be used for business process modeling (BPM). We use BPM to document the current state of a process and the intended future state of our process, focusing on the how the process with sensing technologies could be performed (Cockburn, 2001). This research looks at global

health and public safety initiatives and pairs them with routine use of individuals. The use of UML domain models will serve as our guide and an enabler to advance our research with stakeholders both in information communication technology (ICT) scientists and the end user.

The research at hand looks at the critical environmental factors that could influence societal well-being and could align with global warming, public safety, and crisis response. Our concept model (Figure 2) uses interoperability with lean resources that can interface with sensing technologies via API. The concepts identified serve as a guide and enabler to enhance communication with stakeholders to promote a deeper understanding of ICT usage behavior (Figure 3) and sensor data needs. It should be noted that the focus is on routine day-to-day use.

Figure 2. Local public health community landscape

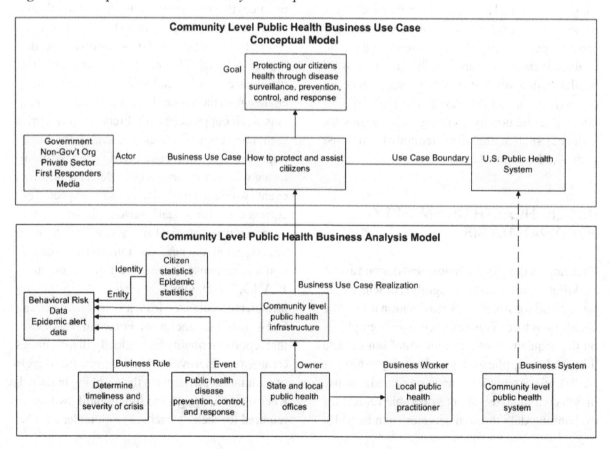

Figure 3. Concept model for ICT usage and societal well-being (Larman, 1998; Fowler, 1997)

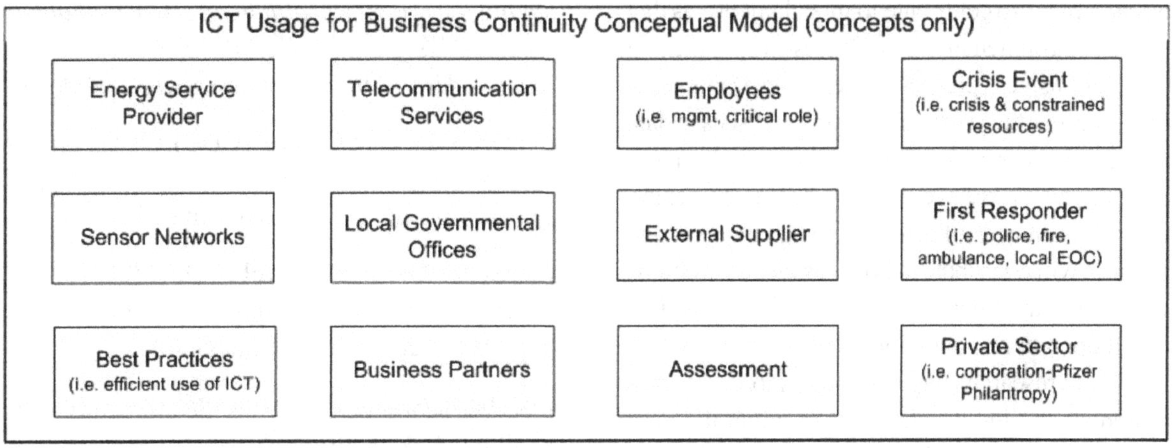

ICT Usage for Business Continuity Conceptual Model (concepts only)			
Energy Service Provider	Telecommunication Services	Employees (i.e. mgmt, critical role)	Crisis Event (i.e. crisis & constrained resources)
Sensor Networks	Local Governmental Offices	External Supplier	First Responder (i.e. police, fire, ambulance, local EOC)
Best Practices (i.e. efficient use of ICT)	Business Partners	Assessment	Private Sector (i.e. corporation-Pfizer Philantropy)

The user who needs personalized environmental data is the primary actor and supports the initiatives (business processes) set forth by the organization with which they are aligned. An actor as defined by Fowler (1997) is a role the user plays with respect to the system; the user being a human or system. For purposes of our research, the system is the mobile ICT infrastructure used to acquire sensor data. Telecommunication services (i.e. telecomm service, wireless, satellite) and sensor networks are considered supporting actors who play a critical role in data delivery and strengthen our targeted goal to ensure societal well-being and contribute to global warming with improved ICT usage behavior.

The primary objectives with respect to ICT usage behavior and response readiness for the alignment of business continuity and sustainability from the business process model (BPM) objectives are identified below (Table 1).

CONCLUSION

This paper presents a vehicle for discussion and feedback on the alignment of sensor data and sustainable business processes as they relate to ICT usage behavior and response readiness. While our focus initiates from crisis management where resources are constrained for a short time and until normalcy is reached, we posit that these efforts

Table 1. Societal well-being to sustainability alignment – an example

Primary Objectives of Societal Well-Being	BPM Summary and Personal (User) Goals
Assessment	Sustainable Protocols
• Monitor usage statistics, energy and sensor network statistics	• Assess critical system usage/capabilities
• Diagnose and investigate problems and large resource loads	• Obtain sensor data (alerts, warnings)
Policy Development	Sustainable Usage
• Inform, educate, and empower people with alternate energy options	• Educate Citizens
• Develop policies for compliance of critical business processes	• Educate Private and Public Sector
• Develop policies and plans that initiate alternate plans in crisis that constrain resources	• Promote Awareness

should become part of day-to-day sustainable activities and can also contribute to global initiatives that tackle environmental issues. Our research presents a need for business process models (BPM) that carefully address the needs and benefits of sensor data along with the interdependencies between lean ICT and people. We believe effective ICT usage behavior and awareness of alternate energy options and services will contribute to both sustainable systems for day-to-day use and empower individuals when called upon to response during times of crises.

Future research suggests updates to the models presented and the development of initial use case diagrams that highlight ICT usage behavior and instances where sensor data can be acquired and paired with day-to-day activities. This analysis is a first step in aligning societal well-being through sensor readings at the source of origin.

Our approach with lean ICT allows for resilience when resources are constrained, facilitates integration of new sensors including human ones and can also be introduced in real-time mode during crisis response. A critical item that must be addressed in parallel is the environmental factors that can lead to data inaccuracy and imprecision when measuring physical phenomena (Bychkovskiy, 2003).

REFERENCES

Bush, G. W. (2001). *Executive order establishing office of homeland security.* Retrieved from http://www.whitehouse.gov/news/releases/2001/10/20011008-2.html

Bush, G. W. (2002). Executive order on critical infrastructure protection. *Communications of the ACM.*

Bychkovskiy, V., Megerian, S., Estrin, D., & Potkanjak, M. (2003). A collaborative approach to in-place sensor calibration. In F. Zhao & L. Guibas (Eds.), *Proceedings of the 2nd International Conference on Information Processing in Sensor Networks* (LNCS 2634, p. 556).

Center for Disease Control (CDC). (2002). *Protecting the nation's health in an era of globalization: CDC's global infectious disease strategy.* Retrieved from http://www.cdc.gov/globalidplan/

Center for Disease Control (CDC). (2011). *About the tracking program.* Retrieved from http://ephtracking.cdc.gov/showAbout.action

Center for Disease Control (CDC). (2011). *National public tracking program.* Retrieved from http://www.cdc.gov/nceh/tracking/

Cockburn, A. (2001). *Writing effective use cases.* Reading, MA: Addison-Wesley.

Cooper, G. F., Dash, D. H., Levander, J. D., Wong, W. K., Hogan, W. R., & Wagner, M. M. (2004). Bayesian biosurveillance of disease outbreaks. In *Proceedings of the 20th Conference on Uncertainty in Artificial Intelligence* (pp. 94-103).

Fowler, M. (1997). *UML distilled: Applying the standard object modeling language.* Reading, MA: Addison-Wesley.

Gantenbein, R. E. (2004, September 1-5). Establishing a rural telehealth project: The Wyoming network for telehealth. In *Proceedings of the IEEE 26th Annual International Conference on Engineering in Medicine and Biology* (pp. 3089-3092).

Gomez, E. A. (2008). Connecting communities of need with public health: Can SMS text-messaging improve outreach communication? In *Proceedings of the 41st Hawaii International Conference on Systems Science* (p. 128).

Gomez, E. A. (2010, December). Towards sensor networks: Improved ICT usage behavior for business continuity. In *Proceedings of the SIGGreen Pre-ICIS Workshop.*

Gomez, E. A., & Bartolacci, M. (2007). Beyond email: The need for immediacy in a wireless world. In *Proceedings of the Networking and Electronic Commerce Research Conference*, Riva Del Garda, Italy.

Gomez, E. A., & Bartolacci, M. (2011, May). Crisis management and mobile devices: Extending usage of sensor networks within an integrated framework. In *Proceedings of the 8th International Information Systems for Crisis Response and Management Conference*, Lisbon, Portugal.

Halperin, W., Baker, E. L., & Monson, R. R. (1992). *Public health surveillance*. New York, NY: Van Nostrand Reinhold.

Health Indicators. (n. d.). *Overview of the HIW web services*. Retrieved from http://healthindicators.gov/Developers/Overview

Healthy People. (2010). *Health related quality of life and well-being*. Retrieved from http://www.healthypeople.gov/2020/about/QoLWB-about.aspx

Healthy People. (2010). *Initiative*. Retrieved from http://www.healthypeople.gov/

Institute of Medicine (IOM). (2003). *The future of the public's health in the 21st century*. Washington, DC: The National Academies Press.

Joint Commission on Accreditation of Healthcare Organizations. (2002). *Sentinel events and alerts*. Retrieved from http://www.premierinc.com/safety/topics/patient_safety/index_3.jsp

Kass-Hout, T., & Zhuang, X. (Eds.). (2010). *Biosurveillance: Methods and case studies*. Boca Raton, FL: Taylor & Francis. doi:10.1201/b10315

Larman, C. (1998). *Applying UML and patterns: An introduction to object-oriented analysis and design*. Upper Saddle River, NJ: Prentice Hall.

Levy, B. S. (1996). Toward a holistic approach to public health surveillance. *American Journal of Public Health, 86*(5), 624–625. doi:10.2105/AJPH.86.5.624

McAdams, J. (2006). *SMS for SOS: Short message service earns valued role as a link of last resort for crisis communications*. Retrieved from http://www.fcw.com/article92790-04-03-06-Print

McDonald, M. D. (2002). Key participants in combating terrorism: The role of American citizens and their communities in homeland security. *IEEE Engineering in Medicine and Biology Magazine, 21*(5), 34–37. doi:10.1109/MEMB.2002.1044158

Obama, B. (2011). *President Obama details plan to win the future through expanded wireless access*. Retrieved from http://www.whitehouse.gov/the-press-office/2011/02/10/president-obama-details-plan-win-future-through-expanded-wireless-access

Rice, R., & Katz, J. (2001). *The Internet and health communication: Experiences and expectations*. Thousand Oaks, CA: Sage.

Zerfos, P., Meng, X., & Wong, S. (2006, October 25-27). A study of the short message service of a nationwide cellular network. In *Proceedings of the 6th ACM SIGCOMM Conference on Internet Measurement* (pp. 263-268).

This work was previously published in the International Journal of Interdisciplinary Telecommunications and Networking, Volume 3, Issue 2, edited by Michael R. Bartolacci and Steven R. Powell, pp. 76-84, copyright 2011 by IGI Publishing (an imprint of IGI Global).

Chapter 12

Modeling and Simulation of Traffic with Integrated Services at Media Getaway Nodes in Next Generation Networks

Izabella Lokshina
SUNY College at Oneonta, USA

ABSTRACT

This paper is devoted to modeling and simulation of traffic with integrated services at media gateway nodes in the next generation networks, based on Markov reward models (MRM). The bandwidth sharing policy with the partial overlapped transmission link is considered. Calls arriving to the link that belong to VBR and ABR traffic classes, are presented as independent Poisson processes and Markov processes with constant intensity and/or random input stream, and exponential service delay time that is defined according to MRM. Traffic compression is calculated using clustering and learning vector quantification (e.g., self-organizing neural map). Numerical examples and simulation results are provided for communication networks of various sizes. Compared with the other methods for traffic compression calculations, the suggested approach shows substantial reduction in numerical complexity.

INTRODUCTION

The next generation networks (NGNs) are expected to be packet-based networks that provide various services including traditional telecommunications. The NGNs use broadband transport technologies that enable quality of service (QoS) manage-

DOI: 10.4018/978-1-4666-2154-1.ch012

ment, and in which service-related functions are independent from underlying transport-related technologies. Additionally, the next generation networks offer unrestricted access to different service providers and support generalized mobility that allows consistent and ubiquitous provision of services to users. The NGNs that we consider in this paper can be defined with the following fundamental characteristics, including:

- Packet-based transfer;
- Control functions that are separate for bearer capabilities, call/session, and application/service;
- Service provision that is largely independent from the network;
- Support for a wide range of services, applications and mechanisms, based on service building blocks that include real-time/streaming/non-real-time services, and multi-media;
- Open interfaces;
- Broadband capabilities that provide end-to-end QoS, and transparency;
- Ability to interconnect with legacy networks;
- Generalized mobility support;
- Converged services between Fixed/Mobile;
- Backward compatibility and support for IP based addressing schemes, including for a variety of IP address recognition schemes designed for routing in IP networks; and
- Unrestricted access to different service providers.

Furthermore, the next generation networks, considered in this paper, have to support unified service characteristics as well as convergence of broadcast and telecommunications. The NGNs architecture is layered, including the transport layer and the service layer, with the boundaries that are strictly defined, and with the following interfaces that have to be available:

- User-to-Network Interface (UNI);
- Network-to-Network Interface (NNI);
- Application-to-Network Interface (ANI).

The transport layer provides a connection between the outer NGN elements (such as, for example, the user terminals), and the elements located at the NGN servers (such as, for example, the databases and media gateways), with access that fully depends on the applied technology. For example, fixed access can be provided through the DSL and wired LAN, and wireless access can be provided through the WiFi, WiMAX, and CDMA.

The service layer provides session and other related services and delivery methods, as soon as the media gateway nodes (MGW) represent the above interfaces between the NGNs and other networks. This paper is devoted to modeling and simulation of traffic with integrated services at the media gateway nodes, based on Markov reward models (MRM), using clustering and learning vector quantification, e.g., self-organizing neural map (SOM). Compared with the other methods for traffic compression calculations, this approach provides substantial reduction in numerical complexity.

NEXT GENERATION NETWORKS AND MEDIA GATEWAY NODES

There are various views on next generation networks, which have been introduced by operators, manufacturers and providers, and that have been subject of research (Cochennec, 2002; Fazekas *et al.*, 2002; Radev & Lokshina, 2008; Lokshina & Bartolacci, 2008). The foremost NGN concept is based on integration of currently divided voice and data networks into a simpler and more flexible IP-based network, where the transport, control and service layers are independent and interact via open interfaces. At the same time, all IP networks allow different access options seamlessly integrated with an IP network layer.

Particularly, the next generation networks contain both wired and wireless access networks. The most important NGN requirements include simplicity to provide new services, portability and accessibility through different networks, and support for Quality of Service. A most popular access to the NGNs is based on the media gateways with changing transfer and switching.

The media gateway nodes are often implemented as independent devices; however, they can also be integrated in another system. In the traditional circuit-switched networks, the intelligence is concentrated in parts of the core of the network (e.g., in specific central switches). In the NGN model, the intelligence for transfer and switching is expected to be decentralized, including to the edge of the network.

At this point, the NGN architecture is conceived to achieve an independence of the applications and services from the basic switching and transport technologies. The entire independence of the applications and control mechanisms from the access and transport layers represents the fundamental feature of the next generation network architecture. That can possibly be achieved with migration of the applications and call functions to open platforms, and introduction of common control protocols supporting communication between the control functions and network resources. Particularly, that can be obtained with the gateways providing a conversion between different communication media and providing protocol adaptations.

The NGN with open architecture consists of the next three main layers:

- Connectivity layer;
- Control layer; and
- Application and service layer.

The connectivity layer consists of the following key elements:

- **Multi-Service Core:** the IP-based transport backbone that carries multiple services over high-speed optical links. This part of the network acts as a long haul transport system providing connectivity among geographically distributed nodes, and is shared by such services as, for example, the phone calls, Web sessions, video-conferences, multi-player games, and movies.

- **Gateway Elements:** they are needed to convert the information between different standards and representations.
- **Access Segment:** consists of various different broadband access technologies (e.g., the xDSL, broadband wireless, optical technologies, etc.).

The control layer (e.g., the call control) is clearly independent from the transport (physical) layer, which provides open and programmable interfaces towards the independent application layer that seamlessly mediates between the signaling protocols of different interconnected networks.

The access layer includes the both wired and wireless network technologies. The core transport network might be built around Dense Wave Division Multiplexing (DWDM) transport system.

The media gateways and soft-switches are important parts of the NGN architecture. The media gateways can be employed to interconnect networks based on different representation of the same signal. The multi-service soft-switches are common elements of the control layer, which are able to operate in spite of different protocols they have to mediate between. The soft-switches are designed as software applications that run on the server or switch to manage the MGW switching activities.

The MGW nodes are located at the ends of the next generation network, and consist of the following important components:

- Interface with the networks with circuit switching (e.g., TDM network);
- Interface with the packet networks (e.g., LAN connection);
- Digital Signal Processor (DSP) for signal processing between circuit-switched networks and packet networks.

There are three categories of the media gateway nodes in dependence of their size:

- Small Office/Home Office (SOHO) for small peripheral networks, including voice, VoIP, data and video devices;
- Office, for medium size peripheral networks;
- Provider or carrier grade with high capacity in terms of simultaneous sessions and aggregate bandwidth.

BANDWIDTH SHARING MODELS WITH PARTIAL OVERLAPPED TRANSMISSION LINK AT MEDIA GATEWAY NODES

The partial overlap of the bandwidth sharing model is defined in the following way: traffic of service i obtains part of the bandwidth equal to $r_i m_i$ bandwidth units, and the rest of traffic classes concur for sharing the rest of the link capacity $C - n_1 m_1 - n_2 m_2$ bandwidth units, where C is the whole capacity. The input traffic is described as traffic of service i, if it has m_i existing items in reserved capacity of $r_i m_i$ bandwidth units, or in sharing capacity of $C - n_1 m_1 - n_2 m_2$ bandwidth units; otherwise connection is blocking and lost (Balsamo *et al.*, 2001; Radev & Lokshina, 2007b).

In this way, the schemes for access in full sharing and full separating of the traffic flows can be introduced as particular cases of the partial overlap scheme, where for $r_1 = r_2 = 0$, full sharing is obtained, and for $r_1 m_1 + r_2 m_2 = C$, full separating is obtained (Gross & Harris, 1998).

The obtained function for retranslation with the partial overlapped transmission link for $(n_1, n_2) \in \Omega$ can be presented according to (1).

$$\alpha_1(n_1, n_2) = \begin{cases} 0, & \text{if } m_1 > C - n_1 m_1 - \max\{n_2, r_2\} m_2 \\ 1, & \text{if } m_1 \leq C - n_1 m_1 - \max\{n_2, r_2\} m_2 \end{cases}$$

$$\alpha_2(n_1, n_2) = \begin{cases} 0, & \text{if } m_2 > C - \max\{n_1, r_1\} m_1 - n_2 m_2 \\ 1, & \text{if } m_2 \leq C - \max\{n_1, r_1\} m_1 - n_2 m_2. \end{cases}$$

$$(1)$$

The partial overlapped link (POL) is introduced in this paper as a particular case of the partial overlap scheme. The basic idea consists in the fact that there is a traffic service class, for which some part of bandwidth is reserved, e.g., $r_1 > 0$ and $r_2 = 0$. At the same time, this class is not concurring for the rest of the free capacity, e.g., $m_1 = 0$, $m_2 > 0$.

For example, such service class can be represented by the constant bit rate (CBR) service class at ATM, the administrative channel at GSM channels, etc. In this way, a hybrid approach is developed for the full separating scheme, such as the CBR service class, and the full sharing scheme, such as the variable bit rate (VBR) service class. This approach helps to determine the optimal capacity of the broadband connection channels shared by two or more traffic classes and services, such as, for example, the CBR traffic class for voice transfer, the VBR traffic class for compressed video data, and ftp transfer.

Continuous Times Markov Chain (CTMC) is used for modeling a transmission link with integrated services, where single channel for bandwidth sharing policy is defined as Markov reward model (Rácz *et al.*, 2003; Radev & Lokshina, 2007b). Models of the traffic classes with guarantee of bandwidth, such as conversational, adaptive and elastic traffic classes, are presented as stationary stochastic processes developed with the network of three parallel queues in the space of states, as shown in Figure 1.

If different traffic classes, such as conversational CBR traffic classes, adaptive VBR traffic classes, and elastic traffic classes with available bit rate (ABR) are observed at the same time, then there is no chance to obtain the equivalent sharing of the bandwidth capacity. Even if an assumption is made that it is possible to separate a certain bandwidth capacity for the VBR class, still there is no chance to make it possible for the ABR class, because of the following reasons:

Figure 1. Bandwidth sharing in wireless system

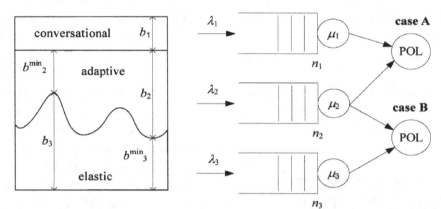

- ABR class doesn't provide the same Quality of Service as VBR or CBR classes;
- Bandwidth offered for ABR calls has a high variation in dependence of the link overload.

When considering the bandwidth sharing policy of the link capacity, the approach of full sharing should be avoided, and an alternative method has to be developed. As soon as ATM transmission link obtains the traffic classes without strong restrictions on quality of service, the bandwidth sharing of the link capacity, which increases the throughput and decreases the blocking probability, must be properly used.

The input parameters of the model consist of set of arrival rates $(\lambda_1, \lambda_2, \lambda_3)$ and departure rates (μ_1, μ_2, μ_3), the bandwidths (b_1, b_2, b_3), and the throughput constraints $(\tilde{\theta}^{\min}, \hat{\theta}^{\max})$. The state model is uniquely described by the triple (n_1, n_2, n_3), where n_1 is the number of states in the constant (conversational) flows, n_2 is the number of states in the adaptive flows, and n_3 is the number of states in the elastic flows. In order to obtain the performance measurement, the CTMC's generator matrix Q and the bandwidth sharing policy are defined, so that the link capacity C is divided into two parts:

a common part and a part reserved for the only two traffic classes. As shown in Figure 1, there are two cases in the POL bandwidth sharing policy: Case A (e.g., partial overlap between the constant and adaptive traffic flows) and Case B (partial overlap between the adaptive and elastic traffic flows).

Case A demonstrates the partial overlapped transmission link between the constant and adaptive traffic flows. The only case of the partial overlapped link between the CBR and VBR traffic classes is considered in the paper. However, the number of traffic classes considered in this model of the sharing policy can vary. When the number of traffic classes is more than one, it leads to a considerable system complexity increase, and the space of possible states should increase as well.

Let us assume that the system consists of ATM channel with the bandwidth capacity *C, Mbps*. The calls of two traffic classes arrive in the system as independent Poisson processes with exponential service times. The service times are defined according to MRM (Rácz *et al.*, 2003; Radev & Lokshina, 2008).

The following assumptions are made:

- VBR calls always use the maximum allowable bandwidth, which is a value that is less than or equal to the bandwidth of b_2. At the same time, this value is equal to a free capacity for CBR classes;

- All VBR traffic flows share the bandwidth in equal parts, e.g., the newly arrived calls and in-progress calls are compressed to the same values, as if they haven't been assigned their peak bandwidth. If during the newly arrived call the bandwidth is less than b_2^{min}, then the last call is not admitted to the system, and next, it is blocked;
- VBR call management is ideal, e.g., the time for adapting the system to new widths of the bandwidth after the newly arrived calls is infinitesimal.
- The actual residency time for VBR calls depends not only on the quantity of transferred data, but also on obtained bandwidth for VBR.

The following parameters are used to define this balance:

- Instant throughput of VBR call at the moment t is determined as a discrete random variable $\theta(t) = \min[b_2, (C-n_1 b_1)/n_2]$;
- The throughput of VBR call during the retranslation of x quantity of data for continuous random variables is determined as $\theta_x = x/T_x$.

The following additional parameters are introduced in the proposed model of the partial overlapped transmission link:

- B_1^{max}: The maximal allowed blocking probability for CBR class;
- b_2^{min}: The minimal allowed bandwidth for VBR class;
- θ^{min}: The minimal allowed throughput for VBR class;
- ε: The threshold, which determines value of θ^{min}.

These assumptions let us develop a CTMC, which state is determined as $i=(n_1, n_2)$, where n_1 is the number of broadband CBR calls, and n_2 is the number of VBR calls. The partial overlapped link capacity is divided into two parts: a common part capacity of C_{COM}, and a separate part capacity of C_{VBR} only for VBR calls. The constraints that define the number of calls are presented in (2),

$$n_1 \leq C_{COM}; \quad N_{VBR}.b_2^{min} \leq C_{VBR}; \quad n_2 \leq N_{VBR}, \tag{2}$$

where N_{VBR} is the maximal number of VBR calls.

These constraints are guaranteed for VBR calls in difference with CBR calls, where the maximal number of N_{VBR} calls is limited, and in this way, the new arriving VBR calls get protected. If there are too many of new arriving VBR calls in the transmission link, the throughput θ is decreased to θ^{min}, and must be regulated via N_{VBR}.

The generator matrix Q is built in such a way that the only transitions between the neighboring states are allowed, where q_{ij} is a depicted transition from state i to state j. There are four possible transitions between states, therefore two equations describe the newly arrived calls, and other two describe their services.

The compression between traffic classes can be obtained according to (3),

$$r_i = \min(n_2, \frac{C - b_1.n_1}{b_2}) \tag{3}$$

where $r_i.b_2$ is a common width of the bandwidth for VBR calls, when the system is at the state i. The partial overlapped transmission link model is completely defined with the following two input parameters: N_{VBR} and C_{COM}. The purpose is to minimize the blocking probability of CBR calls and to determine minimal throughput of VBR calls, under criteria for required quality of service.

Case B demonstrates the partial overlapped transmission link between the adaptive and elastic traffic flows. This model of the sharing policy is used in mobile communications as soon as modeling the media gateways with integrated traffic flows with variable bit rate of the real-time working sources is required.

There are two basic reasons to apply the proposed model in the networks with integrated services. The first reason is related to the quality of the elastic traffic flows with constant width of bandwidth, as the bandwidth that is busy with the elastic flow depends on current load of the transmission link, and on management and control algorithms in the network nodes. The second reason is related to the blocking of the elastic flows, where the service is complete even if the available bandwidth is very limited during the newly arrived calls.

For many services the actual residency time of the elastic flows depends on the throughput that the flow obtains. For example, a file transfer protocol (ftp) session would last longer if the throughput decreased. The traffic service class that is standard for GSM mobile communications doesn't change the bandwidth sharing policy model, because part of the link capacity δ, which is obtained for this class of provider services is extracted from the link capacity dedicated to three basic service classes. In this way, the scheme of full separating is used for the service class, e.g., the link capacity is left equal to C-δ.

The following possibilities can be considered for the proposed scheme of the partial overlapped transmission link:

- If there is enough bandwidth, then all traffic flows will obtain their necessary peak bandwidth, and the second and the third traffic classes will obtain, respectively, b_2 and b_3 bandwidth units.
- If providing compression of the bandwidth is required, e.g., if $n_1 \cdot b_1 + n_2 \cdot b_2 + n_3 \cdot b_3 > C$-$\delta$,

then the bandwidth compression is organized in such a manner that the bandwidth is equally shared between the adaptive and elastic traffic classes up to the moment when a constraint for the minimal possible value of the one of two classes is achieved.

- If providing additional compression of the bandwidth is required for the newly arriving calls, then the class that can allow additional decrease of the width of the bandwidth decreases it up to the moment when the minimal constraint for that class is achieved. After that, the newly arrived calls are rejected.

The described rules demonstrate that not only adaptive flows but elastic flows always obtain their maximum possible bandwidth as well, which is less than:

- Requirements to the peak bandwidth for b_2 and b_3;
- Equal part of the bandwidth left from the adaptive and elastic traffic flows for the constant traffic flows.

The compression of ABR calls is required to achieve a better throughput. The particular feature of the ABR classes is related to their bit rate that changes over time, e.g., vary in a certain interval that makes them appropriate for compression. If the compression is applied with a higher than the maximum possible value, then it leads to the total loss of information. The case with VBR and ABR traffic service classes is used, with the calls arriving as independent Poisson processes and introduced to the model as Markov processes designed for pure birth with constant intensity or random input stream. The time of service delay is exponential.

The time of service delay can be determined using MRM. All adaptive and elastic traffic flows share proportionally the available bandwidth among themselves, i.e., the newly arrived flow and in-progress flows will be squeezed to the same compression values. After that, if a newly arriving flow decreases the flow bandwidth below the minimal accepted value, and is not admitted to the system, then it is blocked and lost.

The compression of traffic classes can be presented via a common width of the bandwidth $r_i.b_i$ for ABR calls, as shown in (4).

$$r_i = \min\left(n_i, \frac{C - b_1.n_1}{b_i}\right), \quad i = 2, 3 \quad (4)$$

The proposed model is based on MRM, in which only transitions between the neighbouring states are allowed, and possible state transitions are described with the nonzero transition rates according to (5),

$$
\begin{aligned}
q\left(n_1, n_2, n_3 \to n_1 + 1, n_2, n_3\right) &= \lambda_1 \\
q\left(n_1, n_2, n_3 \to n_1, n_2 + 1, n_3\right) &= \lambda_2 \\
q\left(n_1, n_2, n_3 \to n_1, n_2, n_3 + 1\right) &= \lambda_3 \\
q\left(n_1, n_2, n_3 \to n_1 - 1, n_2, n_3\right) &= n_1 \cdot \mu_1 \\
q\left(n_1, n_2, n_3 \to n_1, n_2 - 1, n_3\right) &= n_2 \cdot \mu_2 \\
q\left(n_1, n_2, n_3 \to n_1, n_2, n_3 - 1\right) &= n_3 \cdot \rho_3(n_1, n_2, n_3) \cdot \mu_3
\end{aligned}
$$
$$(5)$$

where the first three equations represent the state transitions due to the call arrivals, while the second three equations represent the transitions due to the call departures. The $n_3 \cdot \rho_3(n_1, n_2, n_3) \cdot \mu_3$ quantity denotes the total bandwidth of the interactive flows when the system is in a state (n_1, n_2, n_3), and their compression is denoted as ρ_3.

CALCULATIONS OF TRAFFIC COMPRESSION IN MODELS WITH PARTIAL OVERLAPPED TRANSMISSION LINK

The bandwidth sharing policy is completely determined in the partial overlapped transmis-

sion link models with specifying the following parameters: the link capacity is divided into two parts: a common part of C_{COM} and a separated part of C_{VBR} only for VBR calls, where the total link capacity is equal to $C = C_{COM} + C_{VBR}$. These two parameters guarantee a certain blocking probability for CBR service classes and throughput for VBR service classes.

The blocking probability is guaranteed through proper determination of a common part of the link capacity C_{COM}. If VBR load of the link changes, then adjusting the parameter N_{VBR} to the maximal number of ABR calls is required to keep the same throughput. The input parameters are determined with the following two steps:

Determine the minimal required capacity for CBR traffic class, which guarantee the blocking probability B_1 for CBR traffic class according to (6),

$$\min\left\{C_{COM} : B_1 \leq B_1^{\max}\right\} \quad (6)$$

- Determine the maximal number of simultaneous VBR calls in the system with the minimal blocking probability.

The partial overlapped single transmission link model is considered with the middle-size link capacity of $C_1 = 60$ units and the small capacity of $C_2 = 50$ units. The assumptions are made that the system has three traffic classes: adaptive, elastic and constant; and the last traffic class is not included in the subject of the future analysis.

The partial overlapped transmission link is introduced as the bandwidth sharing policy, and part of the bandwidth is reserved for constant bit rate traffic class. At the same time, this traffic class doesn't compete for the rest of free capacity. The policy is a hybrid between the full bandwidth separating scheme considering CBR class, and the full bandwidth sharing scheme considering VBR class.

Figure 2. The space of feasible states for the link capacity of $c_1=60$ units

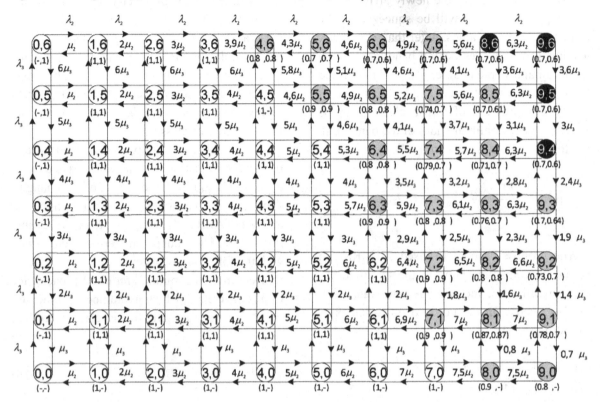

The space of possible states for the link capacity of $C_1=60$ is presented in Figure 2, where the maximal number of adaptive VBR traffic classes is $N_2=9$, and the maximal number of elastic ABR users is $N_3=6$. The peak bandwidth requirements for the traffic flows are $b_2=8$ and $b_3=5$, respectively. The flows are defined with the minimum accepted bandwidth, which set to $b_2^{min}=5.6$ (e.g., 70% of the maximal), and $b_3^{min}=3$ (e.g., 60% of the maximal).

According to the assumptions, the model has 70 feasible states, and for 27 of them the traffic flow is compressed below the peak bandwidth for adaptive and elastic traffic classes specified by (b_2, b_3). Each circle in the Markov chain represents one of the states; the adaptive and elastic flows are depicted with numbers, respectively. The compression is shown in brackets under each of 70 feasible states, which is calculated for both traffic classes (r_2, r_3).

The circles are colored according to the their compression value, where the white ones represent states with no compression, the grey ones represent states with compression and with no losses, and the black ones represent states that the rule for the minimal available bandwidth is applied ($r_2^{min}=0.7$ and $r_3^{min}=0.6$). Because of Markov chain complexity the transitions λ_2 and λ_3 are not presented as there is no change in tendency of their behavior.

More accurate results are obtained through the actual compression developed with μ_2 and μ_3. When defining the size of the generator matrix Q, the calculation of each state capacity without compression is required: $C=N_2.b_2+N_3.b_3$.

For states with capacity obtained greater than present, the compression is applied in the following way:

- If $C<N_2.b_2+N_3.b_3$, then $r_2=r_3$ is a level of compression for both traffic flows, as soon

as the minimal constraint for adaptive and elastic traffic classes is applied ($b_2^{\min} / b_2 \leq r_2 \leq 1$ and $b_3^{\min} / b_3 \leq r_3 \leq 1$).

- If more compression is required and the minimal constraint still is not true for both traffic flows, then the traffic class with newly arrived calls, which needs additional compression, is compressed up to the point when the minimal constraint becomes true.

The calculations of the service coefficients are made through multiplying the number of flows with the coefficients of compression for the corresponding states. For the link capacity of $C=50$ units 35 feasible states are obtained, and for 16 of them the traffic flow is compressed below the peak bandwidth for adaptive and elastic traffic classes, based on the proposed rule for the minimal throughput.

The clustering procedure is used with the steady-state distribution calculated in stochastic node network with Discrete Times Markov Chain (DTMC), which provides capabilities to resolve the tasks described below, in the following conditions:

- The topology and structure of Markov chain are known in advance, e.g., the number and shape of the cluster classes are determined. In that case the task of adjusting the target cluster classes' size cannot be resolved with the conventional stochastic methods. The neural structures with unsupervised learning and clustering algorithms such as Kohonen networks, K-means clustering and Gaussian Mixture models offer the models with a greatly reduced training time. These models, known collectively as Vector Quantifications (VQs), provide capability to present the winning node that represents the same class as a new training pattern (Buhmann, 2002; Webb, 1999; Radev *et al.*, 2006).

With two-layered learning neural structure one can successfully estimate the probability density function, the occupancy distribution, and the rare event probability of DTMCs and MRMs (Radev & Lokshina, 2007a, 2007b).

- The topology of Markov chain is unknown, e.g., the number of cluster classes only is determined in advance. The performance of Markov chain mapping should correspond to possible classes of events that generate the data.

Heuristic classification procedure for the clustering procedure with Markov chain mapping is introduced in this paper. Two-layered Kohonen neural network is developed that minimizes empirical loss function when the class centers are determined (Radev & Lokshina, 2007a, 2007b).

CALCULATIONS OF TRAFFIC COMPRESSION WITH CLUSTERING AND SELF-ORGANIZING NEURAL MAP

Minimum-squared-error algorithm is used to solve the clustering problem. Let's assume that a dataset $\mathbf{x}=(x_1,\ldots,x_n)$ of points is given in some Banach space, which partitions the data into k clusters (e.g., disjoint groups), so that some minimizing empirical loss function (Lokshina & Bartolacci, 2008) can be written according to (7),

$$D(x) = \frac{1}{n} \sum_{j=1}^{k} \sum_{i}^{n_j} \left(\left\| x_{ij} - s_j \right\| \right)^2 \qquad (7)$$

where

$$x_{ij} \in C_j, \qquad x_{ij} = x_i I_{\{x_i \in C_j\}}, \qquad \sum_{j=1}^{k} n_j = n$$

when the dataset points belong to a d-dimensional Euclidean region $(d \geq 2)$, C_j denotes the j-th cluster, n_j denotes the number of point x_i in C_j.

The centroid (with the same expected value) has been partitioned into d clusters with n_j elements; and the mean vectors s_j are given in (8).

$$s_j = \frac{1}{n_j} \sum_{x_i \in C_j} x_{ij} \qquad (8)$$

On the other hand, as of (7) and (8), the patterns are moved from one cluster to another only if such move improves the criterion function $D(x)$, which is known as (9),

$$D(x) = \sum_{j=1}^{k} \sum_{i}^{n} y_{ij} \left(\left\| x_{ij} - s_j \right\| \right)^2 \qquad (9)$$

where y_{ij} is the indicator of $\{x_i \in C_j\}$ - $y_{ij} = I_{\{x_i \in C_j\}}$ and $n_j = \sum_{j=1}^{n} y_{ij}$.

The mean vectors and the criterion function are updated after each pattern move. Like hill-climbing algorithms in general, similar approaches can guarantee local (but yet not global) optimization. Different initial partitions and sequences of the training patterns can lead to different solutions.

The goal of clustering is to partition the sample set of points into k (not necessarily equal) clusters C_j, such that (7) is minimized. In other words, it is necessary to specify the set of centroids $\mathbf{S} = \{s_1, \dots, s_k\}$ and the corresponding partitions $\{C_j\}$, which minimize (7). The definition also combines both the encoding and decoding steps in vector quantification. Clustering methods with the loss function are called minimum-variance methods.

Most well-known clustering and vector quantification methods update a set of centroids $\mathbf{S}$, starting from some initial set $\mathbf{S}_0$ and using iterative, typically gradient-based procedures that are multi-extremes and depend on the initial value $\mathbf{S}_0$ in the gradient-based procedures, then they converge to a local minimum rather than global minimum (Kohonen, 1997).

We can associate with the clustering used with an n dimensional discrete distribution $f(\mathbf{x};\mathbf{p})$ with independent marginal $f_m(x_m;\mathbf{p}_m)$, $\mathbf{p}_m = (p_{m1}, \dots, p_{mk})$, $m = 1, \dots, n$, and so that each $f_m(x_m;\mathbf{p}_m)$ represents a discrete k-parameter of probability density function (PDF) with masses at points $x_m = 0, 1, \dots, k-1$. Note that for $k=2$, $f_m(x_m;\mathbf{p}_m)$ is reduced to Bernoulli PDF. It is crucial to recognize that each generation based on $f(\mathbf{x};\mathbf{p})$ partitions the set samples into k clusters C_j.

The clustering procedure consists of advanced calculations of discriminating hyper-planes W_j for the pair-wise discrimination of k classes $k^* = k.(k-1)/2$, and later, the prediction of the dataset sample $\mathbf{x}$, e.g., the classification procedure. The linear regression is applied that for to discriminate the k^*-dimensional vector V_{nj}, formed for each class.

For example, we can assume that $W_j(\mathbf{x})$ is a regression function, as given in (10).

$$W_j = w_0 + w_1 x_1 + \dots + w_n x_n \qquad (10)$$

Then, we can determine belonging of the pattern $\mathbf{x}$ to pair-wise classes C_1 and C_2.

In that case, the j-th component of the vector V_{nj} is equal to 1 $(\mathbf{x} \in C_1)$ only if $W_j(\mathbf{x}) > 0$; and it is equal to -1 $(\mathbf{x} \in C_2)$ only if $W_j(\mathbf{x}) < 0$; and, finally, it is equal to 0 in all other cases. Using discriminating functions, vector function sw can be defined for each pattern $\mathbf{x}$, as it is shown in (11).

$$sw : X_j \rightarrow \{1, 0, -1\}^{k^*}$$
$$sw(\mathbf{x})_j = sign\left(W_j(\mathbf{x})\right) \qquad (11)$$

For each class C_j, function $s_j(\mathbf{x})$ can be presented as (12):

$$s_j(\mathbf{x}) = \sum_{j=1}^{k^*} V_{nj} \cdot sw(x)_i \qquad (12)$$

The pattern x is uniquely classified with discriminating hyper-planes W_j ($j = 1,\ldots, k^*$) into class C_j only if $s_j(\mathbf{x})=k-1$, i.e., with respect to $k-1$ hyper-planes, which discriminate the class C_j from the other $k-1$ classes. Then, the pattern $\mathbf{x}$ is placed in its half-space that belong to class C_j ($V_{n,j}$ and $W_j(\mathbf{x})$ have the same sign for all $V_{n,j} \neq 0$). If the pattern $\mathbf{x}$ is not uniquely classified, then the Euclidian distances of $\mathbf{x}$ to all these hyper-planes W_j are calculated, and $\mathbf{x}$ is assigned to the class with minimum distance.

The Kohonen's network algorithm provides a tessellation of the input space into patches with corresponding code vectors (Kohonen, 1997). It has an additional feature that the centers are arranged in a low-dimensional structure (usually a string, or a square grid), such that nearby points in the topological structure (the string or grid) map to nearby points in the attribute space.

We should highlight that self-organizing vector quantifications, which conserve topographic relations between centers, are particularly useful in communications, because noise added to the code vectors may corrupt the representation on some level; and the topographic mapping ensures that a small change in the code vector is decoded as a small change in the attribute space, and therefore, a small output change. Self-organizing neural networks (SOM) have layered neural competitive structure, which can learn how to detect regularities and correlations in different input patterns. The neural maps learn both the distribution and the topology of the input vectors, to be capable to recognize neighboring clusters in the attribute space (Radev *et al.*, 2006).

The Kohonen learning rule is used when the winning node represents the same class as a new training pattern, while a difference in class between the winning node and a training pattern causes the node to move away from training pattern by the same distance. In training, the winning node of the network which is nearest node in the input space to a given training pattern, moves towards that training pattern, while dragging with its neighboring nodes in the network topology. This leads to a smooth distribution of the network topology in a non-linear subspace of the training data (Lokshina & Bartilacci, 2008).

The traffic compression example that is shown in Figure 2, represents the link capacity of C=50 units and 35 feasible states, where only for 16 of them the traffic flow is compressed below the peak bandwidth for adaptive and elastic traffic classes based on the proposed rule for minimum throughput; and the traffic compression is calculated with learning vector quantification.

The distribution of arrivals is considered, where 700 independent and identically distributed stochastic values of traffic flows are generated on both inputs of the competitive layer of the neural network. These stochastic values simulate the behavior of the products $N_1.b_1$ and $N_2.b_2$, and they are generated with Gamma and lognormal distribution functions, respectively.

Preliminary vector quantification is developed according to the algorithms described in Lokshina and Bartolacci (2008). Preliminary vector quantification is demonstrated in Figure 3, where the independent arrivals are evenly distributed as 10 hits to 70 target classes. The high concentration of arrivals in inner classes can explain the fact that there is no blocking obtained for the given capacities of C_1=60 and C_2=50 units.

The main purpose of analysis is to achieve optimal throughput and reduce the blocking probability with compression for arrivals, which are out of the link capacity and, evidently, not admitted to the system. The developed compression technique with (r_1, r_2) creates the new grouping of arrivals after rejecting the classes that don't belong to compression. These classes with losses after compression are numbers 10, 20, 30, 40, 50, 60, 69 and 70, as soon as their probability mass function (PMF) has low values, as is demonstrated in Table 1.

For the link capacity of C_1=60 units the blocking probability after compression is about 0.3%, which can be used for the high priority data trans-

Figure 3. Preliminary vector quantification preceding traffic compression

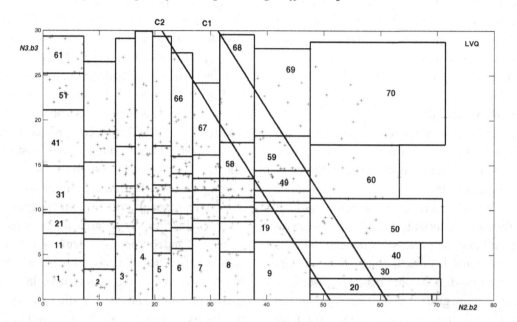

Table 1. Simulation results: arrivals lost following traffic compression

cluster N°		10	20	30	40	50	60	69	70	Total
without compression	C_2=50	10	7	9	8	9	10	9	12	74
with compression		9	7	2	0	1	2	1	7	29
without compression	C_1=60	9	7	5	5	5	10	9	12	62
with compression		0	0	0	0	0	0	0	2	2

fers; while for the link capacity of C_2=50 units the blocking probability after compression is about 4.1%, which is a significant increase, and this link should be used for the low priority data transfers (e.g., e-mails, or ftp files).

CONCLUSION

This paper is devoted to modeling and simulation of traffic with integrated services at the media gateway nodes for the next generation networks, based on Markov reward models. The bandwidth sharing policy with partial overlapped transmission link is considered. The traffic compression is calculated using clustering and learning vector quantification, e.g., self-organizing neural map. Numerical examples and simulation results are provided for communication networks of various sizes.

Modeling and simulation of traffic with integrated services at media gateway nodes for next generation networks, based on Markov reward models, compared with the other methods for traffic compression calculations, shows substantial reduction in numerical complexity.

REFERENCES

Balsamo, S., De Nitto, V., & Onvural, R. (2001). *Analysis of queueing networks with blocking.* Boston, MA: Kluwer Academic.

Buhmann, J. (2002). *Data clustering and learning: Handbook of brain theory and neural networks* (2nd ed.). Cambridge, MA: MIT Press.

Cochennec, J. Y. (2002). Activities on next generation networks under global information infrastructure in ITU-T. *IEEE Communications Magazine*, 98–101. doi:10.1109/MCOM.2002.1018013

Fazekas, P., Imre, S., & Telek, M. (2002). Modeling and analysis of broadband cellular networks with multimedia connections. *Telecommunication Systems*, 19(3-4), 263–288. doi:10.1023/A:1013821901357

Gross, D., & Harris, C. (1998). *Fundamentals of queueing theory.* New York, NY: John Wiley & Sons.

Kohonen, T. (1997). *Self-organizing maps* (2nd ed.). Berlin, Germany: Springer-Verlag.

Lokshina, I., & Bartolacci, M. R. (2008). Effective assessment of mobile communication networks performance with clustering and neural modeling. In *Proceedings of the 7ᵗʰ Annual Wireless Telecommunications Symposium*, Pomona, CA (pp. 9-16).

Rácz, S., Tari, A., & Telek, M. (2003). A distribution estimation method for bounding the reward measures of large MRMs. In *Proceedings of the International Conference on Numerical Solution of Markov Chains. Urbana (Caracas, Venezuela)*, IL, 341–342.

Radev, D., & Lokshina, I. (2007a). Clustering and neural modelling for performance evaluation of mobile communication networks. *Journal of Electrical Engineering*, 58(3), 152–160.

Radev, D., & Lokshina, I. (2007b). Modeling of media gateway nodes for next generation networks. In *Proceedings of the 6ᵗʰ Annual Wireless Telecommunications Symposium*, Pomona, CA (pp. 1-8).

Radev, D., & Lokshina, I. (2008). Modeling and simulation of traffic with compression at media gateways for next generation networks based on Markov rewards models. In *Proceedings of the Industrial Simulation Conference*, Lyon, France (pp. 199-209).

Radev, D., Lokshina, I., & Radeva, S. (2006). Evaluation of the queuing network equilibrium based on clustering analysis and self-organizing map. In *Proceedings of the 13ᵗʰ Annual European Concurrent Engineering Conference*, Athens, Greece (pp. 59-63).

Webb, A. (1999). *Statistical pattern recognition.* London, UK: Arnold.

This work was previously published in the International Journal of Interdisciplinary Telecommunications and Networking, Volume 3, Issue 3, edited by Michael R. Bartolacci and Steven R. Powell, pp. 1-14, copyright 2011 by IGI Publishing (an imprint of IGI Global).

Chapter 13

A Complete Spectrum Sensing and Sharing Model for Cognitive Radio Ad Hoc Wireless Networks Using Markov Chain State Machine

Amir Rajaee
University of Texas at San Antonio, USA

Mahdy Saedy
University of Texas at San Antonio, USA

Brian Kelley
University of Texas at San Antonio, USA

ABSTRACT

This paper presents the Cognitive Radio framework for wireless Ad Hoc networks. The proposed Cognitive Radio framework is a complete model for Cognitive Radio that describes the sensing and sharing procedures in wireless networks by introducing Queued Markov Chain method in spectrum sensing and Competitive Indexing Algorithm in spectrum sharing part. Queued Markov Chain method is capable of considering waiting time and is well generalized for an unlimited number of secondary users. It includes the sharing aspect of Cognitive Radio. Power-law distribution of node degree in scale-free networks is important for considering the traffic distribution and resource management thus we consider the effect of the topology on sensing and sharing performances. The authors demonstrate that CIF outperforms Uniform Indexing (UI) algorithm in Scale-Free networks while in Random networks UI performs as well as CIF.

DOI: 10.4018/978-1-4666-2154-1.ch013

1. INTRODUCTION

In current communication networks, the average spectrum utilization is between 15% and 85%. Cognitive Radio (CR) is a solution to increase the spectrum utilization and ultimately the network capacity leading to generating new revenue streams with higher quality of service. With increasing demand for higher capacity in wireless networks due to the rapid growth of new applications such as multimedia, the network resources such as spectrum should be used more efficiently to fulfill the need for both quantity and quality of service. This implies an optimum resource management (Si, Sun, Yang, & Zhang, 2010; Toroujeni, Sadough, & Ghorashi, 2010). Spectrum is one of the most challenging network resources which need to be carefully consumed. Cognitive Radio Networks (CRN) are supposed to efficiently use idle portions of the spectrum (resource grid). There are many techniques to sense the idle spectrum channels and manage them to increase the networks efficiency. We introduce a spectrum sensing model based on air signal energy detection and study different probabilities for fault and correct detection in Section 3. The information obtained from sensing part will be used to optimally share the idle resources and fulfill the purpose of Cognitive Radio considering different aspects of primary and secondary users and of course network topology in Section 4.

The works done in spectrum sharing has faced some challenges and can be categorized as centralized spectrum sharing vs. distributed spectrum sharing, and cooperative spectrum sharing vs. non-cooperative spectrum sharing. Spectrum sharing can also be considered from inter or intra network perspective as either one or two operators share the resources. On the other hand, the network topology and the user distribution are determining factors that directly affect the network state of being either overloaded or underloaded. CRs can be employed in many applications. CR using dynamic spectrum access can alleviate the spectrum congestion through efficient allocation of bandwidth and flexible spectrum access. It provides additional bandwidth and versatility for rapidly growing data applications. Moreover, a CR network can also be implemented to enhance public safety and homeland security. A natural disaster or terrorist attack can destroy existing communication infrastructure, so an emergency network becomes indispensable to aid the search and rescue. CR can also improve the quality of service when frequency changes are needed due to conflict or interference, the CR frequency management software will change the operating frequency automatically even without human intervention. Additionally, the radio software can change the service bandwidth remotely to accommodate new applications. As communication networks tend to become more social-like networks, Ad hoc networks and in particular power-law distributed networks i.e., scale-free networks are proposed in this paper to be considered for developing spectrum sharing technique then a new method for sharing the spectrum is proposed and proved to have the optimum performance in increasing the network capacity. At the end the results are presented and compared.

2. NETWORK TOPOLOGY

The network topology is one of the main factors in considering the traffic flow and resource management in telecommunication networks. There are different ad hoc topologies like random and scale-free discussed in network theories each presenting certain characteristics.

A. Random Topology

There are classes of networks where the nodes are attached to the network in a random way meaning that the number of connections of nodes has a normal distribution. The degree (number of links to the node) distribution of nodes in such networks is a Gaussian type distribution.

B. Scale-Free Topology

In 1999, A. L. Barabasi, and R. Albert (BA) proposed a scale-free network model based on a mechanism of growth with preferential attachment characterized with power-law distribution of the nodes degree (Si, Sun, Yang, & Zhang, 2010). Scale-free networks are robust to random attacks (node removal) and very well describe the nature of real world networks where there are always few nodes with much higher degree called hubs. Each new node enters the network initially with ability to have m links to existing nodes. The probability to connect to an existing node is dependent on the degree of that node meaning that the new node gets connected most probably to nodes with higher degree. The degree distribution for this model is a power-law distribution. The probability of a node to have a degree d_i is given by

$$P\{d_i\}=d_i^{-\alpha} \tag{1}$$

where $2<\alpha<+\infty$.

The distribution tail shows the nodes with highest degree called hubs. Here we use scale-free properties to better sense network traffic and manage the resources. Since hubs have the highest degrees amongst the nodes and because of their many connections, they have big impact on overall behavior of the network (Si, Sun, Yang, & Zhang, 2010). Consider a cluster of having N_0 nodes and some newcomers tends to attach to this network. The newcomer starts to scan its neighborhood in a radius of r_x which is determined by minimum satisfactory bitrate. There will be some existing nodes within r_x from which only one node is selected to be connected based on scale-free algorithm and that is the node with highest degree and of course the one with the best link quality as in Figure 1.

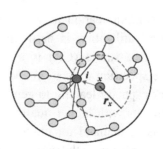

3. SPECTRUM SENSING

Consider a situation where we target sub-channel and the received signal is sampled at f_s. When the primary user is active, we define two hypotheses:

$$\begin{cases} H_1 : y(n) = x(n) + u(n) \\ H_0 : y(n) = u(n) \end{cases} \tag{2}$$

The channel is presume to be AWGN channel with $E[|u(n)|^2]=\sigma_u^2$ and the signal variance is $E[|s|^2]=\sigma_s^2$. We define two probabilities, P_d and P_f. P_d is the probability of detection under hypothesis H_1 and P_f is the probability of false alarm (energy detected) under hypothesis H_0 while there is no signal in current subcarrier. The higher the P_d, the less the amount of interference with primary users and the less P_f, the higher capacity we can allocate to secondary users (Lee & Akyildiz, 2008).

A. Energy Detection

In order to detect the RF energy in the certain subcarrier for a given primary user, the secondary user samples on- the-air signal constructs the following metric to decide on the presence of the active users in targeted subcarrier (Liang, Zeng, Peh, & Anh, 2008).

$$T(y) = \frac{1}{N} \sum_{n=1}^{N} |y(n)|^2 \tag{3}$$

N is the maximum integer not greater than τ f_s, where τ is the sensing time interval and for notation simplicity, we assume $N = \tau$ fs. Under H_0, the metric T(y) is a random variable whose PDF has a Chi-Square distribution If we choose the detection threshold as $\in$, P_f is then given by:

$$P_f(\in, \tau) = P\,r(T(y) > \in |\,H_0) = \int_{\in}^{\infty} p_0(x)dx \tag{4}$$

For Circular Symmetric Complex Gaussian (CSCG) noise:

$$P_f(\in, \tau) = Q((\frac{\in}{\sigma_u^2} - 1)\sqrt{\tau f_s}) \tag{5}$$

P_d will then be:

$$P_d(\in, \tau) = P\,r(T(y) > \in |\,H_1) = \int_{\in}^{\infty} p_1(x)dx \tag{6}$$

For large N, the first and second statistical moment of T(y) are:

$$y_1 = (\gamma + 1)\sigma_u^2 \tag{7}$$

where $\gamma = \dfrac{\sigma_u^2}{\sigma_n^2}$

$$\sigma_1^2 = \frac{1}{N}[E \mid s(n) \mid^4 + E \mid u(n) \mid^4 \\ -(\sigma_s^2 - \sigma_u^2)^2 + 2\sigma_s^2\sigma_u^2] \tag{8}$$

For complex-valued PSK and CSCG:

$$P_d(\in, \tau) = Q((\frac{\in}{\sigma_u^2} - \gamma - 1)\sqrt{\frac{\tau f_s}{2\gamma + 1}}) \tag{9}$$

Threshold for a pre-specified P_d:

$$(\frac{\in}{\sigma_u^2} - \gamma - 1)\sqrt{\frac{\tau f_s}{2\gamma + 1}} = Q^{-1}(\bar{P}_d) \tag{10}$$

Invers Q-function of P_f will be as:

$$Q^{-1}(P_f) = (\frac{\in}{\sigma_u^2} - 1)\sqrt{\frac{\tau f_s}{2\gamma + 1}} \tag{11}$$

$$P_f = Q(\sqrt{2\gamma + 1}Q^{-1}(P_d) + \sqrt{\tau f_s}\gamma) \tag{12}$$

For the pre-specified P_f:

$$P_d = Q(\frac{1}{\sqrt{2\gamma + 1}}(Q^{-1}(P_f) - \sqrt{\tau fs}\gamma)) \tag{13}$$

Later in Section 4 when we discuss spectrum sharing, we will use P_d and P_f.

B. Queued Markov Chain Model for Spectrum Analysis-Access (Sensing)

In this section, we consider group of secondary users that tend to access idle licensed spectrum portions in an opportunistic way. The process of sensing and decision making on the spectrum allocation is analyzed with a Markov Chain Process with a *queue state* in which, primary user has the spectrum and secondary users are waiting to access once it is released by primary user. This model is for one subcarrier and can be generalized for a resource block with multiple subcarriers.

The primary user PU is licensed to operate in the spectrum. The PU traffic is modeled as Poisson random process with arrival rate λ_p and departure rate μ_p. The secondary users are indexed by i (i=1,2,...,N) and modeled with Poisson random process as well; the arrival rate of SU is λ_i and departure rate is μ_i. All users are connected to Cognitive Radio (CR) Control Unit (CU) that controls CR network. To avoid interference with Primary Users (PU), once the PU starts using the subc hannel, CU forces the Secondary Users (SU) to leave and if SUs still need to access the subchannel, CU puts them in Queue State (Sq) to find another band or wait for PU to release the subchannel. Here we assume that spectrum band

Figure 2. Queued Markov chain state machine without sharing

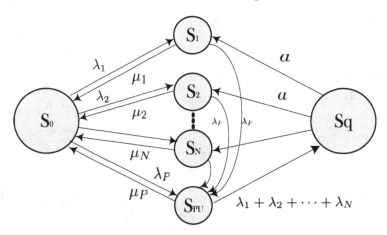

is not allowed to be shared with two or more users concurrently (Yao, Feng, & Miao, 2010; Ghosh, Cordeiro, Agrawal, & Rao, 2009; Wang, Wong, & Ho, 2010).

(N+3)-state Markov Chain State Machine is illustrated in Figure 2, where state S_0 means that the spectrum band is idle.

Without the loss of generality, the probability of transition from S_0 to S_i (*i*=1,2,...,N) is proportional to λ_i's. S_0 is the state when the subchannel is idle. Inversely, the transition probability from S_i to S_0 is μ_i. If SU's are operating in subchannel, and PU shows up, SU's have to leave the subchannel. The probability of S_i to S_{PU} is proportional to arrival rate of primary user. If PU is active in the subchannel by itself i.e., S_{PU}, the transition probability from S_{PU} to S_q is proportional to summation of SU's arrival rates, $\lambda_1 + \lambda_2 + ... + \lambda_N$, because when they request to

access, due to PU's priority, they will be forced to be put in queue. In S_q state, as soon as PU leaves the subchannel, the user state changes to either of S_i whose total probabilities are proportional to departure rate of PU that are considered a_i.

The set of equations based on the Markov Chain Model is as follows:

$$\Pi \mathbf{H} = 0 \tag{14}$$

$$\Pi_0 + \Pi_1 + \Pi_2 + ... + \Pi_N + \Pi_Q + \Pi_P = 1 \tag{15}$$

where H is the matrix that characterizes the transition state of the Markov chain, and $\Pi = [\Pi_0, \Pi_1, \Pi_2,..., \Pi_N, \Pi_Q, \Pi_P]$ is the state probability vector for $S_0, S_1, S_2,..., S_N, S_q, S_{PU}$ respectively.

The H matrix of our model is as shown in Box 1.

So the set of equations from (16) will be:

Box 1.

$$[\Pi_0,\Pi_1,...,\Pi_N,\Pi_Q,\Pi_P] \begin{bmatrix} -(\lambda_1+\lambda_2+...+\lambda_N+\lambda_P) & \lambda_1 & ... & \lambda_N & 0 & \lambda_P \\ \mu_1 & -(\mu_1+\lambda_P) & ... & 0 & 0 & \lambda_P \\ \vdots & \vdots & & \vdots & \vdots & \vdots \\ \mu_N & 0 & ... & -(\mu_N+\lambda_P) & 0 & \lambda_P \\ 0 & a_1 & ... & a_N & -(a_1+...+a_N) & 0 \\ \mu_P & 0 & ... & 0 & \lambda_1+\lambda_2+...+\lambda_N & -(\lambda_1+\lambda_2+...+\lambda_N+\mu_P) \end{bmatrix} = 0 \tag{16}$$

$$-\Pi_0 \left(\sum\nolimits_{i=1}^{N} \lambda_i + \lambda_p \right)$$
$$+\sum\nolimits_{i=1}^{N} (\Pi_i \, \mu_p) + \Pi_p \, \mu_p = 0$$

$$\Pi_0 \, \lambda_i - \Pi_i (\mu_i + \lambda_p) + \Pi_Q \, a_i = 0$$

$$-\Pi_{Q.} \left(\sum\nolimits_{i=1}^{N} a_i \right) + \Pi_{p.} \left(\sum\nolimits_{i=1}^{N} \lambda_i \right) = 0$$

$$\lambda_p . \left(\sum\nolimits_{i=1}^{N} \right) - \Pi_{p.} \left(\sum\nolimits_{i=1}^{N} \lambda_i + \mu_p \right) = 0$$

With which all Π entries will be calculated with known entries of **H**. With more complex systems, computer aided solution is required to above set of equations. This has been done for specific **H** in simulation section.

4. SPECTRUM SHARING

In communication networks, there are always unused resources due to mismanagement or the traffic usage pattern. However, it is possible to share the inactive resources between different portions of the network and in some cases share them with other service providers. Therefore, implementing spectrum sharing would highly improve the spectrum utilization efficiency and reduce the request

blocking rate (Grade of Service). We assume that the network topology is mainly scale free which has ad hoc properties plus that nodes are preferentially distributed mostly around hubs; different portions of the network are then categorized as clusters which have access to certain part of the resources (Toroujeni, Sadough, & Ghorashi, 2010; Si, Sun, Yang, & Zhang, 2010; Lee & Akyildiz, 2008; Liang, Zeng, Peh, & Anh, 2008; Yao, Feng, & Miao, 2010; Ghosh, Cordeiro, Agrawal, & Rao, 2009; Wang, Wong, & Ho, 2010; Akyildiz, Lee, Vuran, & Mohanty, 2010). In this paper we deal with a case where clusters are overloaded and needs extra resources for providing acceptable quality of service. Figure 3a shows the infrastructure of scale free inter-network interaction. Each cluster is defined with a cluster hub and a range of operation. Clusters A, B, and C each of NA, NB, and NC active users are initially planned to operate in separate allocated resource blocks (AU, AD), (BU, BD), and (CU, CD), respectively as in Figure 3b. Index U represents the uplink and D is for downlink communications. If a new user attaches to cluster A and all of the resources in resource block A are busy, the hub node in cluster A i.e., HA, tries to see if there is available idle resources in neighboring clusters. The resource elements in this portion are in the form of REU and RED. For instance, REU(i,j) represents the

Figure 3. (a) Resource Grid allocated to clusters, (b) Scale-Free Network

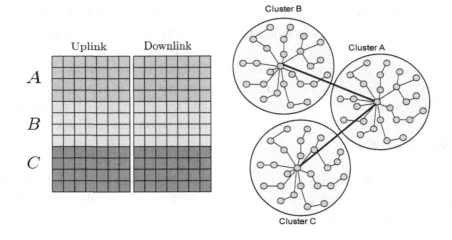

resource element for uplink at ith subcarrier and jth time slot. Now consider a new user willing to attach to cluster A by sending a request to the cluster hub, HA. We presume at the time of request all of the resources of cluster A are occupied. The new user is not blocked at this stage like conventional communication networks. Instead, HA starts to sense and search for potential available resources in neighboring clusters like B, and C. if the resource is available in either neighboring clusters for more than a limited period of time, it will be granted to cluster A and finally allocated to the new user. There is a process to consider associated criteria for releasing and granting the resources from other clusters to requesting clusters. In real world applications the hubs in clusters are distinguished mainly with their degree which is the number of active links either terminated to or originated from these hub nodes. Hubs are basically supposed to have access to as many resources as the number of active links connected to them. This leads to initially interrogating the more populated clusters as opposed to handshaking with less important (lower degree) nodes. The degree of the nodes is then considered in evaluating the merit for a specific hub. The Merit function determines the merit value for each requesting node given a certain set of available resources in granting cluster. One of the main factors in merit function is that the idle resources in neighboring clusters are not idle for unlimited time. As opposed, based on the number of active users in granting clusters, there is an average number of requests coming from user side which leads to always updating a request queue. This queue will be monitored at the time of releasing the idle resource to make sure that there is no potential demand from local cluster for the idle resource.

C. Cognitive Parameters

The hubs are presumed to be Cognitive Enabled in order to be able to sense unoccupied channels. A channel is said to be unoccupied if the instan- taneous radio frequency (RF) energy (plus noise) in this channel, is less than the certain interference limit. The probability that the channel is available for a period of time is greater than a threshold pth, These measures can be evaluated by the CR node through monitoring the traffic pattern. The Interference is measured using Carrier to Interference ratio ($\frac{C}{I}$) for each subcarrier. The probability of a channel to be available for a certain period of time is predicted by looking up the traffic profiles both in real time and the traffic history.

Because the network has a scale-free topology, the requesting users/nodes are characterized with their degree. d_i is the degree of ith node in a cluster. The degree information of the nodes is also communicated along with the request or obtained from network statistics. Nodes with higher degree have higher priority. This information is known in the local cluster and there is no need for global information broadcast (Barabasi & Albert, 1999; Si, Ji, Yu, & Leung, 2010).

Another criterion for granting the network resources to requesting users is the Costumer Classification (Q_i). Each of these new users has a specific service profile with different QoS like gold, silver and bronze. When a resource is reported to be available, it's now time to see which users of what level of quality (priority) have requested the resource. There are users with different subscription profiles which enables the decision making process directed based on the required QoS from user side. Another level of priority is also defined for emergency and security cases which dominate all incoming requests.

Signal to Noise Ratio (SNR) is important parameter that is considered for the users. Finally, the interrogated CR node from neighboring cluster reports the available channels with a set of information ($\frac{C}{I}$), p to the overloaded cluster. At the requesting cluster, d_i, Q_i are used to classify the users for granting borrowed resources.

D. Merit Function

For the reported available resources to be granted to requesting users/nodes in a fairly optimum way there needs to be a function that considers the Cognitive Parameters to calculate the merit for users/nodes.

E. Channel Indexing Function

Let R denote the set of available resources reported by the CR node. (Hub node in neighboring cluster(s)). The Channel Indexing Function (CIF) is meant for indexing the elements in R based on received Cognitive Parameters $R = (r_1, r_2, ..., r_L)$ Where $r_i = \varnothing_i((\frac{C}{I})_i, p_i)$ for every available channel. Then CIF operates on R to generate X

$$\Phi(R) = X \qquad (17)$$

$X = \{x_1, x_2, ..., x_L\}$ is the output of Channel Indexing Function, Φ, which consists of sorted performance indices for all available channels. U is the list of cognitive parameters collected from requesting users/nodes from requesting clusters. $U = \{u_1, u_2, ..., u_L\}$ where $u_i = \psi_i(SNR_i, Q_i, d_i)$. Then the Merit Function is applied to calculate the merit value for requesting users.

$$\psi(U) = M \qquad (18)$$

where M is the set of merit values for all requesting users i.e., $M = \{m_1, m_2, ..., m_L\}$. $\varnothing_i$ and ψ_i are CIF and Merit functions operating on each resource and user respectively and can be defined as:

$$\varnothing_i = w_p P_i + w_I (\frac{C}{I})_i \qquad (19)$$

$$\psi_i = w_D d_i + w_Q Q_i + w_S SNR_i \qquad (20)$$

All ω parameters are set according to technical and commercial constraints. We define the SNR for target node i as the summation of uplink (UL) and downlink (DL) SNRs.

$$SNR_i = SNR_i = SNR_i^{UL}$$
$$+SNR_i^{DL} = \frac{P_x / N_i}{r_x^2} + \frac{P_i / N_x}{r_x^2} \qquad (21)$$

P_x and P_i represent the transmit power of the newcomer node x and target node i, respectively. N_x and N_i are the noise power at newcomer and target node receivers.

F. Competitive Indexing Algorithm

M and X are matched against each other to grant the best performing channel to the users/nodes with highest merit values because the users with higher value are those who require better quality of service and are in urgent need of resources to either use them or distribute them amongst their neighbors. Then the second best resource is granted to the second user with highest merit. This process goes on until either there is no request from overloaded cluster or comes a new request from the local cluster which leads to filling up the request queue in the local cluster (Rajaee, Saedy, & Sahebalam, 2010).

Cluster A sends its request to the neighboring clusters. Matrix X is a set of performance indices for all available resources (channels) in neighboring clusters where X_i is the performance index for ith available resource element which is a function of cognitive parameters for ith subcarrier like interference and availability probability and other potential cognitive parameters that can be defined/measured as well.

Φ and ψ are determined based on network statistics, measurements, topology and operators commercial strategies. Based on the proposed competitive algorithm for granting available resources, X and M are sorted in descending order

Figure 4. Queued Markov chain state machine with sharing

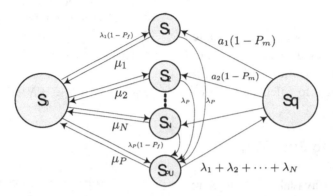

and the winning channel which is the top indexed one in X is granted to the user with highest merit value at the top of M. this process goes on until all the demands from cluster A(or all requesting clusters) are supplied. This algorithm gives optimum performance in terms of the increased capacity in the network compared to uniform allocation of resources (without indexing) in response to incoming request.

To evaluate the performance for different algorithms, We define a resource sharing performance index, $\Upsilon = M_X^T$, which is maximized based on rearrangement inequality for proposed competitive indexing algorithm. Since X and M are sorted, we can write:

$$x_1 \geq x_2 \geq,\ldots,\geq x_L \; ; \; m_1 \geq m_2 \geq,\ldots,\geq m_L \qquad (22)$$

Let $\sigma_k(X)$ and $\sigma_l(M)$ be any arbitrary permutation of X and M. The rearrangement inequality states that for sorted matrices X and M:

$$\sum_{i=1}^{L} x_i m_i \geq \sum_{i=1}^{L} \sigma_k(x_i)\sigma_l(m_i) \qquad (23)$$

$$\Upsilon_{opt} = \sum_{i=1}^{L} x_i m_i \geq \sum_{i=1}^{L} \sigma_k(x_i)\sigma_l(m_i) = \Upsilon_{rand} \qquad (24)$$

Υ_{opt} is the performance index of the proposed competitive algorithm. Υ_{opt} can be defined as different known parameters like total increased capacity if X and M are defined appropriately.

G. Sharing Incorporated in Markov Chain Model

Now, we put sharing part in our Markov Chain Model as shown in Figure 4. Because of each futures the SU's have, their chance to get spectrum from S_q is different based on our specific sharing policy shown by a_i's. Based on different policies we considered for allocating the spectrum with SU's, the a_i's will be different. For example, in Scale-Free Ad Hoc networks, degree is an important feature we can consider for SU that user with greater degree has more priority to get access. In our simulation, we assumed different traffic for each SU. The arrival rates are the same and departure rates are different. The first user has more priority than others in our simulation and the departure rates descending from first to last SU. So, the chance of first user is more than others, because of its less departure rate and its higher priority, and for second user is more than third user, only because of its less departure rate, and so on. As we see, if we increase the number of users, P_q (probability of being in queue) is decreasing and P_{PU} (the probability that only the primary user requested to access) decreasing. At the End, we compared the probability of SU's when number of users is 3. The only difference will be in P_i's that in sharing method, the chance of user with priority is more (user 1) and the difference between other users is because of their different traffics.

Here $P_d = 1 - P_m$.

In this model, imperfect spectrum sensing consisting of false alarm, P_f and miss-detection, P_m is also considered. If the idle channel is reported as busy, the current state does not change. So $(1 - P_f)$ is multiplied by arrival rates. In addition, when a channel is busy but reported as idle, the state S_q will not be allowed to go to S_i's. As a result, a_i x $(1 - P_m)$ is considered as a transition probability.

5. SIMULATION

A. Simulation Results for Sensing

an SU with greater degree has more priority to get access to subchannel. In our simulation, we assumed different traffic models i.e., λ_i for each SU. We simulated a scenario with 2,3,4 and 5 secondary users and one primary user over one subchannel. $\lambda_p = 2$, $\mu_p = 4$ and $\lambda_i = 3$. The rest of parameters are assigned as in Table 1.

The first user is assumed to have higher degree in our scale-free model resulting in having greater parameter a_1, which means it will have better chance than other users in the queue. The departure rates are assigned to SU's in descending order. Therefore, the probability of S_1, Π_1 will be greater than other secondary users because of its less departure rate and higher priority. Also, Π_2, the probability of S_2 is more that others' since its departure rate is smaller than others' and so on. As we see in Figure 5, if we increase the number of secondary users, Π_Q (probability of S_Q) is increasing and Π_P (the probability of S_{PU}) is decreasing. At the End, we compared the probability of

Table 1. Simulation traffic parameters

No.	μ_i	a_i
1	2	2
2	4	0.5
3	6	0.5
4	8	0.5
5	10	0.5

Figure 5. State probabilities using competitive sharing for different number of users

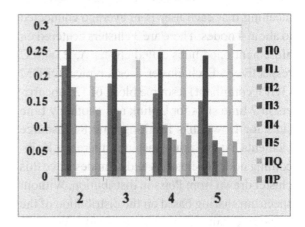

SU's when number of SU's is 3. The only difference between results with *uniform sharing* (all users have equal allocation chance) and *competitive sharing* (users are prioritized) will be in Π_j (j=1,2,...,N). In our simulation we assumed only first SU has higher priority in *competitive sharing*. As shown in Figure 6, Π_1 in *competitive sharing* is greater than Π_1 in *uniform sharing*.

B. Simulation Results for Sharing

The network structure used in the simulation is a scale-free topology with N= 100 nodes and

Figure 6. State probabilities using competitive and uniform sharing N=3

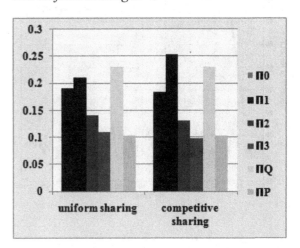

three hubs each of 19, 17, and 15 links. The average degree of nodes in this network is 3.7 meaning that each user is in average connected to about 4 nodes. There are 3 clusters centered on aforementioned hubs called cluster A, B, and C respectively. Each cluster uses a typical OFDM (3GPP compliant) resource block of 12 subcarriers in 7 time slots for Uplink and another 7 time slots for Downlinks resulting total 84 resource elements. At each time instance, there are K_i incoming nodes to cluster A and K_o nodes leave this cluster drawn from Poisson distribution. Without spectrum sharing based on the distribution of the requests coming from the user's side, cluster A may get overloaded. The capacity and load measures are simulated for cluster A without having the chance to borrow resources from neighboring clusters; Figure 9 demonstrates the performance of Uniform Indexing and Competitive Indexing. Our method in Spectrum Sharing outperforms the Uniform Indexing.

As we can see in Figure 7, the network gets saturated after a certain time and all resource elements will be occupied and the capacity tends to zero. Figure 8 shows the cluster load and net request for two different incoming and outgoing

Figure 7. Load and net request in cluster A

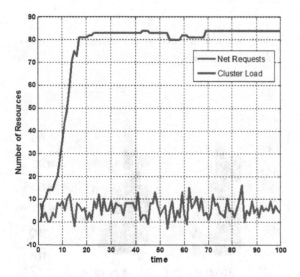

Figure 8. Load with different traffic in cluster A

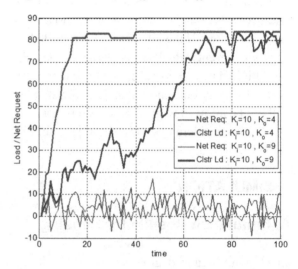

traffic. To avoid user blocking, CR nodes start to search to find idle resources in neighboring clusters. If the found resources are allocated uniformly to requesting nodes, the   .

Capacity will increase to some extent like purple line in Figure 9, but the optimum algorithm i.e., competitive indexing will outperform any *uniformly random allocation* scheme as green curve in Figure 9. Depending on the distribution of incoming request from users, the capacity increase will be different. Figure 9 shows a case where K_i=10 and K_o=4.

Because the degree is directly proportional to the indexing performance, and in Scale-Free networks, the difference between users degree is prominent as a key factor, the outperformance of CIF compared to UI algorithm is much higher than when we apply CIF in Random Network shown in Figure 10 and Figure 11.

CONCLUSION

In this paper, we presented the Cognitive Radio framework for wireless Ad Hoc networks. The complete model describes the sensing and sharing

Figure 9. Capacity improvement in cluster A

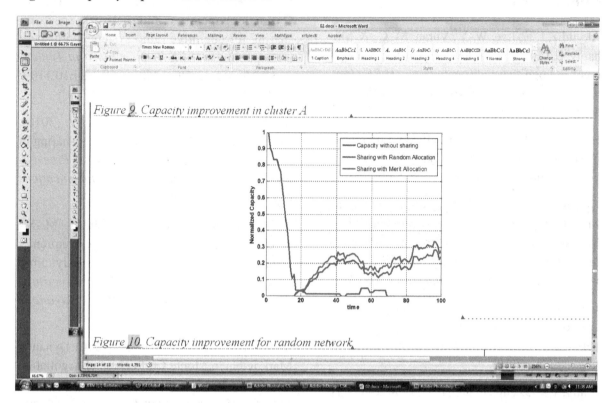

procedures in wireless networks in general but the simulation results are for random and scale-free topologies. We introduced *Queued Markov Chain* method in spectrum sensing and *Competi-* *tive Indexing Algorithm* in spectrum sharing part. We demonstrate that our proposed *Competitive Indexing Algorithm* outperforms the *Uniform Indexing Algorithm*.

Figure 10. Capacity improvement for random network

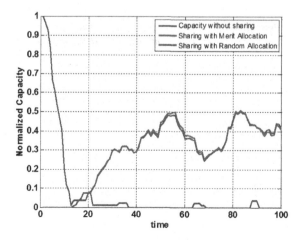

Figure 11. Capacity improvement for scale-free network

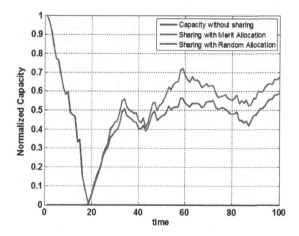

REFERENCES

Akyildiz, I. F., Lee, W.-Y., Vuran, M. C., & Mohanty, S. (2010). A survey on spectrum management in cognitive radio networks. *IEEE Communications Magazine, 46*, 40–48. doi:10.1109/MCOM.2008.4481339

Barabasi, A. L., & Albert, R. (1999). Emergence of scaling in random networks. *Science, 286*, 509–512. doi:10.1126/science.286.5439.509

Ghosh, C., Cordeiro, C., Agrawal, D. P., & Rao, M. B. (2009). Markov chain existence and hidden Markov models in spectrum sensing. In *Proceedings of the IEEE International Conference on Pervasive Computing and Communications* (pp. 1-6).

Lee, W.-Y., & Akyildiz, I. F. (2008). Optimal spectrum sensing framework for cognitive radio networks. *IEEE Transactions on Wireless Communications, 7*(10).

Liang, Y.-C., Zeng, Y., Peh, E. C. Y., & Anh, T. H. (2008). Sensing-throughput tradeoff for cognitive radio networks. *IEEE Transactions on Wireless Communications, 7*(4).

Rajaee, A., Saedy, M., & Sahebalam, A. (2010). Competitive spectrum sharing for cognitive radio on scale-free wireless networks. In *Proceedings of the International Wireless Communications and Mobile Computing Conference*, Istanbul, Turkey.

Si, P., Ji, H., Yu, F. R., & Leung, V. C. M. (2010). Optimal cooperative internetwork spectrum sharing for cognitive radio systems with spectrum pooling. *IEEE Transactions on Vehicular Technology, 59*, 1760–1768. doi:10.1109/TVT.2010.2041941

Si, P., Sun, E., Yang, R., & Zhang, Y. (2010). Cooperative and distributed spectrum sharing in dynamic spectrum pooling networks. In *Proceedings of the Wireless and Optical Communications Conference* (pp. 1-5).

Toroujeni, S. M. M., Sadough, S. M.-S., & Ghorashi, S. A. (2010). Time-frequency spectrum leasing for OFDM-based dynamic spectrum sharing systems. In *Proceedings of the 6th Conference on Wireless Advanced* (pp. 1-5).

Wang, X. Y., Wong, A., & Ho, P. (2010). Dynamic Markov-chain Monte Carlo channel negotiation for cognitive radio. In *Proceedings of the IEEE INFOCOM Conference on Computer Communications* (pp. 1-5).

Yao, Y., Feng, Z., & Miao, D. (2010). Markov-based optimal access probability for dynamic spectrum access in cognitive radio networks. In *Proceedings of the 71st IEEE Conference on Vehicular Technology* (pp. 1-5).

This work was previously published in the International Journal of Interdisciplinary Telecommunications and Networking, Volume 3, Issue 3, edited by Michael R. Bartolacci and Steven R. Powell, pp. 15-28, copyright 2011 by IGI Publishing (an imprint of IGI Global).

Chapter 14

A Novel Approach to Avoid Mobile Phone Accidents While Driving and Cost–Effective Fatalities

H. Abdul Shabeer
Anna University Coimbatore, India

R. S. D. Wahidabanu
Anna University Coimbatore, India

ABSTRACT

This paper presents the results of mobile application which helps in preventing mobile phone accidents to the great extent. An electronic circuit (Transmitter and Receiver block) also designed to detect the driver's mobile phone automatically once he or she starts the vehicle and the circuit will switch OFF and then ON the mobile phone without human intervention with the help of 5 pin relay in order to start the application automatically. The authors further extend the research by comparing the obtained results after installing this application with a recent study of the US National Safety Council, conducted on 2010. The authors also show how far this application helps in reducing economic losses in India.

INTRODUCTION

Mobile phones are essential means of communication when we are away from the office or home and it can be an important security asset in the event of an emergency. However, many studies have shown that driver use of mobile phones increases driving risk (Walsh, White, Hyde, & Watson, 2008; Charlton, 2009; Strayer & Drews,

DOI: 10.4018/978-1-4666-2154-1.ch014

2007). This risk also extends to pedestrians (Loeb & Clarke, 2009; Nasar, Hecht, & Wener, 2008). For example, it is estimated that mobile-phone use for one hour a month increases accident risk by 400–900%. Other studies show that a high percentage of accidents among youngsters are due to mobile phone use (Neyensa & Boyle, 2007). The increased accident risk is due to the fact that drivers using the phone are distracted from their main task, resulting in slower reaction time which leads to accidents.

Distraction

Using a cell phone while driving, whether to talk or to text, is a major distraction that causes car accidents. Dialing and holding a phone while steering can be an immediate physical hazard, but the actual conversations always distract a driver attention. Distraction is broadly classified in to two categories 1) Physical distraction (Visual and Mechanical), and 2) Cognitive distraction.

Physical Distraction

When using a hand-held mobile phone, drivers must take away one hand from the steering wheel to hold and operate the phone (Mechanical Distraction). They must also take their eyes off the road, at least momentarily, to pick up and put down the phone and to dial numbers (Visual Distraction). While using a hand-held phone, the driver always continues to simultaneously operate the vehicle (steer, change gear, use indicators, etc.) with only one hand. Although the physical distraction is far greater with hand-held phones, there is still some physical activity with hands-free systems. Even though they do not need to be held during the call, the driver must still divert their eyes from the road to locate the phone and (usually) press at least one button (Royal Society for the Prevention of Accidents, 2001).

Cognitive Distraction

When mental (cognitive) tasks are performed concurrently, the performance of both tasks is often worse than if they were performed separately, because attention has to be divided, or switched, between the tasks and the tasks must compete for the same cognitive processes. When a driver is using a hand-held or hands-free mobile phone while driving, he or she must devote part of their attention to operate the phone and maintain the telephone conversation, part to operating the vehicle and responding to the constantly changing road and traffic conditions. The demands of the phone conversation must compete with the demands of driving the vehicle safely.

Indian Statistics

Due to this type of distractions, in 2009 nearly one hundred and thirty five thousand people died in Indian road accidents according to National Crime Records Bureau. In India alone, the death toll rose to 14 per hour in 2009 as opposed to 13 in previous year. By 2030, road accidents are projected to become the fifth biggest killer as per Global Status Report on Road Safety (Figure 1).

Many studies highlight the usage of mobile phone and drunk & drive is the major cause for road accidents. In order to observe which is more dangerous either a mobile phone usage or presence of alcohol during driving. We extend our research by collecting special studies and data from various article and research organization. According to their research report by considering parameters like number of accidents, Brake on time, Brake force etc., Almost 90% of the studies proves usage of mobile phone while driving exhibited greater impairment than intoxicated drivers while remain-

Figure 1. Accident deaths year wise in India

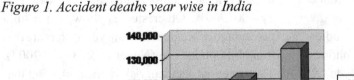
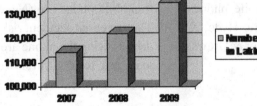

ing 10% studies says effect of both remains same. From these facts, we can conclude that mobile phone usage while driving is the foremost concern, which need to address.

Measures Taken by Government

Initially, almost all the countries around the world bans the usage of hand-held devices while driving by considering the fact of deviation in driver's concentration and their physically incapacitated because of holding the phone to his ear, which slows reaction time and could result in accidents. During that time most of the countries are encouraged to use hand-free devices. Few years later, researchers found even the usage of hand-free device doesn't show any improvements (Ogg, 2006). Later various studies carried out to establish the fact of risk involved in hand-free device. According to recent research from University of Sydney, it's proved that People talking on cell phones while driving are at least four times more likely to be involved in a collision, using a hands-free device does virtually nothing to reduce the risk (Loeb & Clarke, 2009; American Medical Network, 2005). As of results of these types of studies, various governments from all part of the countries ban the hand-free devices too. In New Delhi, India use of cell phones when driving, including use with a hands- free unit was banned from 2001.

In-spite of these ban and various strict laws against cell phone use while driving in India, Drivers are still unlikely to altogether give up using their cell phones while on the road due to various factors like,

- Users don't want to miss any emergency calls like business and personal.
- Calling for help in a Medical Emergency
- Obtaining directions when lost
- Alerting authorities of crimes in progress

REVIEW OF LITERATURE

Researchers and scientists proposed various ways like developing a model or building an application to prevent the usage of mobile phone during driving. But still each has its own demerits.

Japanese patent application JP 10 233836 entitled "On Vehicle Portable Telephone System" discloses a system and method where incoming calls received within a moving vehicle, when the vehicle exceeds a predefined speed limit, are directed to a voice mail system where the caller is invited to record a voice message, before the call is closed. In this Japanese patent application, the objective of emergency calls and risk associated with outgoing call are failed to deal with it.

US Patent "Method for automatically switching a profile of a mobile phone" (Fan & Chiu, 2007) discloses a method of measuring a current environmental noise value and compared with a predetermined noise value to calculate a noise difference. Then switching the profile of the mobile phone based on the value of the noise difference. In this patent User able to get call when struck in traffic signals, Risk of Outgoing call and Emergency call are failed to deal with it.

Another Japanese patent application JA 10 013502 entitled "Portable Telephone Set Used for Vehicle" discloses a system and method where incoming calls received within a moving vehicle, when the vehicle exceeds a predefined speed limit, are blocked by a computer on the vehicle which forces no wireless communications can be carried on. With such a system, the call recipient within the moving vehicle has no chance of knowing that an incoming call has been received. In this Japanese patent application, the objective of emergency calls is not met. Since user knows in advance that he will not get any calls while driving which may encourage the user to drive quickly to reach the destination in short time which in turn increase the risk of accidents.

They are also some of the mobile application exists to prevent mobile phone accidents like PhonEnforcer which automatically turns off the cell phone when the user is driving. This patent pending process enhances driving safety by stopping mobile phone use (Turn Off the Cell Phone, 2010). Even this application increase the risk of accidents as discussed above and chances of missing some important calls are also high.

From the above study, if we implement the above techniques in preventing mobile phone accidents there is high possibility of harsh driving and also there is every chance of missing important calls.

METHODOLOGY

By keeping all the above facts from studies, we proposed a safest application which will significantly reduce the risk of mobile phone accident at the same time the user don't have any stress on missing emergency calls.

Our mobile application comprise various stages (1) Measuring the current speed of the vehicle in mobile phone (2) compare the current speed with predefined threshold speed (3) By capturing the incoming call event and even before the phone rings, we block the call once the speed is beyond a threshold value. (4) Send the message to the caller once the call is disconnected e) Before Step (3) it should check whether the call is 'Emergency'. Emergency call means if the caller is calling from the same number 3 times within the duration of 5mins from the 1st call. (5) User shouldn't be allowed to make any outgoing call irrespective of threshold speed. Figure 2 shows the application setting panel which contains,

- **Driving Profile:** It can be set to either Enable or Disable. Set to enable if the user drives the vehicle.

Figure 2. Application setting panel

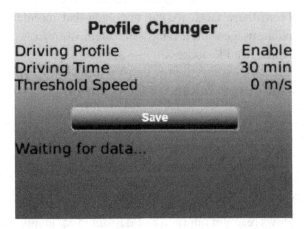

- **Driving Time:** Max. time to reach the destination can be set. Its values are 30min, 45min, 60min and 2hrs.
- **Threshold Speed:** Max. speed above which user will not get call. Its values are 0m/s, 2m/s, 4m/s, 8m/s and 16m/s.

Measuring Current Speed

By using J2ME API GPRSInfo.GPRSCellInfo we get the information of mobile tower through which mobile is connected. These are

- **Mcc:** Mobile country code
- **Mnc:** Mobile network code
- **Lac:** Location area code
- **Cellid:** Current cell id

Once this information is obtained, we will pass this parameter into the loc8.in API which results in JSONObject. After parsing this JSONObjectArray we can get the latitude and longitude of the device. From the obtained data, we can calculate the distance between two corresponding latitudes and longitudes by using the Coordinates API's method. From obtained distance, we can calculate speed by simple calculation refer Figure 3. Once the current speed is obtained it will be compared

Figure 3. Distances & speed with LAT and LONG

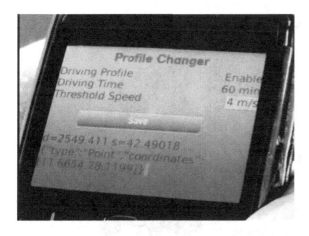

against threshold speed, if it exceeds then it start listening for any incoming call events.

Handling the Incoming Call Event and SMS

To get incoming call event, we need to implement J2ME's PhoneListener API which listens for all act on phone events like incoming call, call disconnection etc. Once our application gets any incoming call event it starts following process to make user uninterrupted while driving based on various conditions.

First application checks for driving profile. If driving profile is disabled then all incoming call will be allowed. Otherwise it will again check

whether it is emergency call (if user has called thrice in 5 mins duration then it is an emergency call) or not.

If it is emergency call, application allows that call. Otherwise disconnect that call and send message to user. The message sent to user is based on various conditions refer to Figure 4:

- If caller calls first time then he receives message "User is on Driving...So please call back after XX minutes and if its emergency call 2 more time continuously" here XX is difference b/w total time of journey (set in setting panel) and time of incoming call.
- If caller calls second time within 5 mins then he receives message "User is on driving... So please call back after XX minutes and if its emergency calls 1 more time continuously"
- If the time interval between 2 calls is more than 5 mins than application will consider it as first call.

Handling the Outgoing Calls Event

To get outgoing call event we need to implement PhoneListener API which listens all act on phone events like call Initiated, incoming call, call disconnection etc. Once the user set Driving profile

Figure 4. Show SMS received by the caller after 1st and 2nd try respectively

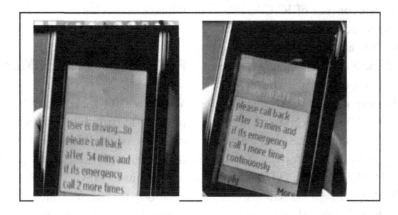

Figure 5. Transmitters and receiver block

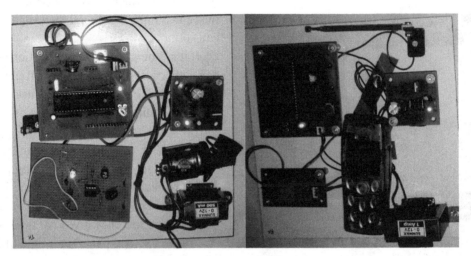

as 'Enable' these application will block the user from making outgoing call irrespective of vehicle speed or threshold speed.

TRANSMITTER AND RECEIVER BLOCK TO SWITCH OFF AND ON OF THE MOBILE PHONE

The circuit is mainly designed to start the mobile application automatically once the driver starts his vehicle. This was due to the fact that there is high chance that user may failed to start the application before he start driving. In order to avoid this, below circuit has been designed.

When driver starts its vehicle by turning the key ON, the supply will be given to mobile phone detection circuit which consists of IC CA3140. This Mobile phone detector circuit can detect the presence of an activated mobile cell phone from a distance of one and-a-half meters. Once RF signal is detected, the circuit will inform you with a flashing LED signal and it will trigger the transmitter circuit which includes ND R433 and ATMAL AT89C52 microcontroller. Signal will be generated by using on-chip oscillator which will be transmitted to ND R433 through TxD/P3.1 of microcontroller. Transmit Flag (TI) will also be set, when a byte has been completely transmitted.

ND R433 will transmit the received signal from microcontroller to IC RXD1 receiver (Figure 5). Bytes of data received from RXD1 passed to ATMAL AT89C52 microcontroller PORT 3.0/ RXD. Controller is programmed in such a way that once the signal is received it will activate a 5 pin relay unit which will turn-off & on the mobile phone within a few seconds. We developed this circuit in order to start a profile changer application only when the mobile phone restarted. These will eliminate the unnecessary memory used by mobile phone and to run this application only on the phone used by the drivers, not on the phone used by the fellow passenger. These profile changer application will automatically end when the driving speed is below the threshold speed for more than 2min by assuming user has completed the task of driving.

PRACTICAL EXPERIMENTS

We carried out our research by installing this application to 10 mobile users who travels frequently to analyses the risk factor involved and we compare these obtained results with risk estimated model carried out by NSC.

From the recent study of National Safety Council, US conducted on 2010, provides a Risk estimation report (National Safety Council, 2010) which explains in details about the risk of using mobile phone while driving by considering various factors. According to the report, the DRIVER POPULATION RELATIVE RISK is determined as the weighted average of the relative risks for the two groups comprising the driving population:

- **Rdp:** $DPo + DPe \times RR$ Where,
- **Rdp:** Driver population relative risk
- **DPo:** Proportion of the driver population not using a cell phone
- **DPe:** Proportion of the driver population using a cell phone
- **RR:** Relative risk of crashing while using a cell phone while driving.

Similarly looking at only the cell phone using portion of the population the EXPOSED DRIVER RELATIVE RISK is:

- **Re:** $DPe \times RR$ Where,
- **Re:** exposed driver relative risk

Using the above two formulas the percentage of crashes involving cell phones or the PERCENT OF EXPOSED DRIVER RELATIVE RISK can be estimated as:

- **Re%:** $Re / Rdp \times 100$

Where,

- **Re%:** Percent of exposed driver relative risk

The percent of exposed driver relative risk is the percentage of crashes involving cell phone use.

Assumptions

- **% Cell phone drivers at any given time:** 30%

- **% Non-cell phone drivers at any given time:** 70%
  Relative risk when using a cell phone while driving = 4

DRIVER POPULATION RELATIVE RISK

- **Rdp:** $DPo + DPe \times RR$
- **Rdp:** $.70 + (.30 \times 4)$
- **Rdp:** 1.90

EXPOSED DRIVER RELATIVE RISK

- **Re:** $DPe \times RR$
- **Re:** $.30 \times 4$
- **Re:** 1.20

PERCENT OF EXPOSED DRIVER RELATIVE RISK

- **Re%:** $Re / Rdp \times 100\%$
- **Re%:** $1.20 / 1.90 \times 100\%$
- **Re%:** 63.15%

Therefore, 63% of the crashes involve a driver using a cell phone. These 63% of crashes includes both the cases i.e., when the user tries to make a call or to attend a call. Thus the probability of one in two cases will be ½.

Percentage of user involve in accidents while making outgoing call (Ro) = 0.5%
Percentage of user involve in accidents while attending incoming call (Ri) = 0.5%

Now if we calculate percent of exposed driver relative risk while making outgoing call (Ro%),

- **Ro%:** $Re\% \times Ro$
- **Ro%:** 63.15×0.5
- **Ro%:** 31.575%

Similarly if we calculate percent of exposed driver relative risk while making incoming call (Ri%) we get 31.575%. Now if we compare these results with the user who has installed this applications, we get the below findings,

Since in our application outgoing call is blocked once the Driving Profile is 'Enabled'. So there is almost negligible chance of user involving in accidents while making call. (Roa% = 0%)

Where, Roa% percentage of user involve in accident while making call after installation of this application (Figure 6):

Now when comes to incoming call, our application eliminate all the calls which is not consider as emergency. To calculate the risk associated with the user after installing this application on mobile while driving. We install the application to 10 people out of which only 1 user got the emergency call while rest of the users also got the call during driving only one time, since our application start sending the SMS to the caller with elapsed time of the user to reach destination. So the caller makes call only after this elapsed time, by that time the user could have reached his destination safely. So out of 10 people only 1 got affected with the probability of risk associated with user after installing this application is 10%.

- **Rd%:** Ro% - Ria%

- **Rd%:** 31.575 – 10
- **Rd%:** 21.575%

Where,

- **Ria%:** Percentage of user involve in accidents while attending incoming call after installing application.
- **Rd%:** Differences in risk involvement of user in accident between before and after installing application (Figure 7).

INDIAN ECONOMIC IMPACT ON ACCIDENTS

As per data registered by the World Health organization, nearly 13 lakh people are known to die each year in road accidents globally. Out of which more than 1.35 lakh people are killed in India. It means, 369 people die every day on Indian roads. According to Indian National Crime Records Bureau at least 14 people die every hour in road accidents when compare to 13 in previous year refer to Tables 1 and 2.

Due to coordinated interagency approaches in developed countries, the situation is improving. However, projections indicate that unless there is a new strong political commitment to prevention, the crash death rate in low and middle-income

Figure 6. Ro% Vs Roa%

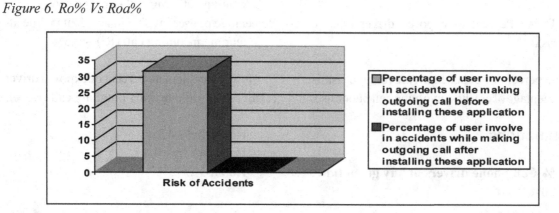

Figure 7. Ri% Vs Ria% and its difference (Rd%)

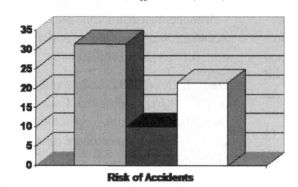

Table 1. Road accidents comparison on 2009 vs. 2008

Road Accident	Daily Statistics		Hourly Statistics	
Statistics	World	India	World	India
Deaths (2009)	3561	369	148	14
Deaths (2008)	3288	314	137	13

Table 2. Total number of accidents in 2008

Road Accident	Daily Statistics		Hourly Statistics	
Statistics	World	India	World	India
Deaths (2008)	3288	314	137	13
Injuries (2008)	136,986	1,275	5,708	53
Death + Injuries	140,274	1,589		

countries will double by 2020, reaching more than 2 million people per year. Road crash injuries impose substantial economic burdens on developing nations like India estimating 3 percent of gross national product.

The estimated cost includes compensation, asset loss, time and energy spent on police, hospital, court cases and Work Loss Costs value productivity losses. They include victims' lost wages and the replacement cost of lost household work, as well as fringe benefits and the administrative costs of processing compensation for lost earnings through litigation, insurance, or public welfare programs like food stamps and Supplemental Security Income. As well as victim work losses from death or permanent disability and from short-

term disability, this category includes work losses by family and friends who care for sick children, travel delay for uninjured travelers that results from transportation crashes and the injuries they cause, and employer productivity losses caused by temporary or permanent worker absence (e.g., the cost of hiring and training replacement workers).

National Safety Council (NSC) estimates that 28% of crashes are caused by a driver using his or her cell phone. Financial losses due to road accidents are close to 3% of our GDP every year as per BBC report. In India, so the total cost of losses due to road accidents was `820 crores per day.

According to NSC, total number crashes due to mobile phone usage while driving every day can be calculated as, 28% of 1589 = 445. Therefore, 445 crash per day due to mobile phone usage.

In India the total cost spent approx. `820 Crores per day for 1589 crashes as per the reports obtained from BBC and NSC. Hence `229.6 Crores spend for crashes involved in mobile phone usage. As discussed earlier, if we install this application in mobile phone while driving the chances of involvement in crashes is 10%

So, 10% of 445 = 44.5 (Approx. 45)

Hence we can save nearly 400 crashes which in turn reduce the economic loss of India to `620cr from `820cr and with improving personal family benefits. Hence nearly ¼th of crashes and economic loss can be reduced.

CONCLUSION

This paper presents a low-cost, non-invasive, small-size system and a profile changer application which helps to detect the driver's use of mobile phone not the phone used by the fellow passenger in the vehicle. It also helps in preventing the road accident due to distraction to a large extent. In addition to this, implementation of above system will also helps in reducing the loss of economy. Though Engineers, researchers or scientist innovated various new technologies, methods or system to prevent road accident, but still road accident continues. To overcome this type of situation all people must educate, realize and give more attention along with newly innovated technology to decrease the rate of road accident.

REFERENCES

American Medical Network. (2005). *Hands-free phones just as dangerous in cars*. Retrieved from http://www.health.am/ab/more/hands_free_phones_just_as_dangerous_in_cars

Charlton, S. G. (2009). Driving while conversing: Cell phones that distract and passengers who react. *Accident; Analysis and Prevention*, *41*, 160–173. doi:10.1016/j.aap.2008.10.006

Dash, D. K. (2009). *India leads world in road deaths: WHO*. Retrieved from http://articles. timesofindia.indiatimes.com/2009-08-17/india/ 28181973_1_road-accidents-road-fatalities-global-road-safety

Ethiopian Review. (2010). *India has the highest number of road accidents in the world*. Retrieved from http://www.ethiopianreview.com/ news/90301

Fan, Y.-F., & Chiu, I.-G. (2007). *U.S. Patent No. 7248835: Method for automatically switching a profile of a mobile phone*. Washington, DC: United States Patent and Trademark Office.

Kent, P. (2008). *Cell phone and automobile accidents on the rise*. Retrieved from http://www. articlesbase.com/automotive-articles/ cell-phone-and-automobile-accidents-on-the-rise-390037. html#ixzz10HgLel8M

Lee, J. D., McGehee, D. V., Brown, T. L., & Reyes, M. L. (2002). Collision warning timing, driver distraction, and driver response to imminent rear-end collisions in a high-fidelity driving simulator. *Human Factors*, *44*, 314–334. doi:10.1518/0018720024497844

Lissy, K. S., Cohen, J. T., Park, M. Y., & Graham, J. D. (2000). *Cellular phone use while driving: Risks and benefits*. Retrieved from http://www. cellphonefreedriving.ca/media/harvard.pdf

Loeb, P. D., & Clarke, W. A. (2009). The cell phone effect on pedestrian fatalities. *Transportation Research Part E: Logistics*, *45*, 284–290. doi:10.1016/j.tre.2008.08.001

McCartt, A. T., Hellinga, L. A., & Braitman, K. A. (2006). Cell phones and driving: Review of research. *Traffic Injury Prevention*, *7*, 89–106. doi:10.1080/15389580600651103

McEvoy, S. P., Stevenson, M. R., McCartt, A. T., Woodward, M., Haworth, C., Palamara, P., & Cercarelli, R. (2005). Role of mobile phones in motor vehicle crashes resulting in hospital attendance: A case-crossover study. *BMJ (Clinical Research Ed.)*, *331*, 20–27. doi:10.1136/bmj.38537.397512.55

Nasar, J., Hecht, P., & Wener, R. (2008). Mobile telephones, distracted attention, and pedestrian safety. *Accident; Analysis and Prevention*, *40*, 69–75. doi:10.1016/j.aap.2007.04.005

National Center for Statistics and Analysis. (2009). *Driver electronic device use in 2008* (Research Report No. []). Washington, DC: National Highway Traffic Safety Administration.]. *DOT HS*, *811*, 184.

National Safety Council. (2010). *Attribute risk estimate model*. Washington, DC: National Safety Council.

Neyensa, D. M., & Boyle, L. N. (2007). The effect of distractions on the crash types of teenage drivers. *Accident; Analysis and Prevention*, *39*, 206–212. doi:10.1016/j.aap.2006.07.004

Ogg, E. (2006). *Cell phones as dangerous as drunk driving*. Retrieved from http://news.cnet.com/8301-10784_3-6090342-7.html#ixzz10HaYIDvx

Raybeck, M. R. (2010). *Information about cell phones & driving*. Retrieved from http://www.ehow.com/about_6402532_information-cell-phones-driving.html#ixzz10HfO3dT3

Royal Society for the Prevention of Accidents. (2001). *The risk of using a mobile phone while driving*. Retrieved from http://www.rospa.com/roadsafety/info/mobile_phone_report.pdf

Strayer, D. L., & Drews, F. A. (2007). Cell-phone-induced driver distraction. *Current Directions in Psychological Science*, *16*, 128–131. doi:10.1111/j.1467-8721.2007.00489.x

SWOV. (2010). *Fact sheet: Use of the mobile phone while driving*. Retrieved from http://www.swov.nl/rapport/Factsheets/UK/FS_Mobile_phones.pdf

Thaindian News. (2008). *Strong link between mobile use and road accidents found*. Retrieved from http://www.thaindian.com/newsportal/world-news/strong-%20link-between-mobile-use-and-road-accidents-%20found_10096231.html#ixzz10Hublokz

Turn Off the Cell Phone. (2010). *PhonEnforcer: Turn off cell phones while driving*. Retrieved from http://turnoffthecellphone.com/

Walsh, S. P., White, K. M., Hyde, M. K., & Watson, B. (2008). Dialling and driving: Factors influencing intentions to use a mobile phone while driving. *Accident; Analysis and Prevention*, *40*, 1893–1900. doi:10.1016/j.aap.2008.07.005

This work was previously published in the International Journal of Interdisciplinary Telecommunications and Networking, Volume 3, Issue 3, edited by Michael R. Bartolacci and Steven R. Powell, pp. 29-39, copyright 2011 by IGI Publishing (an imprint of IGI Global).

Chapter 15
A Study of Speed Aware Routing for Mobile Ad Hoc Networks

Kirthana Akunuri
Missouri University of Science and Technology, USA

Ritesh Arora
Missouri University of Science and Technology, USA

Ivan G. Guardiola
Missouri University of Science and Technology, USA

ABSTRACT

The flexibility of movement for the wireless ad hoc devices, referred to as node mobility, introduces challenges such as dynamic topological changes, increased frequency of route disconnections and high packet loss rate in Mobile Ad hoc Wireless Network (MANET) routing. This research proposes a novel on-demand routing protocol, Speed-Aware Routing Protocol (SARP) to mitigate the effects of high node mobility by reducing the frequency of route disconnections in a MANET. SARP identifies a highly mobile node which forms an unstable link by predicting the link expiration time (LET) for a transmitter and receiver pair. NS2 was used to implement the SARP with ad hoc on-demand vector (AODV) as the underlying routing algorithm. Extensive simulations were then conducted using Random Waypoint Mobility model to analyze the performance of SARP. The results from these simulations demonstrated that SARP reduced the overall control traffic of the underlying protocol AODV significantly in situations of high mobility and dense networks; in addition, it showed only a marginal difference as compared to AODV, in all aspects of quality-of-service (QOS) in situations of low mobility and sparse networks.

INTRODUCTION

Mobile Ad Hoc Networks (MANET) are complex distributed systems comprising wireless mobile devices called MANET nodes that can freely and dynamically self-organize into arbitrary and temporary ad hoc network topologies. In MANETs, nodes internetwork seamlessly in areas with no pre-existing communication infrastructure (e.g., in tactical military networks, disaster recovery environments) providing a new and easily deployed wireless communication medium. Mobile wireless devices otherwise referred to as MANET nodes, within the transmission range connect with one

DOI: 10.4018/978-1-4666-2154-1.ch015

another through automatic configuration and set up an ad hoc network. A MANET node may be a PDA, laptop, mobile phone, and other wireless device mounted on high-speed vehicles, mobile robots, machines, and instruments; thus, the network topology is highly dynamic. The MANET nodes have computational power and routing functionality that allow them to function as sender, receiver, or an intermediate relay node or router.

Initially, MANETs were used primarily for tactical network applications to improve battlefield communications or survivability. More recently, however, the introduction of new technologies such as the Bluetooth, IEEE 802.11, and Hiper-LAN has laid foundation for commercialization of MANET. MANET deployments have begun taking place outside the military domain (Tonguz et al., 2006; Varshey et al., 2000). These recent innovations have generated a renewed and growing interest in the research and development of MANETs.

Mobility has been a major hindrance to the smooth operation of a MANET protocol (Corson & Macker, 1999). It increases link disruption and, consequently, higher network activities, exerting pressure on protocol performance. Increased network operation forces protocols to generate more control packets; thereby increasing the control overhead. Thus, a robust protocol capable of routing effectively within a highly mobile environment and without compromising its inherent attributes is vital for successful deployment of a MANET. In other words, a protocol must maintain information about the speed of the intermediate nodes and use this information to determine a stable routing path with minimal overhead.

The research presented in this paper sought to optimize MANET routing using a new routing mechanism based on node mobility. A popular and widely-employed MANET routing protocol, ad hoc on-demand vector (AODV) (Oliveira et al., 2010), was modified to drop packets when node mobility does not permit a node to form a link for the necessary amount of time. This new rout-

ing protocol is called the Speed-Aware Routing Protocol, referred to as SARP here forth. Network simulator, ns-2.33, was used to implement SARP and design and perform a variety of experiments to ensure that SARP fulfills the need to incorporate speed-awareness in a MANET's route discovery mechanism. In addition, simple empirical simulations similar to those used in Akunuri et al. (2010), Perkins and Royer (1999), Das et al. (2000), Manvi et al. (2010), and Tongus and Ferrari (2006) including random movement and traffic scenarios were run to perform a comparative study to analyze the performance of SARP against the established AODV.

PROBLEM STATEMENT

Mobility in MANETs

The ad hoc and mobile nature of the node imposes a number of restrictions on a MANET. Some of the restrictions are the limited battery power, restricted bandwidth allocation, limited transmission power and hence, limited communication range. This in turn restricts the nodes' involvement in the routing activity. A MANET node should, hence, be utilized in an efficient way with a smart routing mechanism.

Amongst various fields of MANET routing, node mobility has so far grabbed comparatively little research emphasis. The two applications that captured majority of the work that involved node mobility were designing realistic mobility models or the usage of node mobility to improve the link connectivity time. In Xiaojiang and Dapeng (2006), Athanasios (2006), and Aziz et al. (2009), different strategies have been implemented to satisfy different degrees of mobility. Also, much research has been focused on designing competitive mobility models for the simulators; as seen in Bai and Helmy (2007), Bai et al. (2003), Hong et al. (2007), and Hassan et al. (2010).

The dynamic and unpredictable movement of the nodes in a network and the heterogeneous propagation conditions make routing information obsolete; these frequent changes result in continuous network reconfiguration. The random node movements result in frequent exchange of routing packets over the limited networks' communication channels. Mobility also directly impacts the number of link failures within the network. It also causes an increase in network congestion while the routing protocol responds to the topological changes caused by independent node mobility.

Node mobility, coupled with physical layer characteristics, determines the status of link connections. Link connectivity is an important factor affecting the relative performance of MANET routing protocols (Broch et al., 1998; Corson & Macker, 1999; Samarajiva, 2001; Su et al., 2001). From the perspective of the network layer, changes in link connectivity triggers routing events such as routing failures and routing updates. These events affect the performance of a routing protocol, for example, by increasing packet delivery time or decreasing the fraction of delivered packets, and lead to routing overhead (e.g., for route discovery or route update messages) (Chlamtac et al., 2003; Samarajiva 2001; Su et al., 2001).

In Corson and Macker (1999), the impact of mobility on connectivity and lifetime route distributions was explored to isolate breakage from mobility or signal interference; this analysis supports the notion that for small route lifetimes, the link breakage is attributable to packet collisions and intermodal interference, and for longer lasting routes, the breakage is a consequence of node mobility (Tsao et al., 2006; Samarajiva 2001). It can also be stated that larger the amount of data that has to be transmitted between any arbitrary receiver-transmitter pair, the larger would be the impact of node mobility (Su et al., 2001).

When the links break, a large amount of data packets that were being transmitted through those links, are dropped. This reduces the overall throughput of the network. Once the link discon-

nects, the network forces the underlying protocol to repair the broken links or initiate search for new routing paths resulting in a continuous reconfiguration of the network (Gerla & Raychaudhari, 2007; Malek, 2011). The reconfiguration of the network for a routing protocol denotes route maintenance. Route maintenance includes the transmission of routing packets like route disconnections (RERR), route replies (RREP), route requests (RREQ) and possible HELLO packets (i.e., in case of on-demand routing). The cumulative number of routing packets generated is represented by overall control overhead generated by the network. Frequent route disconnections due to high node mobility and frequent topological changes lead to heavy route maintenance; this causes high control overhead which causes high network traffic load. The increase in network traffic load due to node mobility will result in otherwise avoidable resource reservation and bandwidth occupancy; it also increases congestion and contention.

To mitigate the adverse effect of mobility on MANET protocol performance, a new routing protocol, speed-aware routing protocol (SARP) is proposed that promises to dramatically increase the reliability of link routes during the connectivity period. The establishment of routes with unreliable links is a major factor in diminishing the end-to-end performance of established protocols (Chlamtac et al., 2003). This unreliability often causes lapses in the connectivity during the critical period of data packet transmission. Such a loss in connectivity immediately leads to maintenance activities and the subsequent rediscovery of routes, and thus creating excessive overhead and system congestion. Hence, the research proposes the exclusion of unreliable links in the potential routes using the nodes' GPS information. This capability is achieved for reactive protocols by utilizing basic link expiration time (LET) calculation in the route discovery phase. This calculation determines which nodes should participate or remain passive in a potential route. Thus, SARP incorporates speed-awareness in a routing protocol establishing reliable routes and reducing control overhead generated by the underlying routing protocol.

SARP proposes to restrict the formation of unreliable routes resulting from highly mobile intermediate nodes. During a route discovery phase, a node sends out routing packets. When a neighboring node receives this packet, it determines whether a node is too fast to form a reliable route. If the node indeed is too fast, the neighbor rejects the sender node as a potential one-hop link. This method helps is eliminating nodes with high mobility and perhaps more importantly, less reliable routes from the routing activity thereby promising comparatively lower control overhead.

Link Expiration Time

When certain amount of data is required to be transmitted using a MANET, some data is lost due to the handoffs and/or link breakages. To avoid this loss of data, a secure link should be formed; this link must survive the time required to transmit the given data size at a particular data rate supplied by the network. This would ensure the given block of data to be transmitted efficiently. The measure used in this research to represent uninterrupted link time is the link expiration time (LET).

LET between two nodes could be defined as the predicted connectivity time between the nodes (Samarajiva, 2001). In other words, it is the time two nodes are predicted to have an active route without a disconnection. The LET is calculated using the Global Positioning System (GPS) information (Rabbanny, 2002, 1994) of the nodes (Rhim et al., 2009).

In Das et al. (1998), a Zone and Link Expiry based Routing Protocol (ZLERP) was proposed for MANETs. This proactive protocol forms the most reliable links using the received signal strengths from neighboring nodes at periodic time intervals; the determination of which considers node mobility as a key factor. In both Broch et al. (1998) and The ns manual (Fall & Varadhan, 2009), the node mobility was used to predict a connectivity time between two nodes; however, the connectivity times have been used to form backup routes or multicast routing. Nevertheless, the idea of employing the predicted link connectivity time to establish reliable routes initially has not been exploited yet. In Rhim et al. (2009), during the route maintenance phase, the MANET nodes were made capable of predicting the remaining connectivity time with their neighbors in order to avoid disconnections. However, no key progress has been achieved where node mobility was used to establish stable routes.

In Broch et al. (1998), LET was introduced as a statistical derivation to forecast the average distance the relay is within the scope of the nodes. This mobility prediction method utilizes the location and mobility information provided by GPS. Initially, a free space propagation model is used, where the received signal strength solely depends on its distance to the transmitter. It is also assumed that all nodes in the network have their clock synchronized. Therefore, if the motion parameters of two neighboring nodes like speed, direction, radio propagation range are known, the duration of time these two nodes will remain connected can be determined. Assume two nodes i and j within the transmission range of each other. Let (x_i, y_i) be the coordinates of node i and (x_j, y_j) be the coordinates of node j. Let v_i and v_j be the speeds, θ_i $(0 \leq \theta_i)$ and θ_j $(\theta_j \leq 2\prod)$ be the directions of motion for nodes i and j, respectively. Then, the amount of time two mobile hosts will stay connected, is predicted by the formula given by:

$$\text{LET} = \frac{-(ab+cd) + \sqrt{(a^2+c^2)r^2 - (ad-bc)^2}}{a^2 + c^2}$$

(1)

The parameters a, b, c and d are determined using the formulae illustrated by Equations (2), (3), (4) and (5).

Parameter 'a' is the relative velocity of the receiver node with respect to the sender node along Y-axis. It is determined using Equation (2).

$$a = V_r \cos\theta - V_s \cos\theta \qquad (2)$$

'b' is the parameter used to determine the distance of the receiver node from the sender node along X-axis and is determined using Equation (3).

$$b = X_r - X_s. \qquad (3)$$

The third parameter used to determine LET is 'c'. Parameter 'c' is the relative velocity of receiver node with respect to the sender node along Y-axis. Equation (4) gives the formula to determine 'c'.

$$c = V_r \sin\theta - V_s \sin\theta = V_{Yr} - V_{Ys.} \qquad (4)$$

'd' is the distance of the receiver node from the sender node along Y-axis. This parameter is determined using the formula given in Equation (5).

$$d = Y_r - Y_s. \qquad (5)$$

The implementation of SARP will take into account two value of LET, the supply LET denoted by LET_S and the demand LET denoted by LET_D. The LET calculated from Equation (1) is the LET supplied by a potential link and hence, is denoted by LET_S.

When a MANET protocol is required to transfer a given amount of data over the network, it requires each of the links to sustain the transmission of data without a link disconnection. This demand for link connectivity by the network can be measured using the demand LET_D.

Demand Link Expiration Time (LET$_D$)

Consider two nodes i and j are within communication range of each other. Let there be a demand to transfer 'x' KB of data from node i to node j with a packet size of y KB and a rate of z packets per second. Hence, actual size of data transferred between the nodes per second is calculated to be yz Kbps.

For successful data transmission without any link breakage between the nodes, the length of time during which both nodes must have an active link to transfer the 'x' KB through the link is calculated to be:

$$LET_D = \frac{Size\,of\,Data\,to\,be\,Tranmitted(KB)}{Size\,of\,Data\,Transmitted\,per\,Second(Kbps)}\,sec \qquad (6)$$

Hence,

$$LET_D = \frac{x}{yz} + \Delta T \text{ seconds.} \qquad (7)$$

where ΔT seconds is the time-lenience factor to compensate for the delays at the intermediate nodes.

When an intermediate node receives a routing packet, it processes the packet, sets up a forward path and updates its routing table. Depending on the availability of a fresh reverse route in its routing table, it then either floods the network with more routing packets or replies to the source node with a reverse route. This processing at each intermediate node adds to high end-to-end delay in the network. In order to compensate for this delay, a time-lenience factor 'ΔT' is introduced.

The LET_D thus calculated will be the expectant LET a link must last to be considered as a stable route for SARP.

METHODOLOGY

As discussed in the preceding section, node mobility reduces the length of active connectivity within the nominal range thus increasing the potential for link disconnections. The proposed algorithm to reduce the occurrence of such link disconnections, Speed-Aware Routing Protocol (SARP) is based on excluding the nodes that are too fast from inclusion in the route discovery mechanism.

To achieve this functionality the routing protocol drops the packets received from a node that is too fast to maintain an active route.

In Paudel and Guardiola (2009), performance of ad hoc routing protocols AODV, DSDV and DSR was compared against a mobility metric which was designed to reflect the relative speeds of the nodes. This study concluded that in most simulations the reactive protocols (AODV and DSR) performed significantly better than the pro-active protocol DSDV; it also stated that AODV performed better than DSR at higher traffic loads. In addition, the simulations conducted in Mullen and Hong (2005), Das et al. (2000), and Guo et al. (2005) with varying network parameters including mobility levels, multi-path fading and network densities showed that AODV performed better than the other routing protocols in high stress situations of high mobility and fading. Henceforth, this research uses AODV as the underlying routing protocol to implement the Speed-Awareness in the routing algorithm.

SARP Algorithm

In the SARP routing algorithm, when a node receives a routing request (RREQ) or a routing response (RREP), it calculates the link expiration time (LET) of the node with respect to the packet sending node. LET is the parameter that predicts the link disconnection time between two nodes; in other words, it is the time two nodes are predicted to have an active route the ns manual.

The demand LET, LET_D, of a link is determined for a given size of data and transmission rate of the link; it is a limiting factor to identify ineffective routes. When a node receives a routing packet (RREP/RREQ), the supplied LET, LET_S is determined for the sending and receiving nodes. Ideally, when the value of the LET_S is lower than that of the LET_D, the link is predicted to be ineffective for the required amount of time; therefore, the packet is dropped, and the sending node is excluded from further routing activity. Therefore, a node must exclude a packet-sending node from route inclusion unless the condition specified by Equation (8) is satisfied.

$$LET_S \leq (LET_D + \Delta T). \tag{8}$$

The SARP algorithm comprises of the three stages:

1. The determination of node coordinates and velocities,
2. The calculation of LET_S and,
3. The identification and exclusion of unstable links from the routing procedure.

Determination of Node Co-Ordinates and Velocities

When a MANET node receives a routing packet (RREP/RREQ), the packet is transferred from lower network layers to higher node layers. At the medium access layer (MAC) of the packet-receiving node, GPS information of is noted; this includes the spatial coordinates and node spatial velocities of both the sender and receiver nodes.

At a given simulation time 't', the node coordinates and velocities are noted along the three spatial axes, as listed in Table 1.

Table 1. Determination of node coordinates and velocities

	Receiver Node	Sender Node
Node Coordinates	X_r, Y_r, Z_r	X_s, Y_s, Z_s
Node Velocities	V_{Xr}, V_{Yr}, V_{Zr}	V_{Xs}, V_{Ys}, V_{Zs}

Calculation of LET$_s$

The LET of the receiver node with respect to the sender node is determined through each axis. At time 't', the node parameters for both sender and receiver node are given in Table 1.

Since the simulations are performed on grid-frames in ns-2.33, the parameters along the Z-axis are assumed to be zero:

$$Zs = Zr = 0. \tag{9}$$

Similarly the velocities along the Z-axis are zero:

$$V_{Zs} = V_{Zr} = 0. \tag{10}$$

Equations (2), (3), (4) and (5) were substituted values from Equations (9) and (10). The resulting formulae are exemplified in Equations (11), (12), (13) and (14) respectively.

$$a = Vr \, Cos\theta_r - Vs \, Cos\theta_s = V_{Yr} - V_{Ys}, \tag{11}$$

$$b = Xr - Xs, \tag{12}$$

$$c = Vr \, Sin\theta_r - Vs \, Sin\theta_s = V_{Yr} - V_{Ys}, \text{ and} \tag{13}$$

$$d = Yr - Ys. \tag{14}$$

where θ_r and θ_s are the directions of motion of the receiver and sender nodes respectively. The amount of time the nodes are predicted to be in active communication, LET$_s$, is calculated using the formula given by Equation (1).

Identification and Exclusion of Fast-Moving Nodes

Once a node determines its LET$_s$ with respect to the packet-sending node, it can determine whether the packet-sending node is too fast to form a stable route with. The LET$_s$ of the receiver and sender nodes is used to identify the fast moving nodes. This algorithm uses a predetermined value for the demand LET, LET$_D$, as a limiting factor. A node is considered to be fast or travelling in a direction not feasible for effective communication when the LET$_s$ is short of the demand LET$_D$.

$$LET_S \leq (LET_D + \Delta T). \tag{15}$$

For a given link, when the condition given by Equation (15) is not satisfied, the link is treated as unstable and the receiver node considers the sender node too fast for effective communication and hence, dismisses it from further routing processes.

Validation of SARP

Using an ns2.33 all-in-one package, the MAC layer of the AODV protocol was modified to include the speed awareness of SARP within AODV. The functions getLoc() and getVelo() are used to determine a nodes' spatial coordinates and velocities. SARP calculates LET$_s$ based on the formula given in Equation (13). When it is less than the required LET$_D$, the node drops the control packets to ensure that the packet-sending node remains available to participate in further routing activities with the current node. Once the SARP algorithm was implemented, a scenario was simulated to validate the functioning of the SARP. Figure 1 illustrates this scenario.

A set of four nodes has initial spatial coordinates as follows: node 0 (20, 200), node 1 (200, 200), node 2 (220, 200), and node 3 (400, 200). Nodes 0, 1, and 3 travel in one direction at speeds of 5 m/s, 20 m/s, and 5 m/s respectively; however, node 2 travels the opposite direction at 20 m/s. Thus, the nodes at the farthest ends (node 0 and 3) are outside communication range and cannot form a direct route. The nodes between them, nodes 1 and 2, act as intermediate nodes for communication between nodes 0 and 3. At time 1.0 seconds, node 0 tries to connect to node 3, sending out a RREQ. These RREQs are received by intermediate nodes 1 and 2. On receiving the RREQ, nodes 0 and 1 calculate their respective LETs.

Figure 1. Experiment to validate SARP

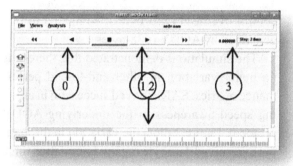 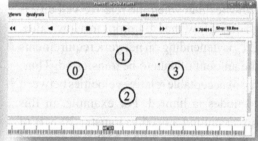

When node 1 receives the RREQ from node 0, it calculates the LET_S using Equation (13). The parameters are calculated:

at simulation time '1.0',

$$a = V_{Xr} - V_{Xs} = V_{X1} - V_{X0} = 0 - 0 = 0,$$

$$b = X_r - X_s = X_1 - X_0 = 200 - 20 = 180,$$

$$c = V_{Yr} - V_{Ys} = V_{Y1} - V_{Y0} = 20 - 5 = 15, \text{ and}$$

$$d = Y_r - Y_s = Y_1 - Y_0 = 0 - 0 = 0.$$

The supply LET is then calculated as

$$LET_S (0-1)$$

$$= \frac{-(0+0) + \sqrt{(15^2 x 250^2) - (0 - 180 * 15)^2}}{(0^2 + 15^2)}$$

$$= 11.57 \text{ seconds.}$$

Similarly, the LET_S of the link 0-2 is calculated to be approximately $LET_S(0-2) = 6$ seconds.

The above scenario was simulated using ns2 and the results are discussed below. Figure 2

Figure 2. Relative velocity between the nodes vs. LET_S

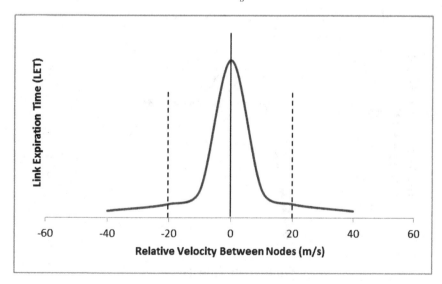

is a graph that shows the how LET_S is affected by the relative velocity of nodes.

It indicates that when the relative velocity is too high or too low, the LET drops to a low value. The LET_D is depending on network requirements or on the amount of data to be transferred. Thus, the range of acceptable relative velocities between the two nodes is limited. For example, in this scenario, assuming a need to transfer 20 MB of data with a packet size of 0.5 MB at a rate of 5 packets per second, using Equation (11), LET_D is calculated as

$$LET_D = \frac{20}{5 * 0.5} + 1 \text{ second} = 9 \text{ seconds.}$$

Thus, to transfer 20 MB for of data through this network without link breakages, two nodes at one-hop distance are expected to be connected for at least 9 seconds. From the plot in Figure 4, at an LET_S of 9 seconds, the relative velocity is 19.28 m/s. Therefore, to transfer the 20 MB of data with no link disconnections, two nodes must have a relative velocity within the range (-19.28 m/s, +19.28 m/s). The relationship between the relative velocity and LET was thus verified, and scenario 1 with LET_D of 9 seconds was then

simulated; and the results are discussed below. The results showed that SARP generated low control overhead (536 bytes); compared to that generated by AODV (638 bytes).

The simulations demonstrated that there was no major variation in other end-to-end performance metrics. SARP proved successful in creating speed awareness in the underlying AODV protocol. Figure 3 plots the variation in throughput of received data bytes against simulation time. Although the average throughput received was almost the same for both protocols, the time at which the peak of throughput occurred showed the difference between the performances of the protocols more clearly.

Node 2 went out of range of node 0 (sender) and node 3 (receiver) at 3.6 seconds. Initially, AODV created route 0-2-3 and began transmitting data at 3.1 seconds, causing an early throughput peak in AODV at 4.6 seconds. When this link broke at 3.6 seconds causing the peak, there was a drop in the throughput until the 4.6s point. Node 0 then began transmitting through a new route, 0-1-3, and throughput stabilized from 6.6 seconds to 9.2 seconds. At 9.2 seconds, the links 0-1 and 1-3 broke and did not generate throughput.

SARP handled this scenario efficiently. While forming an initial route, SARP recognized node

Figure 3. Throughput of received data bytes vs. simulation time

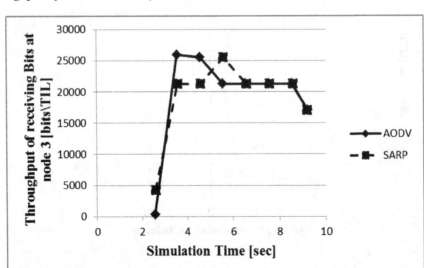

Figure 4. Control overhead generated vs. mobility in networks with 25 nodes

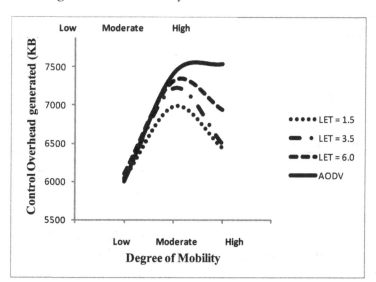

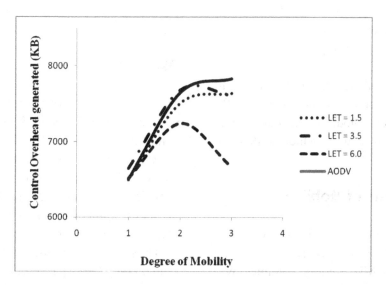

2 as an unstable link with LET above the acceptable limit. Therefore, it formed route 0-1-3, thus maintaining more stable throughput throughout the simulation until the links broke at 9.2 seconds.

SIMULATION DESIGN

Network simulator, Ns 2.33, was used to conduct detailed set of simulation to compare the performance SARP against AODV.

Propagation Channel Specification

All the simulations were performed using the technological specifications of Orinoco IEEE 802.11b wireless card channel (Xiuchao, 2004) for communication and essential network operations.

The parameters for Orinoco 802.11b channel with CCK11 (11 Mbps) were written in NS-2 using OTcl code, as indicated in Table 2.

Table 2. Orinoco 802.11b channel specification

Phy/WirelessPhy set L_ 1.0	;#System Loss Factor
Phy/WirelessPhy set freq_ 2.472e9	;# Channel-13. 2.472GHz
Phy/WirelessPhy set bandwidth_ 11Mb	;# Data Rate
Phy/WirelessPhy set Pt_ 0.031622777	;# Transmit Power
Phy/WirelessPhy set CPThresh_ 10.0	;#Collision Threshold
Phy/WirelessPhy set CSThresh_ 5.011872e-12	;# Carrier Sense Power
Phy/WirelessPhy set RXThresh_ 1.15126e-10	;# Receive Power Threshold
Phy/WirelessPhy set val(netif)	;# Network Interference Type

Achieved Levels of Network Density

The comparative study demonstrates the combined effects of node velocity on the routing protocol performance under sparse, normal, and high network densities and varying traffic densities. A simple flat grid topology measuring 500m X 500m and 700m X 700m was chosen for the simulations. Simulations were performed with 25 and 50 mobile nodes in each topology. By varying the number of nodes per unit area, three different density levels were achieved; they are tabulated in Table 3.

Achieved Levels of Mobility

Mobility was generated using a random waypoint mobility model (RWMM) (Rhim et al., 2009; Bai & Helmy, 2007). CMU "setdest" command was used to generate the communication scenario with random initial placement of nodes within a defined environment. The nodes were set to continuous motion with pause time of 0 seconds.

The simulations were performed under three levels of simulations, low, medium, and high; the level of mobility in the network is determined by the number of nodes travelling at high velocity. Table 4 lists the levels of mobility used in these simulations.

Data Traffic

The traffic pattern was generated using cbrgen routine included in the ns-2.33 following a randomized distribution. Then the number of active routes, that is, the number of active transmitter-receiver (Tx/Rx) pairs, was set to 10 for the 25 nodes scenario and to 20 for the scenario with 50 nodes, initiating communication at different points of time during the simulation.

The source node transmitted 512 bytes of constant bit rate (CBR) packets per second, resulting in a data rate of 256 kbps. This value corresponds to an average of the data rate specified for a high speed vehicle and travel on foot, and it is in accordance with the standard specified by ITU for

Table 3. Achieved network density levels

Grid Dimension (m2)	Number of Nodes	Average Area per Node	Density Level
500 X 500	25	100	Moderate
500 X 500	50	70.7	Dense
700 X 700	25	140	Sparse
700 X 700	50	98.9	Moderate

Table 4. Achieved degrees of mobility

Mobility Type	Node Velocity
Low	80% nodes @ velocity range 0.1 m/s - 3 m/s
	20% nodes @ velocity range 18 m/s - 21 m/s
Medium	50% nodes @ velocity range 0.1 m/s - 3 m/s
	50% nodes @ velocity range 18 m/s - 21 m/s
High	20% nodes @ velocity range 0.1 m/s - 3 m/s
	80% nodes @ velocity range 18 m/s - 21 m/s

Table 5. Simulated values of demand link expiration times

Amount of data to be transferred (MB)	Calculated LET_D (seconds)
1 MB	1.5
5 MB	3.5
10 MB	6.0

multimedia/voice transmission (Muller & Ghosal, 2005). A user datagram protocol (UDP) was implemented at the transport layer, allowing a message to be sent without prior communications to set up a transmission path. It uses a simple transmission model and assumes that error checking and correction is either unnecessary or performed at other layers. A UDP is often used with time-sensitive applications, where, dropping packets is preferred to delayed packets. A transmission control protocol (TCP) can be used alternatively if a reliable stream delivery of packets is desired. This study used UDP to ensure timely delivery of data packets with low network overhead.

Values for LET_D

LET is the decision-making parameter for the implementation of SARP; it accounts for the relative velocity between sender and receiver nodes. The selection of LET is crucial for the analysis of SARP. However, for the simulations performed here, three values of LET_D were selected for each network depending on the amount of data to be

transferred. SARP was analyzed for end-to-end performance using these three LET values. Table 5 gives the calculated values of the LET_D used to simulate SARP for various sizes of data in bytes. Equation (6) is used to calculate LET_D for a network.

Simulations were executed with SARP implementation for three values of LET, 1.5, 3.5, and 6.0 seconds. Simulation time was set to 200 seconds. Each simulation was repeated 10 times with varying traffic routes, traffic sources and traffic receivers, creating a different set of routes for each simulation run.

RESULTS AND DISCUSSION

A comprehensive analysis permitted a comparative study between AODV and SARP protocols, and the results are presented here in terms of graphs and tables. All results discussed here represent an average of the 10 simulation runs for each scenario.

Tracegraph 2.04 (Johansson et al., 1999) was used to extract data from the trace files generated by the simulations. The performance analysis conducted uses four average end-to-end performance metrics: normalized routing load (NRL), packet delivery ratio (PDR), average end-to-end Delay (E2E), and average throughput of the data received. Among these four metrics, NRL was the most significant parameter for measuring the performance of SARP because it focuses on the control overhead generated for each scenario. It is to be noted that the following discussion refers to SARP at a LET value of 'a' as SARP(a).

Normalized Routing Load (NRL)

NRL is defined as the ratio of the amount of control overhead generated to the total number of data bytes successfully transmitted:

$$\text{NRL} = \frac{TotalNumberofControlBytesGenerated}{TotalNumberofDataBytes \, Re \, ceived}$$

In other words, it denotes the useful traffic generated in the network during simulations. This ratio indicates how much traffic was involved in the successful transmission of data. Hence, it is a good measure of the control overhead generated in a network.

1. Networks with 25 Nodes

Figures 4a and 4b compare the control overhead generated by the protocols against various degrees of mobility in 500mX500m and 700mX700m grids, respectively.

The control overhead generated by the protocols in both the networks is similar. However, the NRL generated by the protocols in the smaller yet denser network of 500mX500m grid with 25 nodes, is lower than that of the sparsely-populated network of 700mX700m grid with 25 nodes. IN denser networks, the number of forwarding intermediate nodes is higher thus, forming more number of routes. Hence, the lower NRL in the

denser network is attributed to higher amount of data packets transmitted by the smaller network as compared to the sparser network.

In both networks, AODV and SARP generated similar control overhead at low-moderate mobility but gradual variation was observed with an increase in mobility. This behavior confirmed the initial prediction that SARP would be efficient at moderate-high mobility and would not hinder the functionality of the underlying protocol at low mobility. SARP generated significantly reduced control overhead at high-moderate mobility as compared to AODV. This reduction in the generated control traffic of SARP is a result of the reduced number of fast-moving intermediate nodes. At moderate mobility, SARP generated low control overhead than that of AODV in both the networks. At moderate-high mobility, the protocols show a significant decrease in control traffic, however, SARP still generated low control overhead as compared to that generated by AODV.

The sparse network (i.e., Figure 6d) shows a similar amount of control overhead generated by SARP(1.5) and SARP(3.5) as compared to AODV at various degrees of mobility. This behavior is due to the intended ineffectiveness of SARP in sparse networks.

In the denser network, all three values of LET used in simulating SARP generated a lower control overhead generation as compared to AODV, with increasing mobility. However, there was a slight increase in the control overhead generated with increasing values of LET. This limited increase may be attributable to an increase in control traffic during route discovery phase since the elimination of fast nodes required a longer route discovery. The higher the LET value, the greater the restriction on the acceptable relative velocity of the nodes; the greater the restriction, the greater the possibility of dropped routes. As the number of dropped routes increase, the control overhead generated during route maintenance also increased significantly. This behavior of high value of LET may cause an advert effect on routing by eliminating even useful routes; hence, it suggests that the selection of LET is crucial to ensure that the network is not negatively influenced by incorporating SARP.

The sparser network (i.e., Figure 4b) also showed an increase in the generation of control traffic by SARP with increasing mobility; the only exception was an LET of 6.0 seconds. A node cluster may have formed, which resulted in complicating the routing activity by making two nodes not accessible by cutting of any intermediate nodes. This phenomenon can be understood by considering mechanisms outside the scope of this study.

In most cases, control overhead increased with increasing mobility and the variation was affected most by the values of LET. Nonetheless, SARP performed better than AODV in generating low control traffic at moderate-high mobility; confirming that the SARP algorithm is minimally effective in low mobility scenarios.

Figures 5a and 5b compare the NRL caused by the protocols against various degrees of mobility in 500mX500m and 700mX700m grids, respectively (Figure 5).

As a measure of control overhead, NRL follows the same trend as the control overhead generated. In a similar way, the NRL recorded by the small-

Figure 5. NRL vs. mobility in networks with 25 nodes

(a) 500mX500m Grid

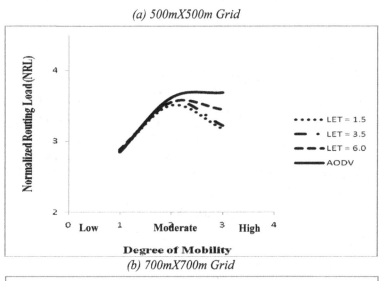

(b) 700mX700m Grid

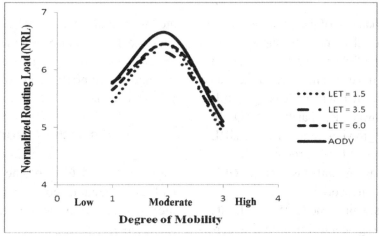

Figure 6. Control overhead generated vs. mobility in networks with 50 nodes

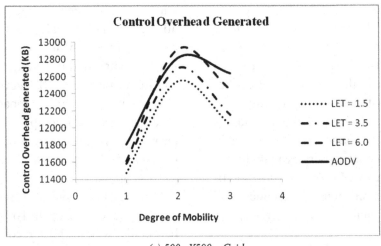

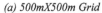

(a) 500mX500m Grid

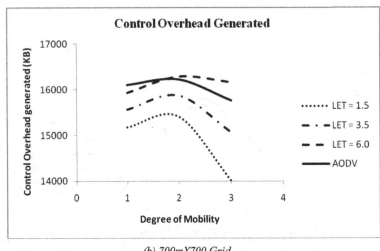

(b) 700mX700 Grid

er network is less than that of the larger network. Also, the trend for data packets successfully transmitted by each of the protocols is observed to be the same. In the dense network, there was a gradual increase in NRL with increasing mobility. SARP caused lower NRL than that of AODV, however, with the growing value of LET, NRL increased. However, all the protocols caused similar NRL; this again confirms that SARP is ineffective in sparse networks.

In the networks with 25 nodes, SARP limits the amount of average control overhead gener-

ated with smart selection of LET. The higher the LET, greater the control overhead generated. With increasing mobility and network density, SARP becomes more efficient in reducing the control traffic generated.

2. Networks with 50 Nodes

Figures 6a and 6b plot the amount of control traffic generated vs. various degrees of mobility in 500mX500m and 700mX700m, respectively.

In 500mX500m network with 50 nodes, both the protocols generated control overhead approximately 3 – 4 MB higher than in the scenario of 25 nodes. This increase can be attributed to greater congestion and intermodal interference in a dense network, considering this is the densest network in these simulations. The moderately dense network of 700mX700m grid with 50 nodes showed a trend very similar to the 500mX500m grid with 25 nodes. This could a result of their almost equal network densities.

Although both protocols showed high control traffic at medium mobility, control traffic dropped significantly when the mobility was high. In high-mobility scenarios, the communicating nodes can be out of range for most of the time during a simulation run. The sender, however, resends routing packets until it reaches the allocated retry limit, which is a MAC layer parameter. If no routes can be established within the maximum retry limit, the sender assumes a permanent link failure and therefore stops sending routing packets.

In general, the amount of control traffic generated increased from low-moderate mobility and decreases from moderate-high mobility. The higher the value of LET employed, higher the control traffic generated; the only exception is the SARP(6.0) in the moderately dense network of 700mX700m grid. This trend is the same as observed in the similar density network, 700mX700m grid with 25 nodes and is explained to be a consequence of network sparseness.

In both the networks, at low-moderate mobility, control traffic generated by the protocols is similar; however, in the dense network, SARP(1.5) and SARP(3.5) have produced slightly lower control overhead than that of AODV. In this network, SARP(1.5) and SARP(3.5) generated significantly less control overhead than AODV at moderate-high mobility.

At moderate mobility, SARP(6.0) generated higher control traffic than AODV in both the networks. This is likely because of the high value of LET that reduced the number of potential intermediate nodes. However, at high mobility SARP(6.0) generated less control traffic than AODV by reducing frequent link disconnections.

At high mobility, the trend remains the same as in the scenario with 25 nodes. AODV generated the highest control overhead and SARP(1.5) generated the least. As expected, with increasing LET the control overhead generated increased in the dense network. However, in the sparser network, all the protocols except SARP (6.0) generated similar amounts of control overhead, further confirming the ineffectiveness of SARP in sparse networks.

Figures 7a and 7b plot the NRL vs. various degrees of mobility in 500mX500m and 700mX700m, respectively. Both the networks have a node population of 50.

In the dense network, all the protocols showed a gradual increase in NRL as mobility increased. However, the performance of the protocols was identical. However, SARP(6.0) recorded highest NRL throughout the varied levels of mobility indicating that this value of LET is too high to effectively implement SARP and hence, a lower value should be appropriate in this scenario.

In the sparser network, at low-moderate mobility, the general trend noted in the networks with 25 nodes was repeated. AODV caused the highest NRL whereas SARP(6.0) reported the least. The amount of control overhead generated increased with increasing values of LET. At moderate-high mobility, all the protocols showed a decrease in NRL, similar to the trend in the 700mX700m grid with 25 nodes. However, at high mobility, AODV generated higher NRL than the other two and SARP(6.0) generated high NRL throughout the simulation despite lower control traffic generation at both low and high mobility. This high NRL could be the result of the reduced number of data packets received by SARP(6.0).

In general, at low-moderate mobility, SARP and AODV performed almost identically in terms of both NRL and control traffic generated. At moderate-high mobility, SARP generated signifi-

Figure 7. NRL vs. mobility in networks with 50 nodes

(a)

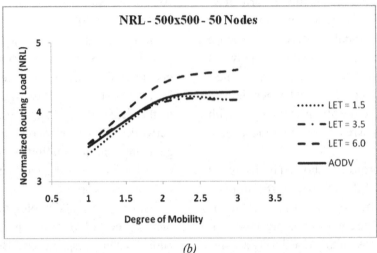

(b)

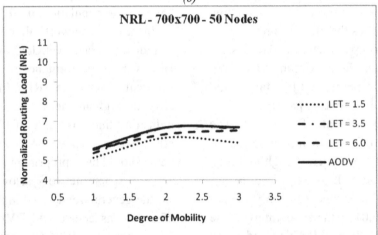

cantly lower control overhead and hence lower NRL than AODV, given an appropriate usage of LET. At high values of LET, the control overhead generated by SARP was higher than that generated by AODV.

The routing loads discussed in this section were larger than that observed in the network topologies described in networks with 25 nodes. This increased routing load can be attributed to high interference and congestion in the scaled-up network. AODV showed insignificant increase in the control overhead and high NRL compared to SARP. Furthermore, NRL increased more under high mobility conditions than in low mobility conditions. The increase in routing load due to mobility can be explained by frequent link updates and by updates to ensure local connectivity through hello packets. The behavior of SARP with respect to control overhead remains similar to that observed in previous scenarios. As expected, SARP again outperformed AODV in generating optimal control traffic.

Packet Delivery Ratio (PDR)

PDR is a significant measure of the rate of successful data transmission within a network. It can be defined as the ratio of the amount of data

Figure 8. Packet delivery ratio vs. mobility

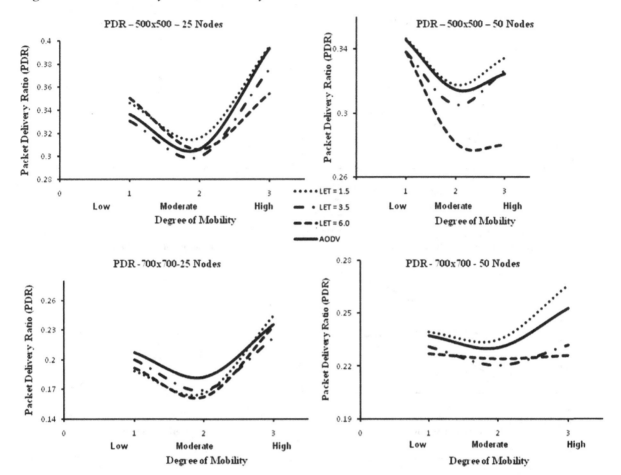

received by an application in the network to the amount of data sent out by the application:

$$PDR = \frac{TotalNumberofDataBytesreceived}{TotalNumberofDataBytesSent}$$

Figure 8 shows graphs for PDR versus mobility under various traffic and movement scenarios.

PDR decreased with an increase in mobility because fewer packets were sent in the high-mobility scenario than in the low-mobility scenario. SARP(1.5) outperformed AODV in all the scenarios that including the least dense network, (i.e., 700mX700 m grid with 25 nodes). In terms of both NRL and PDR, the performance of SARP

in such sparse networks was inconsistence when compared to the other scenarios. SARP(3.5) and SARP(6.0) caused lower PDR than the other two protocols, indicating that the higher the values of LET, the lower the PDR. Hence, the choice of LET is crucial role to the effective functioning of SARP.

Average End-To-End Delay (E2E)

E2E can be defined as the delay that a packet suffers from the time it leaves the sender application to the time it arrives at the receiver application. The average end-to-end delay is the average of such delays suffered by all data packets successfully received within a network; it does not consider

Figure 9. Average end-to-end delay vs. mobility

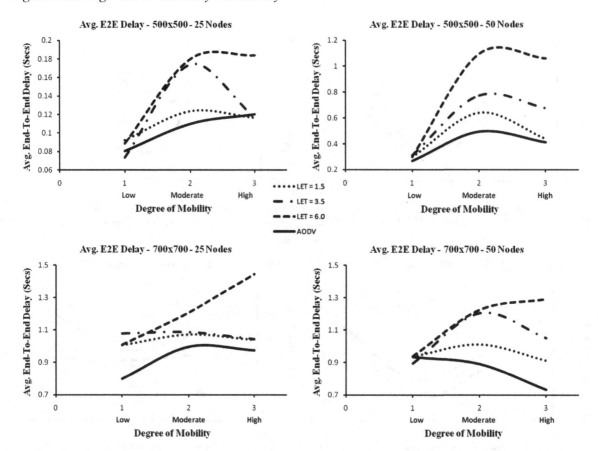

dropped packets. This parameter ensures that the determination of node velocity and the calculation of LET during simulations do not significantly increase the end-to-end delay of the network.

Figure 9 indicates that average end-to-end delay increased with an increase in mobility in case of all the simulation scenarios.

The abnormality of the graphs for a 700mX700m grid with 50 nodes may be due to higher network congestion and increased MAC retries caused by unreliable routes. AODV had less average end-to-end delay than SARP. Both protocols have similar delays at low mobility. In sparse networks, however, AODV had significantly less delay than SARP. Both SARP(3.5) and SARP(6.0) showed a large increase in average end-to-end delay from moderate to high mobility. SARP(1.5) had a slightly greater aver-

age end-to-end delay (about 50ms) than AODV. One can safely conclude therefore that SARP(1.5) did not cause high average end-to-end delay in AODV. Further, this work demonstrated that the value of LET plays a crucial role in the successful realization of SARP.

Average Throughput

Average Throughput of Received Data Packets can be defined as the average of the data rates delivered to all terminals in a network. The maximum throughput is the minimum load in bit/s that causes delivery time (i.e., latency) to become unstable and increase towards infinity. It accurately measures the network performance and confirms that the throughput was not compromised with the implementation of SARP.

Figure 10. Average throughput of received data packets vs. mobility

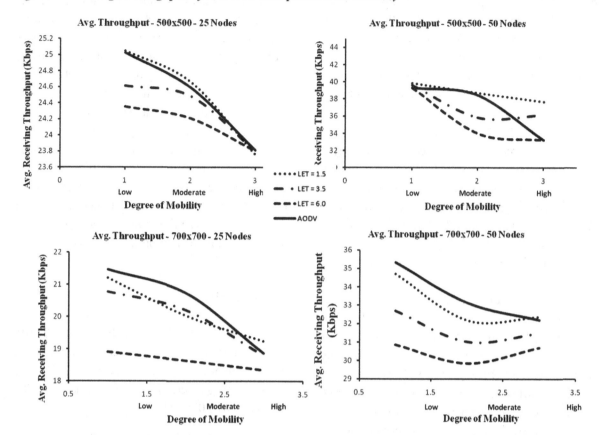

Figure 10 demonstrates that average throughput decreased with increasing mobility, accounting for the relative stability and reliability of routes at lower mobility. Thus, more data packets were successfully delivered to a receiver. SARP(1.5) had higher average receiving throughput than AODV, except in the case of a sparse network of 700mx700m with 25 nodes, in which AODV outperformed SARP. SARP(3.5) and SARP(6.0) once again proved less stable than SARP(1.5) and AODV.

Discussion

The simulations conducted here proved that control overhead generated by both protocols increased with increasing mobility. The overall increase in control overhead and the decrease in the PDR indicate that protocol performance in general degrades with increasing mobility. In addition, the end-to-end delay increases with increasing mobility, as shown in Figure 6. The relationship between the change in NRL and end-to-end delay can be explained in terms of resource utilization. When NRL increases, more network resources and the limited bandwidth are consumed in processing the control overhead. Consequently, the resources needed to process the data traffic become insufficient, causing large number of delayed and dropped packets, significantly reducing the amount of data received, and increasing end-to-end delay.

The use of end-to-end performance metrics to compare the performance of SARP and AODV supports several key conclusions:

a.   SARP(1.5) and SARP(3.5) generate lower control overhead and lower NRL than AODV.

b. SARP(1.5) improves underlying protocol, AODV by generating higher PDR; which confirms more successful data transmission, except in dense networks.

c. SARP(1.5) outperforms AODV by demonstrating higher average receiving throughput, except in dense networks.

d. SARP(1.5) is stable, resulting in only a marginal increase in average end-to-end delay.

e. SARP(3.5) and SARP(6.0) cannot compete with AODV in terms of PDR, average receiving throughput, and average end-to-end delay.

f. With increasing LET, SARP performance degrades.

g. SARP is effective in dense networks.

The outcome of simulation using AODV with two ray ground propagation agrees with findings of Lenders et al. (2006) and Das et al. (2000), indicating that control overhead increases with increasing mobility, whereas PDR decreases. However, since Lenders et al. (2006) and Das et al. (2000) measured mobility in terms of relative velocity and pause time, respectively, rather than in terms of actual speed, no direct comparison is possible. With an appropriate LET, SARP outperforms AODV at moderate-high network density. Comparative study also demonstrated the importance of LET in efficient the SARP routing methodology. Thus, these realistic simulations incorporating numerous variables effectively increase the fidelity of the findings.

CONCLUSION

The results showed that a speed-aware routing algorithm limits the generation of additional control overhead caused by link breakages due to highly mobile nodes. The control overhead generated by the underlying protocol AODV is greater than necessary, and it does not improve data delivery. The simulations conducted here demonstrate that the SARP, which has minimal control overhead, outperforms AODV which generates high control overhead. However, the benefits offered by SARP are heavily dependent on selection of the appropriate LET. The work presented here clearly shows that SARP increases link reliability, decreases control traffic, and shows no deterioration of regular throughput (i.e., number of packets received).

SARP could be further validated by incorporating multi-path fading. In addition, further investigation into the selection of Link Expiration Time for SARP should be warranted. Other methods for encountering and mitigating the effects of node mobility should be explored.

REFERENCES

Akunuri, K., Guardiola, I. G., & Phillips, A. (2010). Mobility adds fast-fading impact study on wireless ad hoc networks. *International Journal of Mobile Network Design and Innovation, 3*(3). doi:10.1504/IJMNDI.2010.038097

Athanasios, B. (2006). A mobility sensitive approach for efficient routing in ad hoc mobile networks. In *Proceedings of the 9th ACM International Symposium on Modeling Analysis and Simulation of Wireless and Mobile Systems*, Terromolinos, Spain.

Aziz, S. R. A., Endut, N. A., Abdullah, S., & Doud, M. N. M. (2009, March). Performance evaluation of AODV, DSR and DYMO routing protocol in MANET. In *Proceedings of the Conference on Scientific and Social Research*.

Bai, F., & Helmy, A. (2007). A survey of mobility models. In *Wireless ad hoc networks*.

Bai, F., Sadagopan, N., & Helmy, A. (2003). IMPORTANT: A framework to systematically analyze the impact of mobility on performance of routing protocols for ad hoc networks. In *Proceedings of the 22nd Annual IEEE INFOCOM Joint Conference* (pp. 825-835).

Bettstetter, C. (2006). Smooth is better than sharp: A random mobility model for simulation of wireless networks. In *Proceedings of the 4th ACM International Workshop on Modeling, Analysis and Simulation of Wireless and Mobile Systems*, Rome, Italy.

Broch, J., Maltz, D. A., Johnson, D. B., Hu, Y. C., & Jetcheva, J. (1998). A performance comparison of multihop wireless ad hoc network routing protocols. In *Proceedings of the ACM/IEEE Conference on Mobile Communications* (pp. 85-97).

Chlamtac, I., Conti, M., & Liu, J. (2003). Mobile ad hoc networking: Imperatives and challenges. *Ad Hoc Networks*, *1*(1), 13–64. doi:10.1016/S1570-8705(03)00013-1

Corson, S., & Macker, J. (1999). *Mobile ad hoc networking (MANET): Routing protocol performance issues and evaluation considerations.* Retrieved from http://www.ietf.org/rfc/rfc2501.txt

Das, S. R., Castaneda, R., Yan, J., & Sengupta, R. (1998). Comparative performance evaluation of routing protocols for mobile, ad hoc networks. In *Proceedings of the IEEE 7th International Conference on Computer Communications and Networks* (pp. 153-161).

Das, S. R., Perkins, C. E., & Royer, E. M. (2000, March). Performance comparison of two on-demand routing protocols for ad hoc networks. In *Proceedings of the IEEE Conference on Computer Communications.*

El-Rabbanny, A. (2002). *Introduction to GPS: The global positioning system.* Boston, MA: Artech House.

El-Rabbany, A. (1994). *The effect of physical correlations on the ambiguity and accuracy estimation in GPS differential positioning* (Tech. Rep. No. 170). Fredericton, NB, Canada: Department of Geodesy and Geomatics Engineering.

Fall, K., & Varadhan, K. (Eds.). (2009). *The ns manual.* Retrieved from http://www.isi.edu/nsnam/ns/doc/ns_doc.pdf

Gerla, M., & Raychaudhari, D. (2007). *Mobility in wireless network.* Rutgers, NJ: Rutgers University.

Gruber, I., & Li, H. (2002). Link expiration times in mobile ad hoc networks. In *Proceedings of the 27th Annual IEEE International Conference on Local Computer Networks* (pp. 743-750).

Guardiola, I. G. (2007). Mitigating the stochastic effects of fading in mobile wireless ad-hoc networks. In *Proceedings of the IIE Annual Conference.*

Guardiola, I. G., & Matis, T. I. (2007). Fast-fading, an additional mistaken axiom of wireless-network research. *International Journal of Mobile Network Design and Innovation*, *2*, 153–158. doi:10.1504/IJMNDI.2007.017319

Guo, S., Yang, O., & Shu, Y. (2005). Improving source routing reliability in mobile ad hoc networks. *IEEE Transactions on Parallel and Distributed Systems*, *16*(4).

Hassan, Y. K., El-Aziz, M. H. A., & El-Rad, A. S. A. (2010). Performance evaluation of mobility speed over MANET routing protocols. *International Journal of Network Security*, *11*(3), 128–138.

Hong, X., Kwon, T., Gerla, M., Gu, D., & Pei, G. (2007, January). A mobility framework for ad hoc wireless networks. In *Proceedings of the ACM 2nd International Conference on Mobile Data Management.*

Johansson, P., Larsson, T., Hedman, N., Mielczarek, B., & Degermark, M. (1999). Scenario-based performance analysis of routing protocols for mobile ad hoc networks. In *Proceedings of the IEEE/ACM Conference on Mobile Communications* (pp. 195-206).

Lenders, V., Wagner, J., & May, M. (2006). Analyzing the impact of mobility in ad hoc networks. In *Proceedings of the ACM/SIGMOBILE Workshop on Multi-hop Ad Hoc Networks: From Theory to Reality*, Florence, Italy.

Malek, J. (2002). *Trace graph program*. Retrieved from http://www.angelfire.com/al4/esorkor/

Manvi, S. S., Kakkasageri, M. S., Paliwal, S., & Patil, R. (2010). ZLERP: Zone and link expiry based routing protocol for MANETs. *International Journal on Advanced Networking and Applications*, 2(3), 650–655.

Mueller, S., & Ghosal, D. (2005, April). Analysis of a distributed algorithm to determine multiple routes with path diversity in ad hoc networks. In *Proceedings of the Conference on Modeling and Optimization in Mobile, Ad Hoc, and Wireless Networks* (pp. 277-285).

Mullen, J., & Hong, H. (2005, October 10-13). Impact of multipath fading in wireless ad hoc networks. In *Proceedings of the 2nd ACM Workshop on Performance Evaluation of Wireless Ad Hoc, Sensor, and Ubiquitous Networks*.

Nikkei Electronics Asia. (2009). *Mobile phone sales fall; Smartphone sales up, Q1*. Retrieved from http://techon.nikkeibp.co.jp/article/HONSHI/20090629/172377

Oliveira, R., Luis, M., Bernardo, L., Dinis, R., Pinto, R., & Pinto, P. (2010). Impact of node's mobility on link-detection based on routing hello messages. In *Proceedings of the Wireless Communications and Networking Conference*.

Paudel, B., & Guardiola, I. G. (2009). *On the effects of small-scale fading and mobility in mobile wireless communication network*. Rolla, MO: Graduate School of Missouri S&T.

Perkins, C. E., & Royer, E. M. (1999). Ad hoc on-demand distance vector routing. In *Proceedings of the 2nd IEEE Workshop on Mobile Computing Systems and Applications*, New Orleans, LA (pp. 90-100).

Rhim, A., & Dziong, Z. (2009). Routing based on link expiration time for MANET performance improvement. In *Proceedings of the IEEE 9th Malaysia International Conference on Communications* (pp. 555-560).

Samarajiva, R. (2001). The ITU consider problems of fixed-mobile interconnection. *Telecommunications Policy*, 155–160. doi:10.1016/S0308-5961(00)00085-9

Su, W., Lee, S.-J., & Gerla, M. (2001). Mobility prediction and routing in ad hoc wireless networks. *International Journal of Network Management*, 3–30. doi:10.1002/nem.386

Tonguz, O. K., & Ferrari, G. (2006). *Ad hoc wireless networks: A communication perspective*. New York, NY: John Wiley & Sons. doi:10.1002/0470091126

Tsao, C.-L., Wu, Y.-T., Liao, W., & Kuo, J.-C. (2006). Link duration of the random way point model in mobile ad hoc networks. In *Proceedings of the IEEE Wireless Communications and Networking Conference* (pp. 367-371).

Varshney, U., & Vetter, R. (2000). Emerging mobile and wireless networks. *Communications of the ACM*, 73–81. doi:10.1145/336460.336478

Xiaojiang, D., & Dapeng, W. (2006). Adaptive cell relay routing protocol for mobile ad hoc networks. *IEEE Transactions on Vehicular Technology*, 278–285.

Xiuchao, W. (2004). *Simulate 802.11b channel within ns-2*. Singapore: National University of Singapore.

This work was previously published in the International Journal of Interdisciplinary Telecommunications and Networking, Volume 3, Issue 3, edited by Michael R. Bartolacci and Steven R. Powell, pp. 40-61, copyright 2011 by IGI Publishing (an imprint of IGI Global).

Chapter 16
A Survey of Cloud Computing Challenges from a Digital Forensics Perspective

Gregory H. Carlton
California State Polytechnic University, USA

Hill Zhou
California State Polytechnic University, USA

ABSTRACT

Computing and communication technologies have merged to produce an environment where many applications and their associated data reside in remote locations, often unknown to the users. The adoption of cloud computing promises many benefits to users and service providers, as it shifts users' concerns away from the physical location of system components and toward the accessibility of the system's services. While this adoption of cloud computing may be beneficial to users and service providers, it increases areas of concern for computer forensic examiners that need to obtain data from cloud computing environments for evidence in legal matters. The authors present an overview of cloud computing, discuss the challenges it raises from a digital forensics perspective, describe suitable tools for forensic analysis of cloud computing environments, and consider the future of cloud computing.

INTRODUCTION

As computing and communications technologies continue to expand rapidly, cloud computing is emerging as a method by which services are provided, often without the knowledge of the users. Contemporary computing environments typically consist of some type of workstation and network connection, and nontechnical computer users may give little consideration as to where applications or data reside. Often, user applications and data exist on remote computing and storage systems while users access their applications and data from the mysterious "cloud" without understanding the physical locations of these devices.

It is not just the nontechnical computer users that are often unaware of the physical computing infrastructure. One of the appeals of cloud comput-

DOI: 10.4018/978-1-4666-2154-1.ch016

ing is the concept that users need not be concerned with the physical location of data, services, or devices in order to utilize these services. Many organizations realize numerous benefits from implementing systems based on cloud computing, as this approach often provides cost savings, increased flexibility, and perhaps higher reliability and lower maintenance.

The increase of cloud computing usage brings an increase in legal matters (i.e., civil lawsuits or criminal investigations) in which data from cloud computing environments are used as evidence. This directly results in an increase in the need to conduct computer forensics examinations from cloud computing environments. Computer forensics examiners must determine the authenticity of the data used as evidence; therefore, it is essential that computer forensics examiners have a thorough understanding of the infrastructure used in cloud computing.

This paper provides an orientation to cloud computing for forensics examiners. We first offer an overview of cloud computing, then a discussion on cloud computing from a forensics perspective, and then we describe forensic tools that are useful in examinations of cloud computing environments. Lastly, we present a view of the future of cloud computing. The following sections address each of these areas, and we are hopeful that this paper will provide useful information to forensic examiners and it will encourage researchers to expand upon the concepts presented.

OVERVIEW OF CLOUD COMPUTING

Most of today's computer users are impacted by cloud computing in some form, and it is becoming an increasingly attractive approach for organizations that seek to transition from in-house data centers to remote, third-party managed data centers (Brodkin, 2009). While the usage of cloud computing is increasing, the concept of cloud computing triggers different perceptions in different people, largely since the nature of cloud computing is not based on a single technology. Instead, it is a combination of many existing technologies, including thin clients, virtualization, online storage, and service oriented architecture (SOA) (Amrhein & Quint, 2009). Similar to the apologue described in John Godfrey Sax's poem, *The Blind Men and the Elephant*, (Saxe), people have different views on the composition and significance of cloud computing yet fail to capture the big picture. For many end-users and managers, the idea of cloud computing is nothing newer than exchanging information and documents through web-based, e-mail services, such as Hotmail or Gmail, and others recognize the concept as an extension of the timesharing model developed in the 1960s (Schneier, 2009). However, from IT professionals' perspectives, these elder computing models hardly resemble the contemporary cloud computing age, as new inventions of virtualization, online collaboration, connectivity, and processor power combine to create our current era of computation. Recognizing the array of technological components comprising cloud computing, the National Institute of Standards and Technology (NIST) offers the following definition: "Cloud computing is a model for enabling convenient, on-demand network access to a shared pool of configurable and reliable computing resources (e.g., networks, servers, storage, applications, services) that can be rapidly provisioned and released with minimal consumer management effort or service provider interaction" (Mell & Grance, 2009).

Prominent enterprises that offer cloud computing services include Amazon, IBM, and Microsoft Corporation. An example is Amazon's Elastic Compute Cloud (EC2) web services, where individuals are provided with an image running a chosen operating system, and users are permitted to utilize this image at their discretion. With this service, users own the virtualization of the operating system image, but not the network infrastructure or any of the supporting hardware. Similarly, Microsoft Azure Services Cloud Platform supports applications that are hosted and run at Microsoft data centers. Additionally, VMware has offered virtualization software for many years, and IBM and Juniper Networks have formed a collaborative partnership delivering cloud computing services. Overall, these cloud services are operated and managed at a data center owned by a service provider that has the ability to host multiple clients while ensuring that the processing power is available.

According to the definition provided by NIST, cloud computing is constrained by five essential characteristics, three service delivery models, and four deployment models (Mell & Grance, 2009). We describe each of these three constraints:

Essential Characteristics of Cloud Computing

The NIST definition of cloud computing includes five essential characteristics. They include on-demand self-service, ubiquitous network access, resource pooling and location independence, rapid elasticity, and measured service. We discuss each of these five characteristics as follows.

On-demand self-service promises that, "A consumer can unilaterally provision computing capabilities, such as server time and network storage, as needed automatically without requiring human interaction with each service's provider" (Mell & Grance, 2009). This characteristic requires the cloud computing service to be agile and user-friendly, as there elements are necessary in order to provide the ability to use cloud computing services for network storage, computation, and online access automatically without human interaction between the users and the cloud service provider. This characteristic represents the factor that is largely responsible for ensuring efficiencies and cost savings opportunities that have helped escalate the popularity of cloud computing services.

The second essential characteristic identified within the NIST definition of cloud computing is ubiquitous network access. With ubiquitous network access, the cloud computing platform ensures

that its "capabilities are available over the network and accessed through standard mechanisms that promote use by heterogeneous thin or thick client platforms (e.g., mobile phones, laptops, and PDAs)" (Mell & Grance, 2009).

The third essential characteristic identified within the NIST definition of cloud computing is resource pooling and location independence. Cloud computing providers pool their computing resources to "serve multiple consumers using a multi-tenant model with different physical and virtual resources dynamically assigned and reassigned according to consumer demand" (Mell & Grance, 2009). Generally, the customers of cloud computing have no control of the physical location of the computing resources, nor do they typically have any knowledge regarding its location; however, they may have the ability to "specify location at a higher level of abstraction (e.g., country, state, or datacenter)" (Mell & Grance, 2009). The resources that are available for pooling and are location independent include processing units, virtual machines, data storage, random access memory (RAM), and network bandwidth, thus allowing any programs or applications to be optimized to meet individual customer's needs (Mell & Grance, 2009). Although these programs and applications are logically grouped together, their physical locations are flexible, allowing the ability to assign devices or components from many geographic areas, as needed for virtual systems.

Rapid elasticity is the fourth essential characteristic within the definition of cloud computing, and it ensures that, "Capabilities can be rapidly and elastically provisioned, in some cases automatically, to quickly scale out and rapidly released to quickly scale in. To the consumer, the capabilities available for provisioning often appear to be unlimited and can be purchased in any quantity at any time" (Mell & Grance, 2009). This characteristic expands the self-service requirement of providing computing environments on demand without human intervention by allowing the cloud computing environment to expand or shrink allocated resources efficiently, promptly, and in some cases, automatically, as necessary.

The last of the five essential characteristics described within the NIST definition of cloud computing is measured service. Measured service provides the ability to monitor, control, and report on the resource usage. With measured service, "cloud systems automatically control and optimize resources use by leveraging a metering capability at some level of abstraction appropriate to the type of service (e.g., storage, processing, bandwidth, and active user accounts)" (Mell & Grance, 2009). Reporting measured services automatically offers the benefit of transparency to service providers and consumers alike.

Primary Service Delivery Models

The NIST definition of cloud computing expands beyond the five essential characteristics by defining three primary service delivery models. These three primary service delivery models include Cloud Software as a Service, Cloud Platform as a Service, and Cloud Infrastructure as a Service. These three primary service delivery models are presented in more detail:

Cloud Software as a Service (SaaS) focuses on the delivery of applications to customers through the usage of thin clients. This capability emphasizes application delivery through web browsers where customers are not concerned with the managing aspects of the cloud infrastructure. Here, the operating systems, servers, storage facilities, and networking infrastructures are transparent to the users. Even application software can be outside of the customers' administrative rights except for "limited user-specific application configuration settings" (Mell & Grance, 2009).

The Cloud Platform as a Service (PaaS) delivery model is one level lower than SaaS, as described above. With PaaS, "the consumer is to deploy onto the cloud infrastructure consumer-created or acquired applications created using programming languages and tools supported by

the provider" (Mell & Grance, 2009). Here, the cloud service provider controls the operating systems, servers, storage systems, and network infrastructure, whereas, the consumer is responsible for managing the deployed applications. The NIST definition also indicates that the consumer might also choose to control the "application hosting environment configurations" (Mell & Grance, 2009).

The lowest of the three cloud computing delivery models is Cloud Infrastructure as a Service (IaaS). This model provisions the consumer with the fundamental computing resources, including network infrastructure, servers, and storage systems while the consumer has the freedom to deploy operating systems and application software. The cloud service provider manages the underlying cloud infrastructure, and the consumer manages applications and operating systems. Additionally, within this delivery model, the consumer might select options to control storage systems and even limited networking components, such as host firewalls (Mell & Grance, 2009).

Deployment Models

Above we described characteristics and delivery models within the NIST definition of cloud computing. The third area within this definition pertains to deployment models. There are four different deployment models provided within the NIST definition of cloud computing, and they are private clouds, community clouds, public clouds, and hybrid clouds.

Private clouds have a single organization focus. The physical existence of the private cloud infrastructure may be either local (i.e., on premise) or remote (i.e., off premise). Also, the organization may manage the private cloud internally or outsource the private cloud's management to a third party service provider. The primary factor is that "the cloud infrastructure is operated solely for an organization" (Mell & Grance, 2009).

A community cloud expands the focus from the single organization of a private cloud to multiple organizations. Using an analogy to describe this deployment model, an intranet is to a private cloud as an extranet is to a community cloud. Several organizations within a logical community share and support the cloud infrastructure. The community determines the mission, policy, compliance considerations, and security requirements. Either an external third party or organizations within the community may manage the community cloud. Also the physical infrastructure of the community cloud may be located within the community (i.e., on premise) or outside of the community (i.e., off premise) (Mell & Grance, 2009).

Beyond the single organization focus of private clouds and the limited organization focus of community are public clouds. The public cloud delivery model is depicted as a cloud infrastructure owned by an organization that provides the cloud to the general public as a service.

Lastly, the fourth delivery model is a hybrid that consists of a combination of two or more of the delivery models described. With a hybrid cloud delivery model, the clouds "remain unique entities but are bound together by standardized or proprietary technology that enables data and application portability (e.g., cloud bursting for load-balancing between clouds)" (Mell & Grance, 2009).

The Value of Cloud Computing

The rapid growth of cloud computing indicates that there are real or perceived value achieved through its adoption. The value of cloud computing is derived from achieving cost efficiency, scalability, accessibility, or improvements in business processes.

The adoption of cloud computing offers the potential to achieve cost efficiency in its ability to allow organizations to implement complex systems by renting or leasing space from an existing infrastructure. Startup usage within an existing cloud infrastructure can began quickly

and relatively expensively. Thus, time and funds are then available for these organizations to allocate toward other resources, such as implementing business processes instead of computer and network installation.

Organizations are typically faced with continuous changes in their environments, and adapting to these changes can be costly. Cloud computing offers these organizations the flexibility to adjust the size of their IT infrastructure according to their needs at short temporal intervals. For example, an organization that faces significant seasonal shifts in sales can obtain a temporary increase in processing power for a short duration relatively easily by utilizing cloud computing services, such as a server farm.

The value of cloud computing is enhanced in two areas related to accessibility. First, the nature of internetworked computing platforms provides an environment where computing resources are reachable. Cloud computing has the ability to accelerate the rate in which technologies and resources are deployed and made available to users. A second benefit of improved accessibility concerns the methods available for individuals with disabilities to utilize resources, and while cloud computing does not specifically address accessible input and output devices, the concept of anytime, anyplace technological resources greatly enhances the likelihood of making resources available to everyone, regardless of disabilities.

Complex organizations, particularly business enterprises, have largely implemented information systems to address key business processes through applications, such as, electronic data interchange (EDI), enterprise resource planning (ERP), customer relationship management (CRM), and supplier relationship management (SRM). These complex, enterprise-wide applications typically expand not just among multiple locations within an enterprise, but to other organizations. The financial and temporal efficiencies afforded by the implementation of cloud computing described above combined with its scalability options yields

Table 1. Challenges of computer forensics in cloud computing infrastructures

Challenges of Computer Forensics in Cloud Computing Infrastructures
Data acquisition challenges
Legal environment challenges
Organizational security issues
Organizational implementation issues
Monitoring real-time traffic
Analysis of virtualization machines
Challenges of progressive complexity

an environment where the delivery of entire business processes can feasibly be rethought and redesigned.

To summarize, usage of cloud computing can provide immediate cost savings to an organization and reduce or virtually eliminate IT infrastructure maintenance by the organization. The benefits gained through shared infrastructure, remote hosting, and dynamic licensing and provisioning are strong enticements for many companies. Also, public cloud implementation facilitates remove the burden of infrastructure maintenance for many organizations.

VIEWING CLOUD COMPUTING FROM A FORENSICS PERSPECTIVE

With all its benefits, cloud computing services also present new challenges and concerns to forensics examiners as they acquire and analyze matters involving cloud computing environments (Rittinghouse & Ransome, 2010). While many components within cloud computing infrastructures possess integral security technologies (e.g., IDS, firewalls, and antivirus software) that have the potential to block external threats, the data available for forensics examiners to collect as evidence is limited and difficult to collect. Audit logging options within cloud computing environments are often not automatically enabled, and forensics examiners frequently are not knowl-

edgeable regarding the infrastructure within cloud computing. We have identified seven challenges to applying computer forensics to cloud computing, as shown in Table 1, and we discuss each of these challenges in the following sections.

Data Acquisition Challenges

Heretofore, computer forensics has typically consisted of acquiring forensic images from one or more computer workstations or servers. These forensic images have traditionally been acquired from a device that is powered off (i.e., static or dead acquisition) (Carlton, 2007). More recently, forensic images have been acquired from a device that is operating (i.e., live or network acquisition). In ideal circumstances, static acquisitions are preferred over live acquisitions, as static acquisitions can obtain a full image of a physical device that is repeatable. Attempting to repeat a live acquisition will yield different results, as changes occur on the suspect device as the acquisition takes place. It usually takes an extended period of time to perform an acquisition; therefore, the changes that occur to the suspect device during a live acquisition can be substantial.

The usage of cloud computing brings new challenges to the traditional approach for obtaining forensically sound data acquisitions. The presence of a cloud infrastructure basically eliminates the opportunity to perform static data acquisitions, especially in implementations where multiple enterprises share the infrastructure. Therefore, the need for a different mindset seems appropriate concerning acquiring evidence from cloud computing environments. The forensic examiner will need to understand the capabilities different providers offer, its characteristics, service models, and deployment models. Based on the basic guidelines from forensic data acquisition principles, examiners should recognize the differences cloud computing environments present compared to traditional acquisitions.

The foremost difference likely to be recognized is that instead of acquiring data from physical devices (i.e., computer hardware), the examiner will acquire machine images, often from virtual servers logically connected to various configurations of enterprise storage solutions. Additionally, instead of acquiring an image in an EnCase (i.e., .E01) format or raw data image (i.e., dd) format, the examiner might choose to format the suspect data as an International Organization for Standardization (i.e., ISO) image.

Advantages of choosing an ISO image when encountering suspect data within cloud computing environments include: it contains the full data collection allocated to the virtual machine, and it is also a convenient and consistent platform from which virtual machines (e.g., WMware, Hyper-V) can boot into a forensics environment. This method limits the use of physical discs, including CD ROM, hard disk devices, floppy disks, and flash drives (Shavers, 2008).

Cloud Computing Legal Environment Concerns

Additional levels of complexity emerge from the legal environment when addressing matters involving cloud computing. Beyond the legal constraints concerning acquiring suspect data from one organization's computing environment, cloud computing frequently results in components of the system infrastructure shared among multiple organizations, and this clearly adds complexity

regarding legal access to the data. Cloud computing also provides opportunities for the computing infrastructure to span internationally, and this produces an additional magnitude of complexity concerning legal access to suspect data.

Organizational Security Issues

While the practice and study of digital forensics and computer security, also known as information security, address different objectives, many elements of both overlap. The essence of cloud computing is based on dispersed, interconnected elements of infrastructure; therefore, compared to traditional, single-site computing environments, the multi-location environment found in cloud computing inherently raises security concerns. In reaching an understanding of cloud computing from a digital forensics perspective, it is important to recognize organizational security issues too. The foundation of information security is comprised of availability, integrity, and confidentiality (Harris, 2008). We discuss each of these three elements of the foundation of information security:

The availability element of information security addresses the need for systems to be available for use at all times. Many components within cloud computing infrastructures are fault tolerant to ensure that the information system will remain functional in the event of a component failure, or perhaps even a failure of multiple components. Organizational performance can be negatively impacted to a large degree if requested information is not readily available (Harris, 2008).

Numerous cloud computing components are charged with ensuring system availability. Domain controllers manage login and logout activities, and they are essential to the availability of the system. Data backup facilities ensure that data manipulation mistakes do not render the system unavailable or damage the integrity of the system, and redundant arrays of independent disks (RAID) are included within the category of data backup or data recovery facilities. Similarly, E-mail servers require zero downtime to ensure effective communications. Also, denial of service and other attacks can result in systems being unavailable; therefore, countermeasures, such as stateful firewalls are needed.

Digital forensics examiners need to be aware of the components utilized within the cloud computing infrastructure. Many of the components described above provide logging functions to document the activities performed by the component, and the potential exists for the digital forensics examiner to retrieve valuable evidence from these logs. Additionally, as with traditional servers, data recovery from RAID is often possible; however, the availability requirement places severe limitations on the methods forensic examiners may use to acquire data. For example, a static data acquisition will not be an acceptable method for a system that has a requirement of being available to users at all times.

Integrity represents the accuracy of data, and to maintain integrity, any unauthorized data modification, whether intentional or unintentional must be prevented. Maintaining integrity within cloud computing environments require that all of the hardware and software data communications components interact correctly. Another concern that increases the risk of losing data integrity stems from organizations' policies regarding user privileges. To reduce this risk, organizations should implement a least privilege policy whereby the data access rights for users are restricted in such a way as to provide them with only the necessary access to perform their authorized functions and no more. Enforcement of a least privilege policy reduces the risk of jeopardizing the data integrity from either intentional or unintentional human data manipulation.

Another area of concern that organizations should consider regarding data integrity addresses modification restrictions for audit logs. In general, only the software processes that create, or append data to audit logs should have the rights to perform these tasks. All users, including administrators, should be restricted from editing or deleting any audit logs. A forensic examiner should consider whether the data contained within audit logs maintain integrity, as without the proper restrictions on user access rights, the potential exists for users or unauthorized applications to create counterfeit logs, modify entries in authentic logs, or destroy authentic logs.

Confidentiality, in the context of information security, ensures "that information is accessible only to those authorized to have access" (Harris, 2008). Breaches in confidentiality occur through different attacks, including social engineering, network sniffing, password cracking, and shoulder surfing. Similar to issues regarding integrity, breaches also occur from users that disclose sensitive information accidentally or intentionally. Encryption of saved or transmitted data is one method that improves confidentiality.

In addition to the need for users of information systems to understand the importance of confidentiality and how to maintain it, digital forensics examiners and IS auditors need to ensure that the data they encounter is maintained in a confidential manner.

The three elements described above, confidentiality, integrity and accessibility are referred to by their acronym, the CIA triad, as shown in Figure 1 (Buenz, 2007). Confidentiality, integrity, and

Figure 1. CIA triad (Buenz, 2007)

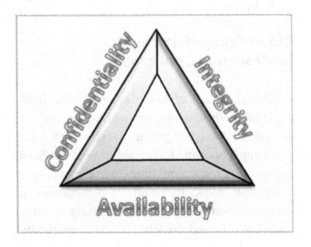

availability work together to secure an organization's perimeter; however, they are sometimes at opposition. For example, to ensure maximum confidentiality, one would reduce the availability of users to access data. Cloud service providers and their clients, including IT professionals and digital forensics examiners, must understand the CIA triad and make the necessary tradeoffs to achieve the proper balance.

In addition to the CIS triad, another set of four elements are associated with the study of information security. These elements are identification, authentication, authorization and accountability (IAAA) (Harris, 2008). IAAA addresses concerns regarding ensuring the validity of the individuals that access information networks by ensuring that those individuals that are genuinely who they claim to be and they are able to access only data in which they are authorized to access. When a user accesses an information system, he or she must provide his or her identity by providing credentials to verify that he or she is not an imposter. The domain controllers within information systems compare users' credentials with control matrices to determine whether to grant access to the system, and if access is permitted, to grant the appropriate rights and privileges to the users. An audit trail, such as access logs, must be maintained to document all of the actions taken by the users.

Organizational Implementation

Organizational implementation of information systems security consists of three types of controls for the enterprise, namely administrative controls, technical controls, and physical controls. Administrative controls build a foundation for a thorough security implementation by developing and publishing high-level, organizational policies, standards, procedures, and guidelines. The areas addressed by the administrative controls include: risk management, personnel screening, security awareness training, and change implementation control procedures. Forensics and auditing func-

tions are more easily performed within organizations in which these controls are established and enforced.

Technical controls consist of implementing and maintaining access control mechanisms, password and resource management, identification and authentication methods, security devices, and the configuration of the infrastructure. Many of these technical controls include rule-based or role-based access controls, and if these controls are applied properly, then much of the activity that occurs will be recorded by logging processes. These logs provide valuable information to auditors or forensic examiners.

Physical controls entail controlling individual access into the facility's premises and access to specific departments within the facility. Physical controls also are used for locking mechanisms on physical computer systems. Physical controls also include removing unnecessary peripherals, such as floppy drives, CD-ROM drives, external disk drives, and USB thumb drives. It addresses protecting the perimeter of the facility, monitoring the facility and systems for intrusion, and environmental controls. Thorough, well-documented physical controls are helpful to forensics examiners too, as the information available, such as logs of physical access, may help associate a suspect with a specific system during a specific time.

Monitoring Real-Time Traffic

In traditional networks, intrusion detection systems, such as sniffers, can be activated in promiscuous mode to capture the network traffic from the network interface cards (NIC) in every device connected to the network. However, the use of promiscuous mode in sniffers does not achieve these results in cloud computing environments. Additionally, the sniffing tools currently available typically lack the ability or capacity to capture large amounts of real-time traffic transmitted through high-speed local area networks (LAN) or wide area networks (WAN). This lack of ability or capacity results in many packets being dropped, ignored, or lost, thus leading to incomplete results when conducting a forensic analysis of network communications.

A technique to successfully capture the network traffic in these situations would utilize a sniffer with the function of a spanning port or network tap. If dynamic host configuration protocol (DHCP) is implemented on the network, DHCP logs can be a valuable source of information to forensics examiners, as these logs provide the internet protocol (IP) addresses to clients' computers assigned by the DHCP service. Furthermore, this information leads to the physical address or media access control (MAC) address for each device that communicates on the network. Also, services that provide domain name system (DNS) records (e.g., Whois) are available to resolve IP addresses by converting the dotted quad numbers into readable text.

Firewall logs also provide a useful source of information in forensic analyses. Many of the firewalls currently available offer more than one service and logging capability. For example, Microsoft Threat Management Gateway (TMG) not only provides basic firewall functionalities to open or block ports and services, it also has both anomaly-based and signature-based intrusion detection systems (Microsoft Corporation, 2009). Information detailing source and destination IP addresses, ports, dates and times can be obtained from the event logs. The placement of firewalls and IDS on networks requires careful consideration. For example, a firewall can be placed at the entrance of an enterprise's network to monitor all incoming and outgoing traffic, it can be placed on the demilitarized zone (DMZ) to watch all internal traffic, and firewalls can be installed on clients' computers to record all users' activities.

Analysis of Virtualization Machines

The adoption of virtualization introduces convenience and cost savings for many organizations;

however, related risks, threats, and vulnerabilities are emerging as the complexity of the network infrastructure rises. New attacks have become more sophisticated, and unsecured hardware and software flaws can pose serious threats, including abuse of privilege, unauthorized use, and denial of service, to name a few. Also contributing to these threats are potential human errors and the misconfiguration of this new technology (Catbird Networks, Inc., 2008).

The security technologies and forensic methodologies are lagging behind the rapid adoption of virtualization. As of the time this study was conducted, the major providers of virtualization are Microsoft's Hyper-V, Citrix XenServer and XenApp, and VMware. These virtual machines are vulnerable to the inherited threats common to physical machines, and they are also at risk to some additional embedded functions, including keystroke logging, shared clipboard and virtual machine backdoors. These functions are open to pass keystrokes, screenshots, and data through channels, possibly covertly, from the host device to the virtual machine and vice versa. Selecting the correct combination of forensic tools to fully capture all of this traffic is challenging, and many current network security devices do not have the ability to log all of the traffic between host devices and virtual machines. This lack of available tools increases the level of difficulty faced by digital forensics examiners at this time. As a result, the potential exists for much evidence to be overlooked.

Challenges of Progressive Complexity

As we have discussed above, cloud computing environments yield higher levels of complexity and risk to security and digital forensics than do more traditional LANs and WANs. While the tools and techniques described can be helpful in obtaining and analyzing evidence when used properly, additional challenges surface when progressively complex layers of data concealment are applied to cloud computing environments. Examples of these concealment technologies include data encryption and steganography.

Matters involving cryptography can be difficult to resolve in isolation, and this is even more evident when enveloped within cloud computing environments. Encryption can be applied rather easily to e-mail communications through available technologies including secure sockets layer (SSL), virtual private networks (VPN) with Internet Protocol Security (IPSec) for encrypting network traffic, and symmetric or asymmetric methods in the public key infrastructure (PKI) for public and private keys. In addition to these readily available tools, determined individuals familiar with cryptanalysis may apply techniques of such a high level of complexity that it is practically impossible for forensic examiner to uncover the actual message from the encrypted data.

There are a variety of products available (e.g., AccessData's PRTK) to forensics examiners to assist in recovering encrypted data, and they utilize numerous attack algorithms, including custom dictionary attacks and brute-force methods. However, one of the biggest obstacles in uncovering encrypted information is work factor. Work factor is the time, effort, and resources necessary to break a cryptosystem. If the work factor is too high, especially in forensic matters where time is of the essence, a successful decryption cannot be accomplished.

Evidence concealed through steganography is arguably more difficult to resolve than matters involving encryption. Whereas data that is concealed through encryption yields an initial set of unreadable data that provides forensic examiners with clues that encryption has been applied, steganography yields usable carrier data files, such as viewable graphic images or audio files. Specialized applications or utilities create covert communication channels by embedding hidden information within carrier files, and since these carrier files are fully functional, it is especially difficult for forensic examiners and security professionals to discover. One helpful technique in recognizing steganography is identifying steganography applications installed or executed on suspect systems; however, sophisticated suspects might choose to embed their payloads on other, non-networked workstations, and then store only the post-concealed data files onto the network.

FORENSICS TOOLS FOR CLOUD COMPUTING ENVIRONMENTS

Above, we discussed the need to consider alternate image formats when acquiring suspect data from cloud computing environments. In addition to selecting alternate image formats for suspect data from cloud computing environments, the forensics examiner must consider forensic analysis tools that are capable of additional complexity. Instead of traditional applications, such as, Guidance Software's EnCase Forensic or AccessData's Forensic Tool Kit (FTK), more elaborate tools that include networking capabilities are necessary, such as EnCase Enterprise, AccessData's SilentRunner, or Computer Online Forensic Evidence Extractor (COFEE). Also, a thorough understanding of data communications, network infrastructure, and access control setup is essential. Armed with this understanding, a forensic examiner will recognize the importance and relevance of audit logs from various products or applications (i.e., firewalls and event viewers).

Despite the challenges described above, utilizing the correct tools and properly implementing control software at the onset of the cloud service, cloud computing environments have the potential to offer a wealth of evidence for forensics examiners. Much information is available by collectively assembling dispersed log files and storing these large log files in a single location for easy data retrieval and discovery. We identified six useful tools for forensic examinations of cloud computing environments, and we discuss each of them:

Symantec Network Access Control

Symantec Network Access Control is not designed particularly for forensic examiners; however, the unified framework that the software provides gives forensic examiners the ability to easily collect detailed and holistic reports on security and system configurations (Symantec Corporation, 2011). Also, this software can be implemented correctly on a cloud computing service network. To do so, it will require that administrative policies, standards and guidelines are enforced and server rooms are physically secured. In addition to protecting the network environment from malicious code (i.e., viruses, worms, Trojans, spyware, etc.), it "provides pervasive endpoint coverage for managed and unmanaged laptops, desktops, and servers existing both on and off the corporate network" (Symantec Corporation, 2011). The software automatically discovers all devices that attempt to connect to the network, including coverage for guests or client access to enterprise LANs, WANs, VPNs, and web applications, and this is beneficial to both security and forensic requirements. Also beneficial to security and forensic purposes, the service enforcement capabilities of Symantec Network Access Control includes gateway, DHCP, NAP, and 802.1X (Symantec Corporation, 2011).

Cisco IOS NetFlow

NetFlow is a Cisco proprietary technology that is used to collect and categorize network traffic. The IOS portion of the product name refers to "Internetwork Operating System," and this connotation helps describe the embedded nature of the product (Cisco Systems Inc., 2011). Unlike enterprise software, such as Symantec Network Access Control, NetFlow is more like an operating system and set of utilities built into supported devices to record all IP traffic passing through the devices.

NetFlow scans the following fields within internet protocol (IP) datagrams: IP address, source IP address and port number, destination IP address and port number, type-of-service byte, and input logical interface. The product also has the ability to capture full data content including data payload, and it can monitor a wide range of packet information. NetFlow is designed primarily to address security concerns, as it can enhance network anomaly and security detection.

As a result of NetFlow's embedded operating system and utility approach, its user interface is more cryptic and less intuitive than similar software offerings. Also this approach prohibits the tool from being implemented directly on an existing server, but instead, it is purchased with qualifying products (Cisco Systems Inc., 2011). The requirement of obtaining NetFlow through available, qualified Cisco products also results in additional expenses associated with purchasing the corresponding hardware devices (Garrison, Lillard, Schiller, & Steele, 2010). However, given these limitations, NetFlow does yield much, low-level data that can be of substantial value to computer forensics examiners, especially in validating data communications.

Guidance Software EnCase Enterprise

EnCase, by Guidance Software, has long been established as a leading software application for computer forensics. The standard product is known as EnCase Forensic, and it operates on Windows-based personal computer workstations. This software application has the capability of creating and analyzing static images using the EnCase proprietary, and *de facto* industry standard, E01 format. Additionally, EnCase Forensic provides extensive analysis and reporting functionality (Carlton, 2008).

Around 2005, Guidance Software added the EnCase Enterprise product their family of forensic applications. In addition to the features available in EnCase Forensic edition, EnCase Enterprise has network capability that allows forensic examiners to conduct a live investigation on a targeted network (Guidance Software, Inc., 2011). EnCase Enterprise uses a component called "Secure Authentication for EnCase (SAFE), which is inserted between the server and the client to ensure data integrity (Bunting & Anson, 2007). Also, a small, enstart.exe program should be installed on clients' workstation prior to an investigation. This provides EnCase Enterprise with the ability to monitor, image, or analyze a live workstation on a network, even covertly if desired.

Recently, Guidance Software introduced a Mobile Enterprise Edition of EnCase that offers a solution to examiners that must travel to physical locations within a network. The mobile version can be installed on small computer workstations, such as a notebook computer. Additionally, Guidance Software now offers the EnCase Field Intelligence Model (FIM) for data acquisition. The Guidance Software product offering provides a wide span of features that enable forensic examiners to acquire, analyze, and report on data in cloud computing environments; however these products come with a relatively expensive purchase price (Bunting & Anson, 2007). Notwithstanding the purchase price, it seems likely to the authors that EnCase Enterprise is an application that is an essential tool for computer forensics examiners that conduct network-based investigations.

NetWitness Investigator

The NetWitness Investigator tool is a Microsoft Windows Server-based application that enables forensics examiners to perform audits and monitor tasks on network traffic. NetWitness Investigator is able to capture network traffic and import packet data into the network forensics application, storing the data into different logical containers, such as, domain controller data, SQL data, DMZ, VPN, etc. The product's components include parsers, feeds, and rules to process live or imported data by decoding network traffic according to metadata values that are predefined by the forensic examiner (EMC Corporation, 2011).

The NetWitness Investigator navigation view is user customizable to enhance the visibility of captured data. The tool also allows forensic examiners to analyze data through various techniques, including: temporal, frequency, transition state, pre-occurrence, historical, traffic behavior, stage, port, statistical, protocol, payload, source linkage, destination, size, correlation, impact, relationship, stylistics, and content analysis. The product also provides the ability to export captured data (Garrison, Lillard, Schiller, & Steele, 2010).

AccessData's SilentRunner

Similar to Guidance Software's product line with the EnCase Forensic application and the EnCase Enterprise application for workstation-based and network-based investigations respectively, AccessData offers their Forensic Tool Kit (FTK) and SilentRunner products. AccessData's applications are designed specifically for forensics purposes, and they include functionality for forensic data acquisition, forensic data analysis, and reporting.

SilentRunner offers similar functionalities as NetWitness Investigator, as both capture, group, analyze, and visualize data. The components are named differently within SilentRunner, as they are called collectors, loaders, database, and analysis workstations (AccessData Group, LLC, 2011). Collectors capture the network traffic through adapters, and loaders import collected data into the database. The database component contains all grouped data for analysis, and the analysis workstations are used to perform testing on the suspect data (Garrison, Lillard, Schiller, & Steele, 2010).

Microsoft Computer Online Forensic Evidence Extractor

Microsoft introduced its Computer Online Forensic Evidence Extractor (COFEE) in 2008, as a tool for the exclusive use of law enforcement agencies. Microsoft made the decision to make COFEE available to law enforcement agencies internationally at no charge. However, shortly after its introduction, the product was leaked to the public in November, 2009 (Nicholas, 2009).

The application "brings together a number of common digital forensics capabilities into a fast, easy-to-use, automated tool" (Microsoft Corporation, 2011). COFEE contains hundreds of functions and commands that can be used to log network activities, collect information, and decrypt passwords. It is a relatively small application with a size of about 15 MB; therefore, it can easily be stored on a portable USB device and deployed on live systems (Microsoft Corporation, 2011).

COFEE is designed to be used in three phases. In first phase, a computer forensics examiner generates a profile that indicates the tasks COF-FEE is to perform. The second phase can be performed by a nontechnical, law enforcement first responder to insert the USB device into a suspect computer and execute the program using the profile defined in the first phase. A forensic examiner performs the third phase by extracting and analyzing the evidence obtained during the second phase (P, 2009).

Acknowledging the public leak of this software, Microsoft indicated that it poses no threat to the viability of the software. As posted on an Internet news site, "In a statement from Vole's Internet Safety Enforcement Team, senior attorney Richard Boscovich said, 'We do not anticipate the possible availability of COFEE for cybercriminals to download and find ways to 'build around' to be a significant concern' (Pullin, 2009). Nonetheless, a counteractive software measure has surfaced named DECAF, and acronym for Detect and Eliminate Computer Acquired Forensics, which is designed to render COFEE ineffective. In discussing DECAF, Zetter explains that, "the program deletes temporary files or processes associated with COFEE, erases all COFEE logs, disables USB drives, and contaminates or spoofs a variety of MAC addresses to muddy forensic tracks" (Zetter, 2009).

THE FUTURE OF CLOUD COMPUTING AND DIGITAL FORENSICS

Today, organizations are pursuing cloud computing technology, as it offers benefits in terms of cost savings, administrative efficiency, and flexibility. According to a Microsoft commissioned survey of 437 information technology decision makers (ITDMs) in United States manufacturing organizations, 59% are planning to implement, or have implemented, cloud-based collaboration tools. Similarly, 57% have implemented or plan to implement cloud-based productivity applications (Hodges, 2011). Cloud computing is clearly emerging as a major driver in the IT marketplace.

Information security systems have already been influenced greatly by the emergence of cloud computing, and forensic analysis is beginning to evolve to adapt to the changes brought by this technology. Network forensics in a cloud computing environment increases the level of complexity regarding performing forensically-sound tasks, and this additional level of difficulty requires forensic examiners to obtain new knowledge, skills, and methodology. Computer forensics examiners must adapt to these challenges since these individuals are viewed by the courts as experts in the field of digital forensics, and in order to maintain expertise within this field, computer forensics examiners must be competent in their understanding of cloud computing technologies, business processes, security policy, and the legal environment of this emerging field.

As with most information technology issues, the technologies available regarding cloud computing emerge rapidly while the relevant laws and forensic methodology lag behind. There is a need for practitioners and researchers within the field of digital forensics to contribute to a better understanding of cloud computing. This better understand should lead to better practices and methodologies. While currently, many professionals familiar with the practice of computer forensics are reluctant to rely on data acquired from logical volumes representing the best evidence in legal matters, as the trend for software application deployment continues to migrate toward cloud computing environments, there will be fewer instances where simple, static data acquisitions of physical volumes are available.

It seems inevitable that forensics examiners will depend more on tools that allow forensically-sound data acquisitions of remote computing environments. Additionally, we anticipate that the volume of data storage and the complexity of network infrastructures will continue to increase. Numerous challenges exist regarding acquiring large amounts of forensically-sound data that includes, not just the logical files, but file slack and unallocated clusters from logical volumes in live systems. Therefore, we urge researchers and developers to become familiar with the challenges of collecting forensic data acquisitions from large installations of live systems in diverse, cloud computing environments. Lastly, we urge computer forensics practitioners to obtain a working knowledge of cloud computing environments and to be aware of the challenges these environments present to digital forensics.

REFERENCES

AccessData Group. LLC. (2011). *SilentRunner sentinel network forensics software*. Retrieved July 25, 2011, from http://accessdata.com/products/cyber-security-incident-response/ad-silentrunner-sentinel

Amrhein, D., & Quint, S. (2009, April 8). *Cloud computing for the enterprise: Part 1: Capturing the cloud*. Retrieved July 18, 2011, from IBM: http://www.ibm.com/developerworks/websphere/techjournal/0904_amrhein/0904_amrhein.html

Brodkin, J. (2009, May 18). *10 cloud computing companies to watch*. Retrieved July 18, 2011, from Network World, Inc.: http://www.networkworld.com/supp/2009/ndc3/051809-cloud-companies-to-watch.html

Buenz, A. (2007, April 11). *Reliable repositories: Using Microsoft Forefront security for SharePoint to defend collaboration* (Microsoft, Producer). Retrieved July 24, 2011, from http://technet.microsoft.com/en-us/library/cc512661.aspx

Bunting, S., & Anson, S. (2007). *Mastering Windows network forensics and investigation*. Indianapolis, IN: Wiley.

Carlton, G. H. (2007). *A protocol for the forensic data acquisition of personal computer workstations* (Doctoral dissertation). Available from ProQuest Dissertations and Theses database (UMI No. 3251043).

Carlton, G. H. (2008). An evaluation of Windows-based computer forensics application software running on a Macintosh. *Journal of Digital Forensics. Security and Law*, *3*(3), 43–60.

Catbird Networks, Inc. (2008, September 27). *Virtualization security: The Catbird Primer*. Retrieved July 25, 2011, from http://www2.catbird.com/news/analyst_reports.php

Cisco Systems Inc. (2011). *Cisco IOS Netflow*. Retrieved July 25, 2011, from http://www.cisco.com/en/US/products/ps6601/products_ios_protocol_group_home.html

Corporation, E. M. C. (2011). *NetWitness Investigator*. Retrieved July 24, 2011, from Netwitness Corporation: http://netwitness.com/products-services/investigator

Garrison, C., Lillard, T. V., Schiller, C. A., & Steele, J. (2010). *Digital forensics for network, internet, and cloud computing*. Amsterdam, The Netherlands: Elsevier Science.

Guidance Software, Inc. (2011). *EnCase Enterprise*. Retrieved July 25, 2011, from http://www.guidancesoftware.com/computer-forensics-fraud-investigation-software.htm

Harris, S. (2008). *CISSP all-in-one exam guide* (4th ed.). New York, NY: McGraw-Hill.

Hodges, C. (2011, April 8). *Cloud computing rains cost savings, productivity benefits*. Retrieved July 2, 2011, from Industry Week Social Media: http://www.industryweek.com/articles/cloud_computing_rains_cost_savings_productivity_benefits_24337.aspx?SectionID=3

Mell, P., & Grance, T. (2009). *The NIST definition of cloud computing*. Washington, DC: U.S. Department of Commerce, National Institute of Standards and Technology.

Microsoft Corporation. (2009). *Microsoft forefront threat management gateway 2010*. Retrieved July 25, 2011, from http://www.microsoft.com/forefront/threat-management-gateway/en/us/default.aspx

Microsoft Corporation. (2011). *Computer Online Forensic Evidence Extractor (COFEE)*. Retrieved July 25, 2011, from http://www.microsoft.com/industry/government/solutions/cofee/default.aspx

Nicholas, D. (2009, November 6). *Siren.gif: Microsoft COFEE law enforcement tool leaks all over the Internet~!* Retrieved July 14, 2011, from http://techcrunch.com/2009/11/06/siren-gif-microsoft-cofee-law-enforcement-tool-leaks-all-over-the-internet/

P, M. J. (2009, November 9). *More COFEE please, on second thought. ..* Retrieved July 14, 2011, from Praetorian Perfect: http://praetorianprefect.com/archives/2009/11/more-cofee-please-on-second-thought/

Pullin, A. (2009, November 11). *Microsoft's not bothered about COFEE leak*. Retrieved July 14, 2009, from The Inquirer: http://www.theinquirer.net/inquirer/news/1561911/microsoft-bothered-cofee-leak

Rittinghouse, J. W., & Ransome, J. F. (2010). *Cloud computing: Implementation, management, and security*. Boca Raton, FL: CRC Press/Taylor & Francis Group.

Saxe, J. G. (n. d.). *Blind men and the elephant*. Retrieved July 18, 2011, from http://wordinfo.info/unit/1?letter=B&spage=3

Schneier, B. (2009, June 4). *Schneier on security: Cloud computing*. Retrieved July 18, 2011, from http://www.schneier.com/blog/archives/2009/06/

Shavers, B. (2008). *Virtual forensics: A discussion of virtual machines related to forensics analysis*. Retrieved July 20, 2011, from http://www.forensicfocus.com/downloads/virtual-machines-forensics-analysis.pdf

Symantec Corporation. (2011). *Network security software solutions: Symantic*. Retrieved July 25, 2011, from http://www.symantec.com/business/network-access-control

Zetter, K. (2009, December 14). *Hackers brew self-destruct code to counter police forensics*. Retrieved July 14, 2011, from Wired Magazine: http://www.wired.com/threatlevel/2009/12/decaf-cofee/

This work was previously published in the International Journal of Interdisciplinary Telecommunications and Networking, Volume 3, Issue 4, edited by Michael R. Bartolacci and Steven R. Powell, pp. 1-16, copyright 2011 by IGI Publishing (an imprint of IGI Global).

Chapter 17
Value Creation in Electronic Supply Chains by Adoption of a Vendor Managed Inventory System

Yasanur Kayikci
Vienna University of Economics and Business, Austria

ABSTRACT

Many strategies have been developed to manage supply chain operations effectively. Vendor Managed Inventory (VMI) system is one of the prevalent strategic tools of the supply side logistics based on the electronic data exchange and business process automation among the suppliers and customers to enhance the competitive advantage. VMI is widely used in different industries including automotive sector. The VMI concept is a continuous replenishment program where suppliers are given access to demand and inventory level of customers and they are fully responsible for managing and replenishing the customer's stock. VMI's extension on customer satisfaction cannot be perceived sufficiently by decision-makers who are responsible to develop and invest in the customer-supplier relationship. This paper presents a path model using the method of Partial Least Squares (PLS) regression to give insight to decision-makers to understand effect of the VMI adoption on customer satisfaction. This paper investigates both determinants of relative factors of successful VMI adoption and the relationship in the supply chain with an empirical automotive industry case. The results show that the collaboration and coordination between customer and supplier and infrastructure of the information-sharing are the important dimensions to add value to the supply chain and to enhance customer satisfaction.

INTRODUCTION

Vendor Managed Inventory (VMI) strategy is one of collaborative applications, in which the supplier manages the retailer's inventory, allows suppliers and retailers to significantly improve supply chain performance. This concept has been around since the late 1980s in the retail industry then spread to other industries including automotive (Wittfeld *et al.*, 2008), but now gradually progressing towards strategic-partnership based forms which is a way to achieve efficiencies across partners in the supply chain. The basic drivers of VMI are to minimize supply chain complexity, to

DOI: 10.4018/978-1-4666-2154-1.ch017

reduce inventory cost for supplier and buyer and to improve service level while sharing information among trading partners (Yao *et al.*, 2007; Waller *et al.*, 1999). Moreover, Internet and new Information and Communication Technologies (ICT) are considered the most important enabler for obtaining and executing VMI relation. VMI has been described as an alternative for the traditional order-based replenishment tool in which the vendors (supplier) has taken the responsibility for making decisions as to the timing and amounts of inventory replenishment while using customer's inventory data (VendorManagedInventory.com, 2009; Southard & Swenseth, 2008). There are different sensitive and timely data that should be shared and agreed in a VMI partnership include inventory levels, sales data and forecasts, order status, promotional activities, production schedules and performance metrics (Angula *et al.*, 2004). All sharing information enables the supplier to make a decision to replenish inventory at the right time and the proper cost level.

The concept of Supply Chain Management (SCM) has been largely fragmented by the trend of outsourcing. Even if outsourcing is obligated strongly for information-sharing, collaboration and transparency between trading partners, especially it can drive huge inventory write-offs in the short product life cycles and cause longer lead times (Kilgore *et al.*, 2002). VMI may take the advantage of benefits to create a demand driven supply chain with easy applications while counterbalancing the drawbacks of outsourcing.

The lack of mutual trust between trading partners, expensive technology investment, personnel training and the uncertainty are difficult handicaps to understand the potential benefits of VMI (Jung *et al.*, 2004; Kaipia *et al.*, 2002). In such an environment, the success of VMI very much depends on how customers perceive and expect this adoption to be adding value to their enterprise.

In this paper, the impact of VMI adoption on end-customer satisfaction is analyzed by developing a path model with the method of Partial Least Squares (PLS) in which the relationship is presented between parameters in ICT infrastructure, customer orientation, cost, confidence, service quality, collaboration and coordination. Parameters for proposed research model were selected by using literature review and interviews. This work is based on an empirical study of both suppliers and their customers from European automotive industry. The rest of the paper is organized as follows: In the next section, a literature review is given. Then the model is described and in the following sections, its assumptions and the result of the experiment respectively are discussed, and the finally in the last section, the result summarizes implications of the study.

REVIEW OF LITERATURE

VMI is not a new collaborative strategy which is reviewed in the literature for both the customer and the supplier side and well documented in terms of its benefits and opportunities for supply chain performance improvements with analytical experiments, simulations and practical case studies in different industries (Southard & Swenseth, 2008; Wittfeld *et al.*, 2008; Simchi-Levi *et al.*, 2008; Yao *et al.*, 2007; Van der Vlist *et al.*, 2007, Dong *et al.*, 2007, De Toni & Zamolo, 2005; Angula *et al.*, 2004; Jung *et al.*, 2004; Yonghui & Raiesh, 2004; Kaipia & Tanskanen, 2003; Disney & Towill, 2003; Zhao *et al.*, 2002; Kaipia *et al.*, 2002; Yu *et al.*, 2001; Achabal *et al.*, 2000; Waller *et al.*, 1999). The majority of research conducted in the benefit of using VMI where mostly indicated on electronic information sharing, transparency and demand visibility in the supply chain. Moreover, it is also indicated that by implementing ICT supported VMI concepts, companies can achieve benefits in terms of reduced inventory levels, improved service levels, increased flexibility and reduced lead-times. Furthermore, the use of advanced information systems increases the level of data integration, level of data utilization and computer system properties and compatibility which are also essential for VMI success (Vigtil & Dreyer, 2006). VMI requires also strategic partnership among trading partners. Management literature defines strategic partnership as stable cross-organizational ties which are strategically important to participating firms. They may take the form of strategic alliances, joint-ventures, long-term buyer-supplier partnerships and other ties (Gulati *et al.*, 2000). Typically, organizations enter into a partnership/alliance in order to: exploit synergies; achieve strategic objectives; develop joint strategies; reduce potential risks, while increasing reward; knowledge sharing, improve returns on scarce resources (Gattorna & Walters, 1996). Therefore VMI is an important collaboration concept in electronic supply chain not only to achieve corporate strategy and also to develop an integrated logistics strategy within participating companies. In addition, high employee involvement and level of logistics integration obtain success and add value to VMI (Kuk, 2004). Finally, a strategic perception model for VMI adoption is not yet available.

METHODOLOGY

The decision of VMI adoption causes considerable changes in the optimizing supply chain performance at the strategic, tactical, and operational levels. At the strategic level, selection of the appropriate concept of VMI is the initial stage in investment. As Figure 1 illustrates, in the main part of the supply chain network, VMI is used as a category of replenishment solution between supply and demand, such as supplier and manufacturers and manufacturers and distributors. Constitution of an integrated cross-organizational information system is important to succeed the success of VMI, consequently all information exchange process between trading partners should be carried out by supporting ICT applications.

A research model is proposed by integrating the key concept of VMI which indicates the perceived success of VMI adoption to obtain customer satisfaction. The proposed model, Figure 2, is based on the conclusion of the studies and findings from literature review and recommendation of consultants/experts.

Figure 1. VMI integrated network system

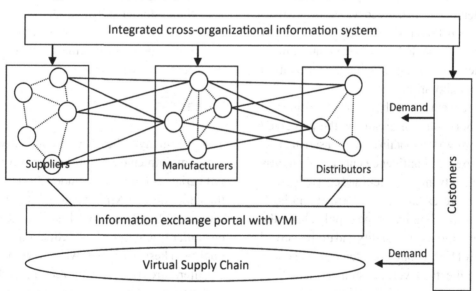

Figure 2. The structural design of the proposed research model

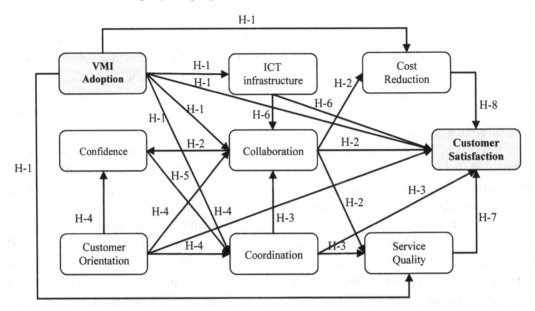

The major building blocks for the path model (structural equation model) have been conceptualized with seven factors: ICT infrastructure, collaboration, coordination, confidence, customer orientation, cost reduction and service quality. The following hypotheses are proposed:

H-1: The higher the VMI adoption success, the greater is the perceived value of VMI to obtain customer satisfaction in respect of cost reduction, ICT structure enhancement, service quality improvement, coordination and collaboration.

H-2: The more extensive the external and internal collaboration, the greater is the perceived value of VMI to obtain customer satisfaction in respect of confidence enhancement, service quality improvement and cost reduction.

H-3: The more extensive the vertical and horizontal coordination in the supply chain, the better is the service quality and the perceived value of VMI to obtain customer satisfaction.

H-4: The higher the level of customer orientation, the higher is the vertical and horizontal coordination, external and internal collaboration

and the better perceived value of VMI to obtain customer satisfaction.

H-5: The higher the confidence, the better is the vertical and horizontal coordination.

H-6: The better ICT structure, the better is the collaboration between trading partners and the more satisfied the customers are.

H-7: The higher the service quality, the higher is the customer satisfaction.

H-8: The more is the cost reduction; the better is the perceived value of VMI to obtain customer satisfaction.

Measures

Test of eight hypotheses were conducted on VMI programs implemented in the automotive industry. All variables (Table 2) included 20 statements on the adoption of VMI in terms of ICT infrastructure, collaboration, coordination, confidence, customer orientation, cost reduction and service quality. These variables were developed based on literature review and recommendation of experts/consultants from the industry. Automotive industry because of its relatively complex

Table 2. Participant list of survey

Respondent	Number of Respondents
Inventory management	8
Process engineer	5
Quality manager	6
Supervisor	7
Plant manager	2
Information system manager	3
Product development members	5
Accounting manager	2
Others	15
Total respondent	53

products is characterized by dedicated logistics networks, the need for lean supply philosophies and a requirement for supplier reliability in terms of component quality and delivery (Kehoe & Boughton, 2001). VMI adoption has become a very necessary tool for the automotive industry especially in the current vague situation to provide an efficient and effective supply network and to enhance better customer satisfaction subsequently. The first hypothesis (H-1) is that is associated with VMI adoption on integrative relationships under demand uncertainty in the automotive industry corresponding internal and external collaboration and vertical and horizontal coordination, a well organized information and communication technology infrastructure, high service quality and cost reduction, which are prerequisite for successful VMI adoption (Simchi-Levi *et al.*, 2008). The second hypothesis (H-2), collaboration is important to improve synchronization and management control, joint optimization, better external integration and a reduction of complexity which has great effect on the logistical service performance; especially external collaboration seems to be necessary which aims at a more direct working partnership than coordination (Barratt, 2004; Stank *et al.*, 2001). Extensive horizontal and vertical coordination, which is pointed to third hypothesis (H-3), is enabler of the relationship to

move towards the shared goal in whole supply chain (Mohr & Spekman, 1994). The vertical co-ordination occurs between supply chain members located at different levels, (between supplier and manufacturer, manufacturer and distributer) of the supply chain and horizontal coordination can be defined as the coordination between different supply chains members located on the same level (between various suppliers/manufacturers mostly coordinated replenishments and standardized information systems); However horizontal coordination is more unusual and difficult to implement than vertical coordination, as partners are first and foremost competitors and conflicts might rise from situations like the *prisoner's dilemma*. The fourth hypothesis indicates customer orientation (H-4). VMI facilitates a stronger customer orientation by suppliers which is seen as a crucial for competitive advantage. By following such an orientation, vendors are likely to play a greater role in driving markets as opposed to reacting to markets. They cannot just rely on data provided by their retailers. Vendors need constantly to assess customer perceptions of the value proposition of their products and services, availability/stock-out, and support services provided (Levy & Grewal, 2000). Confidence is denoted in the fifth hypothesis (H-5) which is a vital ingredient both within the company and the two partners which provides the sound ground for healthy relationship for success (Barratt, 2004; Mohr & Spekman, 1994). The sixth hypothesis (H-6) emphasizes the importance of the information availability. The key to successful VMI adoption is not solely efficient information transfer but timely information availability where the use of information system ensures the visibility (transparency) of item demand, location, status etc. within logistics network. The VMI adoption connects with service quality (H-7) and cost reduction (H-8) which are important to provide better customer satisfaction, was adopted from Kuk's model (2004).

All items included 20 statements (Table 1) on the adoption of VMI to enhance better customer satisfaction in terms of VMI adoption (5 items), collaboration (3 items), coordination (4 items), customer orientation (5 items), confidence (4 items), IT infrastructure (6 items), service quality (5 items), cost reduction (6 items) and customer satisfaction (5 items). These statements are derived after literature reviews. The survey were instructed participants to indicate their answer as "agree" or "disagree" which was measured on a five-point Likert scale of 1 (strongly agree) to 5 (strongly disagree). Twenty-three companies were identified. 53 useful responses are received from the survey which shows 71% response rate. The participant list of survey is shown in Table 2.

The proposed research model was tested with PLS Graph 3.0. PLS is a variance based latent variable structural equations modeling technique and PLS is becoming a major tool in customer satisfaction studies (Jakobowicz & Derquenne, 2007), which uses an estimation approach that places minimal demands on sample size and residual distributions (Chin, 1998). The evaluation of the proposed model fit was conducted in two stages (Chin, 1998; Hulland, 1999). The measurement model is firstly assessed, in which construct validity and reliability of the measures are assessed and secondly, the structural model with hypotheses is tested.

Results

The measurement model was used to assess discriminant and convergent validity of the constructs that were used in the hypotheses and research questions. All constructs which are modeled using reflective indicators. For reliability analysis, Composite Reliability (CR) was assessed. CR values vary from 0.702 to 0.847. After the first test, loading of six items is lower than 0.70 which was suggested as a common minimum level (Chin, 1998). These all items were deleted from the model afterwards the model was estimated again. All loading figures in the new model were equal or greater than 0.711 (Table 1). To test for discriminant validity, the square root of the Average Variance Extracted (AVE) for each latent variable was compared to the latent variable correlation matrix (Hulland, 1999). To possess adequate discriminant validity, the square-root of the AVE for each latent variable should be considerably greater than its correlation to each of the other latent variables. All constructs satisfy this criterion (Table 3). For each construct, the AVE is at least 0.543, which is above than the recommended minimum (threshold value) AVE, 0.50 (Fornell & Lacker, 1981).

The path loadings between the item and its corresponding factor were all positive and significant at $p < 0.001$, and the value of the path loading ranged between 0.50 and 0.847 (Table 3). Composite reliability exceeded 0.70 for all eight factors. AVE exceeds the 0.50 threshold value for all eight factors. These results indicate that construct reliability depends on the measure employed to asses it. Composite reliability and variance extracted present an acceptable picture.

The predictive power of a model is assessed by *Stone-Geisser*'s Q^2 test. Since the developed PLS model represents a theoretical causal model, the cross-validated redundancy measure is chosen and a Q^2 value of 0.244 (The PLS path coefficients are shown in Table 4) is obtained, thus implying that the model has predictive relevance

Table 1. Summary of measurement scales

Factor and item description	Mean	Standard deviation	Item Loading	CR	AVE
VMI Adoption				0.847	0.657
Enhanced better demand visibility	3.905	1.304	0.882		
Enhances the logistics performance (full truck, vehicle fill, less empty running, less cost)	3.612	1.649	0.845		
Enhances the logistics planning (materials, production, capacity and transportation)	3.665	1.802	0.790		
Increases the accuracy of order forecasts*	4.341	1.324	n/a		
Has more advantages to enhance better supply chain efficiency	4.213	1.522	0.865		
Collaboration				0.736	0.587
Enhances better communication between trading partners	3.678	1.345	0.765		
Enhances timely replenishment decision	3.457	1.564	0.834		
Enhances timely shipment decision	4.225	1.624	0.725		
Coordination				0.766	0.610
We collaboratively plan supply chain activities with our business partners ranging from forecasting to shipment planning	4.568	1.545	0.849		
Managerial innovations are transferred among plants within our corporation.*	4.213	1.354	n/a		
Our corporation transfers technological innovations and know-how between plants.	3.664	1.397	0.856		
The choice of information systems standards and technologies is coordinated at the corporate level.	3.908	1.516	0.745		
Customer Orientation				0.728	0.596
We frequently are in close contact with customers	3.889	1.543	0.856		
Customers are actively involved in improving services	4.792	1.679	0.765		
We respond to market and customer needs promptly and flexibly*	3.457	1.843	n/a		
Customers give us feedback on quality and delivery performance.	3.585	1.765	0.711		
Customers frequently share current and future demand information	3.406	1.843	0.867		
Confidence				0.810	0.693
We have corporate level communication on important issues with key suppliers.	3.567	1.567	0.805		
We are willingly to change our premises to be flexible to find a better solution to deal with our suppliers.	4.029	1.653	0.765		
We believe that cooperation with our supplier is beneficial for both parties*	4.346	1.691	n/a		
We give full weight the openness of communications in collaborating with our suppliers.	3.791	1.732	0.753		
ICT Infrastructure				0.734	0.655
VMI can be applied easily in the existing ICT infrastructure	3.267	1.632	0.881		
Is good to ensure better visibility (transparency)	3.589	1.730	0.873		

continued on following page

Table 1. Continued

Factor and item description	Mean	Standard deviation	Item Loading	CR	AVE
Enables readily access data from all applications involved in the project	4.044	1.844	0.716		
Is very easy to use	4.672	1.571	0.871		
Is compatible with trading partner's ICT system*	3.746	1.537	n/a		
Provides accurate and reliable info	3.226	1.759	0.747		
Service Quality				0.702	0.598
reduces the replenishment cycle times	3.745	1.673	0.850		
has effect on the stock rates for retailers	3.497	1.456	0.754		
reduces the frequency of out-of-stocks	4.045	1.539	0.765		
increases the number of delivered on time	4.236	1.743	0.853		
Better prepared for rush orders*	3.743	1.632	n/a		
Cost Reduction				0.711	0.543
VMI reduces substantially on inventory/holding costs.	4.045	1.654	0.879		
VMI has effect on transportation costs as to efficient planning.	3.654	1.743	0.856		
VMI implementation causes high operational costs and implementation costs.	4.567	1.641	0.754		
VMI has effect on overhead costs.	4.054	1.538	0.879		
VMI adoption shrinks transaction cost.	3.235	1.521	0.845		
VMI adoption enables faster inventory turns.	3.456	1.632	0.762		
Customer Satisfaction				0.765	0.603
Our customers are pleased with the products and services we provide them.	4.529	1.456	0.849		
Our customers seem happy with our responsiveness to their problems.	4.780	1.556	0.850		
Customer standards are always met by our plant.	4.024	1.744	0.876		
Our customers have always been well satisfied with the quality of our products over the past three years.	3.654	1.586	0.752		
Our organization satisfies or exceeds the requirements and expectations of our customers.	3.285	1.623	0.765		

SD: Standard Deviation, CR: composite reliability, AVE: Average Variance explained

*All scale items were measured using five-point Likert scales. Respondent were asked to what extent they agreed or disagreed with each statement (1: strongly agree and 5: strongly disagree).

* These items were eliminated from the proposed model, as their loading figures were under 0.70.

(Chin, 1998). Latent variables which show a low influence on explained variance in the ultimate dependent variable. Customer orientation, confidence and collaboration have a low impact or no impact on the overall model estimation, whereas coordination, ICT infrastructure, service quality and cost reduction exhibit an impact. Based on path coefficient, however collaboration, customer orientation and confidence have a significant impact on their internal counterparts and are therefore still of high importance. Furthermore, the model highlights the importance of

Table 3. Discriminant and convergent validity of the constructs

	Latent Variable	1	2	3	4	5	6	7	8
1	Collaboration	0.736							
2	Coordination	0.455	0.766						
3	Customer Orienta.	0.332	0.235	0.668					
4	Confidence	0.345	0.632	0.353	0.810				
5	ICT Infrastructure	0.487	0.543	0.556	0.459	0.734			
6	Cost Reduction	0.443	0.236	0.256	0.723	0.351	0.702		
7	Service Quality	0.258	0.345	0.491	0.704	0.554	0.453	0.711	
8	VMI Adoption	0.134	0.279	0.019	0.201	0.104	0.123	0.357	0.847

Average Variance Explained (AVE) of variables, bold numbers on diagonal.
Square of correlations between latent variables, numbers below diagonal.

Table 4. Statistics for the measurement model

	Latent Variable	q^2 value		Latent Variable	q^2 value
	Initial Q^2	0.244	H-5	Confidence	-0.001
H-1	VMI Adoption	0.028	H-6	ICT Infrastructure	0.042
H-2	Collaboration	0.003	H-7	Service Quality	0.016
H-3	Coordination	0.021	H-8	Cost Reduction	0.027
H-4	Customer Orientation	-0.001			

All correlations are significant at $p < 0.001$.

coordination, the relationship between service quality and cost reduction and the direct impact of ICT infrastructure through VMI adoption on customer satisfaction.

CONCLUSION

The primary purpose of this study is to gain insight into VMI adoption in the automotive industry which allows the benefit to obtain better customer satisfaction. VMI is as an effective supply chain program based on the information shared between trading partners has grown over the years. VMI lets companies reduce overhead by shifting responsibility for managing and replenishing inventory to suppliers (Emigh, 1999). The benefits of VMI consist of cost reductions, service enhancements and creating transparency in the supply chain. The importance of VMI is a trustful relationship between the involved parties. Some automakers already possess close relationship to their suppliers which is a good place to implement VMI. However, the growing complexity of autos, long product development times as well as comparatively long product life cycles complicate the situation in the automotive industry.

The result of this study shows that the VMI adoption with contribution of ICT infrastructure, service quality, cost reduction and coordination can enhance better customer satisfaction. It means that with these relationships customers can perceive the essence of VMI adoption into their supply chain systems, as customer satisfaction highly depends on fulfillment of customer expectation by using efficient supply chain techniques like VMI. Automakers should make some arrangements in order to be ready for VMI adoption. But taking everything into consideration, it offers tremendous potential for the auto industry and seems to be feasible.

ACKNOWLEDGMENT

The author wishes to thank the "COMET K2 Forschungsförderungs-Programm" of the Austrian Federal Ministry for Transport, Innovation and Technology (BMVIT), the Austrian Federal Ministry of Economics and Labour (BMWA), Österreichische Forschungsförderungsgesellschaft mbH (FFG), Das Land Steiermark and Steirische Wirtschaftsförderung (SFG) for their financial support. In addition, the author wants to extend her gratification to the anonymous reviewers for their time, assistance, patience and thoughtful insights during research gatherings.

REFERENCES

Achabal, D., Mcintyre, S., Smith, S., & Kalyanam, K. (2000). A decision support system for vendor managed inventory. *Journal of Retailing*, *76*(4), 430–454. doi:10.1016/S0022-4359(00)00037-3

Angulo, A., Nachtmann, H., & Waller, M. A. (2004). Supply chain information sharing in a vendor managed inventory partnership. *Journal of Business Logistics*, *25*(1), 101–120. doi:10.1002/j.2158-1592.2004.tb00171.x

Barratt, M. (2004). Understanding the meaning of collaboration in the supply chain. *Supply Chain Management: An International Journal*, *9*(1), 30–42. doi:10.1108/13598540410517566

Chin, W. W. (1998). The partial least squares approach to structural equation modeling. In Marcoulides, G. A. (Ed.), *Modern methods for business research* (pp. 295–336). Mahwah, NJ: Lawrence Erlbaum.

De Toni, A. F., & Zamolo, E. (2005). From a traditional replenishment system to vendor-managed inventory: A case study from the household electrical appliances sector. *International Journal of Production Economics*, *96*(1), 63–79. doi:10.1016/j.ijpe.2004.03.003

Disney, S. M., & Towill, D. R. (2003). The effect of vendor managed inventory (VMI) dynamics on the bullwhip effect in supply chains. *International Journal of Production Economics*, *85*(2), 199–215. doi:10.1016/S0925-5273(03)00110-5

Dong, Y., Xu, K., & Dresner, M. (2007). Environmental determinants of VMI adoption: An exploratory analysis. *Transportation Research Part E, Logistics and Transportation Review*, *43*(4), 355–369. doi:10.1016/j.tre.2006.01.004

Emigh, J. (1999). Vendor-managed inventory. *Computerworld*, *33*(34), 52.

Fornell, C., & Larcker, D. F. (1981). Evaluating structural equation models with observable variables and measurement error. *JMR, Journal of Marketing Research*, *18*(1), 39–59. doi:10.2307/3151312

Gattorna, J. L., & Walters, D. W. (1996). *Managing the supply chain: A strategic perspective*. London, UK: Palgrave Macmillan.

Gulati, R., Nohria, N., & Zaheer, A. (2000). Strategic networks. *Strategic Management Journal*, *21*(3), 203–215. doi:10.1002/(SICI)1097-0266(200003)21:3<203::AID-SMJ102>3.0.CO;2-K

Hulland, J. (1999). Use of partial least squares (PLS) in strategic management research: A review of four recent studies. *Strategic Management Journal*, *20*(2), 195–204. doi:10.1002/(SICI)1097-0266(199902)20:2<195::AID-SMJ13>3.0.CO;2-7

Jakobowicz, E., & Derquenne, C. (2007). A modified PLS path modeling algorithm handling reflective categorical variables and a new model building strategy. *Computational Statistics & Data Analysis, 51*(8), 3666–3678. doi:10.1016/j.csda.2006.12.004

Jung, S., Chang, T. W., Sim, E., & Park, J. (2004). Vendor managed inventory and its effect in the supply chain. In D. K. Baik (Ed.), *Proceedings of the Third Asian Simulation Conference on Systems Modeling and Simulation: Theory and Applications* (LNCS 3398, pp. 545-552).

Kaipia, R., Holmström, J., & Tanskanen, K. (2002). VMI: What are you losing if you let your customer place orders? *Production Planning and Control, 13*(1), 17–25. doi:10.1080/09537280110061539

Kaipia, R., & Tanskanen, K. (2003). Vendor managed category management—An outsourcing solution in retailing. *Journal of Purchasing and Supply Management, 9*(4), 165–175. doi:10.1016/S1478-4092(03)00009-8

Kehoe, D., & Boughton, N. (2001). Internet based supply chain management: A classification of approaches to manufacturing planning and control. *International Journal of Operations & Production Management, 21*(4), 516–525. doi:10.1108/01443570110381417

Kilgore, S., & Orlov, L. M. (2002). *Balancing supply and demand.* Cambridge, UK: TechStrategy Research/Forrester Research.

Kuk, G. (2004). Effectiveness of vendor-managed inventory in the electronics industry: Determinants and outcomes. *Information & Management, 41,* 645–654. doi:10.1016/j.im.2003.08.002

Levy, M., & Grewal, D. (2000). Supply chain management in a networked economy. *Journal of Retailing, 76*(4), 415–429. doi:10.1016/S0022-4359(00)00043-9

Mohr, J., & Spekman, R. (1994). Perfecting partnerships. *Marketing Management, 4*(4), 52–60.

Simchi-Levi, D., Kaminsky, P., & Simchi-Levi, E. (2008). *Designing and managing the supply chain: concepts, strategies & case studies.* New York, NY: McGraw-Hill.

Southard, P. B., & Swenseth, S. R. (2008). Evaluating vendor-managed inventory (VMI) in non-traditional environments using simulation. *International Journal of Production Economics, 116*(2), 275–287. doi:10.1016/j.ijpe.2008.09.007

Stank, T. P., Keller, S. B., & Daugherty, P. J. (2001). Supply chain collaboration and logistical service performance. *Journal of Business Logistics, 22*(1), 29–48. doi:10.1002/j.2158-1592.2001.tb00158.x

Van der Vlist, P., Kuik, R., & Verheijen, B. (2007). Note on supply chain integration in vendor-managed inventory. *Decision Support Systems, 44*(1), 360–365. doi:10.1016/j.dss.2007.03.003

VendorManagedInventory.com. (2009). *Definition.* Retrieved May 5, 2011, from http://www.vendormanagedinventory.com

Vigtil, A., & Dreyer, H. C. (2006). Critical aspects of information and communication technology in vendor managed inventory. In *Proceedings of the APMS Conference* (pp. 443-451).

Waller, M. A., Johnson, M. E., & Davis, T. (1999). Vendor-managed inventory in the retail supply chain. *Journal of Business Logistics, 20*(1), 183–203.

Wittfeld, K., Helferich, A., & Herzwurm, G. (2008). Vendor-managed inventory as a knowledge-intensive service in procurement logistics in the automotive industry. *VIMation Journal, Knowledge, Service & Production. IT as an Enabler, 1,* 17–25.

Yao, Y., Evers, P. T., & Dresner, M. E. (2007). Supply chain integration in vendor-managed inventory. *Decision Support Systems*, *43*(2), 663–674. doi:10.1016/j.dss.2005.05.021

Yonghui, F., & Rajesh, P. (2004). Supply-side collaboration and its value in supply chains. *European Journal of Operational Research*, *152*(1), 281–288. doi:10.1016/S0377-2217(02)00670-7

Yu, Z., Yan, H., & Cheng, T. C. E. (2001). Benefits of information sharing with supply chain partnerships. *Industrial Management & Data Systems*, *101*(3), 114–121. doi:10.1108/02635570110386625

Zhao, X., Xie, F., & Zhang, W. F. (2002). The impact of information sharing and ordering co-ordination on supply chain performance. *Supply Chain Management*, *7*(1), 24–40. doi:10.1108/13598540210414364

This work was previously published in the International Journal of Interdisciplinary Telecommunications and Networking, Volume 3, Issue 4, edited by Michael R. Bartolacci and Steven R. Powell, pp. 17-27, copyright 2011 by IGI Publishing (an imprint of IGI Global).

Chapter 18
Distributed and Fixed Mobility Management Strategy for IP– Based Mobile Networks

Paramesh C. Upadhyay
Sant Longowal Institute of Engineering & Technology, India

Sudarshan Tiwari
Motilal Nehru National Institute of Technology, India

ABSTRACT

Hierarchical Mobile IP (HMIP) reduces the signaling delay and number of registration messages to home agent (HA) by restricting them to travel up to a local gateway only. It uses centralized gateways that may disrupt the communications, in the event of a gateway failure, between a gateway and the mobile users residing with underlying foreign agents (FAs) in a regional network. Dynamic mobility management schemes, using distributed gateways, proposed in literature, tend to circumvent the problems in HMIP. These schemes employ varying regional network sizes or hierarchy levels that are dynamically selected according to call-to-mobility ratio (CMR) of individual user. In reality, this information cannot be readily available in practice. Also, any unusual alterations in CMR values may hamper the system performance. This paper proposes a new mobility management strategy for IP-based mobile networks, which is independent of individual user history. The proposed scheme uses subnet-specific registration areas and is fully distributed so that the signaling overheads are evenly shared at each FA. The scheme provides a viable alternative to dynamic mobility management schemes for its simplicity, performance, and ease of implementation.

INTRODUCTION

During the last decade, the mobile communication and Internet technologies have gained tremendous popularity among the users throughout the world. This has led towards the convergence of mobile communications and Internet technologies together so as to achieve their fullest advantages. It is forecast that, by the year 2015, the amount of total mobile user traffic will have a 23-fold increase as compared to the amount of present traffic and the multimedia traffic will account for 90% of the total user traffic (Otsu, Umeda, & Yamao, 2001).

DOI: 10.4018/978-1-4666-2154-1.ch018

Mobile IP (MIP) (Perkins, 1997, 2002), an IETF standard, has laid the foundation stone for IP mobility. It incorporates three additional entities in existing IP network: mobile host, home agent (HA), and foreign agent (FA). The two mobility agents, HA and FA, are used to handle the movement of the mobile hosts in the network. For a mobile host, its HA is fixed whereas the FA changes during its movement from one subnet to another. Therefore, each MH is assigned two addresses, namely a permanent address or home address which is assigned by the HA, and a temporary address or care-of-address (CoA) which is assigned by its visiting FA. Whenever a mobile host moves to a new FA, it obtains a new CoA from the current FA advertisement message. It is mandatory for the host to register its new location information with its HA to facilitate correct data delivery. The HA maintains the current mobility binding of each host, which is one-to-one mapping between home address and CoA of a mobile host. The data packets from a correspondent host, a fixed host on the Internet, are first intercepted by the HA of the mobile host, which, after encapsulation, tunnels them at the last registered CoA of the host. The FA with designated CoA decasulates the packets and forwards them to intended recipient.

Mobile IP provides an elegant solution for macro-mobility. However, it is not suitable for micro-mobility. In Mobile IP, packets addressed to a mobile host are always routed via its home agent. This results in a sub-optimal path for packet routing. This is known as triangular routing problem. Mobile IP incurs heavy registration cost for users with high mobility. This cost may become very significant as the number of mobile hosts increases. Moreover, if a mobile host roams far away from its HA, the signaling delay becomes longer. MIP does not support paging as well. Therefore, for a dormant mobile host, most of the effort to keep its location information up-to-date at its HA is of no use. Only the last location information is needed to establish the data exchange session. Paging is required to save wireless resources and

battery life of mobile hosts. MIP also suffers from triangular routing problem which causes longer handoff delay and, hence, the higher packet loss.

In this paper, we propose a new mobility management strategy for IP-based mobile networks. This strategy employs distributed architecture that helps in sharing the gateway responsibilities uniformly on each FA. In the proposed scheme, an FA can act in three modes of operation namely, as a gateway FA (GFA), a regional FA (RFA), or simply as a FA, as explained. Thus, the scheme employs three-tier hierarchy in a regional network or registration area (RA). It considers a subnet or FA-specific RA size, which is fixed for all users registering the current FA either as a GFA, or an RFA. Therefore, the RA sizes are independent of call and mobility patterns of individual user, and can be easily implemented. This scheme is named as Distributed and Fixed Hierarchical Mobile IP, abbreviated as DFHMIP. The proposed scheme uses overlapping RAs, which helps in mitigating the ping-pong (or zig-zag) effect near the RA boundaries.

Following this section, the paper is alienated in four major sections. First we provide state-of-art scenario in the field of IP-based mobility management. Then we describe the network architecture and an algorithm to implement DFHMIP. An analytical model is developed to compute signaling cost for DFHMIP scheme. We present the performance evaluation of DFHMIP and make its comparisons with DDHMIP and DHMIP. Finally, the paper is concluded.

RELATED WORKS

Micro-mobility management schemes for IP-based networks can be broadly classified in two categories: centralized and distributed schemes. Centralized schemes employ a dedicated mobility agent for gateway functionality. HAWAII (Ramjee, La Porta, Thuel, Vardhan, & Salgarelli, 1999; Ramjee, Varadhan, Salgarelli, Thuel, Wang, &

La Porta, 2002), Cellular IP (Campbell, Gomez, Wan, Kim, Turanyi, & Valko, 1999; Valko, 1999), and Hierarchical Mobile IP (HMIP) or Mobile IPv4 Regional registration (Johnson & Perkins, 2002; Gustafsson, Jonsson, & Perkins, 2005) form the basis for micro-mobility protocols. A survey of IP mobility management schemes with their comparisons can be found in Campbell and Castellanos (2001), Campbell, Javier, Kim, Wan, Turanyi, and Valko (2002), Saha, Mukherjee, Misra, Chakraborty, and Subhash (2004), Akyildiz, Xie, and Mohanty (2004), Reinbold and Bonaventure (2003), Henderson (2003), and Upadhyay and Tiwari (2010).

In Hierarchical Mobile IP (HMIP), the FAs, in a regional network, access Internet through one common gateway, called Gateway Foreign Agent (GFA). When a mobile host moves from one regional network to another, it registers with HA with the address of GFA as its CoA. When the MH moves from one FA to another within the same regional network, it registers locally with its GFA only. The packets destined for the mobile host are intercepted by the HA, and are then tunneled to the GFA of the host. The GFA re-tunnels these packets to the current FA of the host, which eventually forwards them to the recipient. Though HMIP reduces the signaling registration cost and transmission delay by localizing registration traffic, but this centralized network structure is neither robust nor flexible, and the processing overhead on GFA is heavy.

TeleMIP (Das, Misra, & Agrawal, 2000) is similar to Hierarchical Mobile IP with a GFA replaced by a new logical entity, the mobility agent (MA). The authors have compared the scheme with Mobile IP, HAWII and Cellular IP with respect to the performance metrics such as message update latency, handoff delay, and packet loss during message update and have shown that TeleMIP architecture supports intradomain mobility more efficiently than other existing approaches. IDMP (Misra, Mcauley, Datta, & Das, 2001; Das, McAuley, Datta, Misra, Chakraborty, & Das,

2002) is an extension to the base intradomain protocol used in TeleMIP. It offers intradomain mobility by using multi-CoAs. It is designed as a standalone solution for intradomain mobility and does not assume the use for global mobility management. Helmy, Jaseemuddin, and Bhaskara, (2004) have proposed a multicast-based micro-mobility protocol.

The biggest disadvantage of these micro-mobility protocols is that they employ centralized gateways, and therefore, in the event of a gateway failure, communications from/to mobile users, residing with underlying FAs, may get disrupted. In addition, the burgeoning number of mobile users may impose a heavy traffic burden on the centralized gateway, resulting in poor system performance (Xie & Akylidiz, 2002).

To combat the drawbacks of centralized micro-mobility management schemes, distributed schemes have been proposed in the literature. In distributed schemes, the gateway functionality is evenly distributed among all the mobility agents in the network. These schemes can be further classified in two categories: fixed and dynamic. In distributed and static schemes, the domain size for each user remains fixed, whereas the domain size in distributed and dynamic schemes depends on the mobility and packet arrival pattern of individual user.

Distributed and dynamic location management scheme for MIP (Xie & Akylidiz, 2002) is an extension of HMIP and is more flexible in the sense that there is no strict regional network boundary for a mobile host. Hereinafter, this scheme will be referred as DDHMIP. This scheme uses two-level hierarchy in a regional network and is adaptive to mobility and traffic patterns for each user. Each FA can function either as an FA or a GFA depending on the user mobility. Thus, unlike the centralized schemes, the GFA identity is not fixed. After entering a regional network, the first FA acts as a GFA of the host. The optimal number of FAs for a regional network is computed based on the up-to-date call and mobility patterns of the host, same way as in Xie, Tabbane, and Goodman (1993). Due to its dynamic nature, this scheme provides an optimal solution for each user in terms of total signaling cost, as the optimal number of FAs underlying a GFA can be adjusted from time to time. The traffic loads are evenly distributed on each FA and the system robustness gets improved. Wang, Chen, and Ho (1997) present a Mobile IP extended with routing agents. The routing agents have been used to handle local mobility of the mobile hosts in a dynamic two-level hierarchical architecture. Authors have used movement-based analysis in their scheme.

DHMIP (Ma & Fang, 2004) attempts to reduce the location update signaling traffic by registering the new CoA of the mobile host to the previous FA. In this scheme, an optimal hierarchy level threshold is set for each user based on its current traffic load and mobility pattern. The optimal threshold value is computed in a way similar to Xie, Tabbane, and Goodman (1993). The MH registers with its HA only when this threshold is reached. Since, the HA is not aware of the user's up-to-date location information, the packets are sent to the last FA registered at the HA. Thereafter, the packets are tunneled along the FA hierarchy to the mobile terminal. Thus, the location update traffic can be localized. The authors have developed analytical model for DHMIP using forwarding pointer method as in Jain and Lin (1995). Yu, Yu-mei, and Hui-min (2005) have compared signaling transmission cost, packet delivery cost and total cost of HMIP, DDHMIP and DHMIP. They found that DHMIP scheme performs better than other two with regard to signaling transmission cost, but the packet delivery cost increases with the hierarchy threshold as a packet needs to travel several FAs along the hierarchy before arriving at the mobile host.

The authors in Pyo, Li, and Kameda (2003a, 2003b, 2005) have suggested a dynamic scheme for MIPv6. In this scheme, a domain can be configured dynamically and mobility agent functionality is distributed among all access routers

(ARs), similar to GFA functionality on each FA as in Xie and Akylidiz (2002). In order to configure a domain dynamically, each AR produces the list of routers located at certain distance threshold. The distance threshold is determined based on the mobility and packet arrival pattern of individual user. The list of ARs, called a domain list, indicates the domain size of an AR. Each router in a domain list is identified by an IPv6 address. For domain detection, an AR periodically sends entire domain list via its router advertisement messages. Each AR can act as a domain mobility agent (DMA) for ARs in the domain list. Each mobile host maintains the domain list of its DMA in its buffer to detect that it has moved out of its current domain. Huang, Feng, Liu, Song, and Song (2004) have proposed an algorithm for domain construction. The domain size is not fixed and it is re-organized on the basis of handover frequency between two neighboring ARs. According to this proposal, the ARs having high handover frequency lie in the same domain. Xu, Lee, and Thing (2003) proposed a local mobility agent (LMA) selection algorithm that uses mobility pattern of individual user to form an LMA tree on the movement path of a mobile host. A stable LMA in the tree, which is nearest to the mobile host, is selected as new LMA of the host. A chain anchor scheme for IP mobility management is proposed in Bejerano and Cidon (2003). The first FA in the chain acts as anchor for the rest of FAs in chain. The authors have developed an algorithm to determine the length of the chain.

SYSTEM DESCRIPTION

Network Architecture

Similar to MIP, DFHMIP (Upadhyay & Tiwari, 2004, 2005, 2006) also uses two mobility agents, viz. Home Agent (HA) and Foreign Agent (FA), to make user mobility possible from one subnet to another in the network. Each mobile host is permanently registered with a subnet where it started its mobility services first time. This subnet is called the home subnet of the host. Any other subnet in the network is a foreign subnet for the host. The HA, located at the home subnet, assigns a permanent IP address to the MH, called its home address. Similarly, when a mobile host leaves its home subnet and enters a foreign subnet, the FA, located at the foreign subnet assigns a temporary address, also known as care-of-address (CoA) of the host. The proposed scheme is based on having neighbors' information, wherein each FA is assumed to be aware of the IP addresses of its neighboring FAs. The FA_x is said to be the neighbor of FA_y if a mobile host moves from area covered by FA_x to another area covered by FA_y. This means that the neighbors are located at a single hop-distance only. The neighboring information of each FA is assumed to have been embedded in the FA advertisement messages. This assumption does not have significant impact on data rate in the network (Zhang, Jaehnert, & Dolzer, 2003). Also, unlike the scheme proposed in Pyo, Li, and Kameda (2003a, 2003b, 2005) wherein the entire domain list is broadcast in the coverage area of an AR through the advertisement messages, in our proposed scheme, each FA broadcasts the list of neighboring FAs only in its coverage area. An MH, residing in the coverage area of the FA, listens to the advertisement messages, and uses it solely to decide if a registration is necessary.

An FA, in this proposal, can concurrently work in three modes of operation: a Gateway FA (GFA), a Regional FA (RFA), or simply as an FA. The FA acts as a GFA whenever a mobile host requests it to perform a registration with its HA. This registration is termed as a macro-registration. Having performed a macro-registration, a mobile host can freely roam from one subnet to another until the next macro-registration is requested with the visiting FA of the mobile host. At this moment, the visiting FA of the mobile host becomes its new GFA. The group of the subnets, wherein a mo-

bile host can move without a macro-registration, constitutes a registration area, called as macro-registration area (MacRA). The GFA resides at the center of a MacRA. Thus, any two consecutive MacRAs overlap with each other. The size of a MacRA is subnet or FA-specific rather than the user-specific.

A MacRA consists of a number of FA-specific smaller regions. These FA-specific regions are called micro-registration areas (MicRA). In fact, each FA is surrounded by its neighboring FAs, with the FA being at the center. This FA together with its neighboring FAs constitutes a MicRA. The center FA acts as an RFA for itself as well as for the neighboring FAs. Thus, the size of MicRA depends upon the number of neighboring FAs only. When a mobile host leaves a MicRA, the visiting FA becomes its new RFA, and a new MicRA is formed. This makes two successive MicRAs overlapping with each other. When a mobile host changes its MicRA, it registers with its GFA via its RFA. This registration is called a micro-registration. A host can move in its MicRA without any registration with the GFA. Thus, the movement of a mobile host within MicRA is transparent to its GFA. Based on the algorithm given, a mobile host can decide to perform a macro, micro, or a local registration.

When a mobile host registers an FA as an RFA, the registered FA and its neighboring FAs simply act as FAs for the host. While moving from one FA to another within the MicRA, the host registers with its RFA locally. This registration is referred as a local registration. Note that when a host registers an FA as a GFA, at that moment, the registered FA behaves like the GFA, RFA, and the FA for the mobile host. Similarly, if the host registers an FA, other than the GFA, as its RFA, the registered FA acts as an FA for the host, as well.

Thus, a macro-registration is performed at MacRA level, a micro-registration at MicRA level, and a local registration at subnet level. Thus, the scheme uses three level hierarchies in the network, the GFA being at the highest level, RFA at the intermediate level, and the FA at the lowest level. The size of a MacRA or a MicRA for all the mobile hosts using same FA as a GFA or a RFA, respectively, is fixed. The traffic load in the network is evenly distributed at each FA. Therefore, the proposed scheme has been given a name: Distributed and Fixed Hierarchical Mobile IP, abbreviated as DFHMIP.

Each FA maintains three visitor lists namely, GFA visitor list, RFA visitor list and FA visitor list. These visitor lists are used to keep location records of each mobile host at its GFA, RFA and an FA, respectively. The packets intended for a mobile host are first intercepted by the HA, which then tunnels them to the GFA. The GFA forwards these packets to the RFA, which forwards them to the mobile host via itself or through one of its neighboring FAs, if necessary.

DFHMIP employs overlapping MacRAs and MicRAs, which helps in mitigating the ping-pong (or zigzag) effect near the two adjacent MacRA or MicRA boundaries. Here, the ping-pong effect refers to several boundary crossings that occur back and forth across two successive MacRAs or MicRAs. It is undesirable from both, the user's quality perception and the network load view points. The formation of overlapping MacRAs and MicRAs with host movement is shown in Figure 1.

Figure 1. Formation of overlapping macro- and micro-registration areas

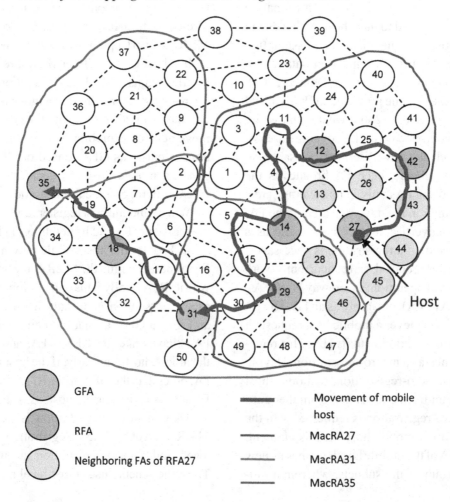

PROPOSED ALGORITHM

To implement DFHMIP, each mobile host maintains three buffers R_1, R_2 and R_3, for storing the IP addresses from GFA, RFA and FA, respectively. Now onwards, these IP addresses are referred as an identifier of a GFA, RFA and an FA, respectively. A GFA identifier consists of IP addresses of GFA and its neighboring FAs. Similarly, the RFA identifier includes the IP addresses of RFA and its neighboring FAs, and the FA identifier is the IP address of the visiting FA and those of its neighboring FAs. A mobile host receives these identifiers from advertisement messages from its GFA, RFA and FA, respectively. When a mobile host enters a foreign subnet, it uses following steps to perform a macro-registration, a micro-registration, or a local registration:

1. The host listens to the FA identifier from the agent advertisement messages from the visiting FA, and stores in its buffer R_3.

2. The visiting FA becomes the GFA and RFA of the mobile host. In this particular case, the GFA, RFA, and FA identifiers are same. Therefore, the FA identifier, in buffer R_3, is copied into the buffers R_1 and R_2, i.e.,

$$R_1 \xleftarrow{\quad identifier \quad} R_3 \, ; \; R_2 \xleftarrow{\quad identifier \quad} R_3$$

3. At each subnet crossing, the mobile host computes $R_1 \cap R_2$ and $R_2 \cap R_3$.

4. IF $R_2 \cap R_3 \geq 4$, as shown in Figure 3, and $R_1 \cap R_2 \neq X$, then local-registration occurs. Here, X represents a null set.

5. IF $R_2 \cap R_3 < 4$ and $R_1 \cap R_2 \neq X$, a micro-registration occurs. At this moment, visiting FA becomes, the new RFA. Therefore, the FA identifier, in buffer R_3, is copied to buffer R_2, i.e.,

$$R_2 \xleftarrow{\quad identifier \quad} R_3$$

6. ELSE IF $R_1 \cap R_2 = X$, then macro-registration takes place.

7. Go To step 1.

ANALYTICAL MODEL AND SIGNALING COSTS

DFHMIP uses random-walk model for the host mobility. In random-walk mobility model, a mobile host can move to its neighboring subnets with equal probability (Akyildiz & Ho, 1995). The analytical model for micro-registration and macro-registration are developed using continuous-time Markov chains. It is assumed that the subnet-residence time of a mobile host follows an exponential distribution with mean $\dfrac{1}{\lambda_m}$, and the packet arrival rate for a host follows Poisson distribution with mean λ_a.

A MacRA, in proposed DFHMIP, can be considered as consisting of four layers namely, layer 0, layer 1, layer 2, and layer 3, as shown in Figure 2. Layer 0 consists of a single FA only, and lies in the interior most vicinity of a MacRA. This FA is referred as a GFA. A mobile host performs a macro-registration in two situations. First, when it moves to an FA at layer 4 and second, when it moves to an FA at layer 3 and finds that it has changed its MicRA. Under both the circumstances, the visiting FA becomes the new GFA of the host, and a new MacRA is formed. A mobile host experiences a change of MicRA or RFA at an FA in layer 3 only when it's current RFA is located at either layer 1 or layer 2. This is due to the fact that, in proposed DFHMIP, the layer 3 FAs in a MacRA can act as general FAs only. The number of RFAs, and hence MicRAs, in k^{th} MacRA, $N_{RFA}^{(k)}$, is given by:

$$N_{RFAs}^{(k)} = \sum_{i=0}^{2} n_i^k \tag{1}$$

where, n_i^k Number of FAs at i^{th} layer in k^{th} MacRA; ($i = 0, 1 \, and \, 2$).

Similarly, number of FAs in l^{th} MicRA of kth MacRA, denoted as $N_{FAs}^{(l,k)}$, can be written as:

Figure 2. Layered network architecture for DFHMIP

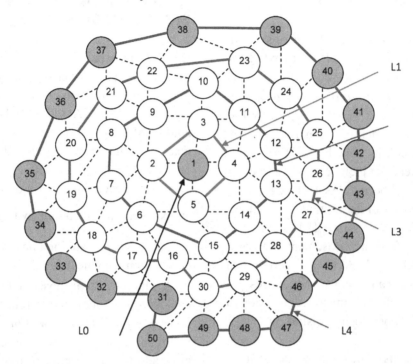

$$N_{FAs}^{(l,k)} = \sum_{i=0}^{1} n_i^k \qquad (2)$$

The transition probability $p_{i,j}^{(k)}$ of a mobile host that it has moved from i^{th} layer to j^{th} layer in k^{th} MacRA can be given as:

$$p_{i,j}^{(k)} = \begin{cases} 1 \\ \dfrac{1}{n_i^{(k)}} \displaystyle\sum_{l=1}^{n^{(k)}} \dfrac{n_{i,j}^{(l,k)}}{\varphi_l^{(i,K)}} \end{cases}$$

$$\begin{aligned} & i = 0, j-1 \\ & 0 \le i \le 3, j - (i-1),(i+1) \qquad (3) \\ & otherwise \end{aligned}$$

where, $\eta_i^{(k)}$ Number of FAs in i^{th} layer of kth MacRA; $\eta_{i,j}^{(l,k)}$ Number of FAs at j^{th} layer which are neighbors of lth FA at i^{th} layer of kth MacRA; $\varphi_l^{(i,k)}$ Number of FAs which are neighbors of lth FA at i^{th} layer of kth MacRA.

The transition rate, $q_{i,j}^{(k)}$, of a mobile host from i^{th} layer to j^{th} layer, in k^{th} MacRA, depends upon the subnet crossing rate of the host, and is given as:

$$q_{i,j}^{(k)} = \lambda_m p_{i,j}^{(k)} \qquad (4)$$

MICRO-REGISTRATION MODEL

The micro-registration model for l^{th} MicRA in k^{th} MacRA is shown in Figure 3. This model uses two states namely, s_0, and s_1. The state s_0 represents the RFA, and the state s_1 can be any of its neighboring FAs, in MicRA. A mobile host can stay in any of these two states without a micro-registration. When the host enters in a higher state, it performs a micro-registration, and the higher state becomes the state s_0. Therefore, $q^{*(l,k)}_{1,0}$ is expressed as:

Figure 3. Markov chain for micro-registration

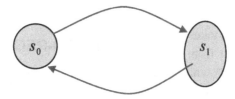

$$q *^{(l,k)}_{1,0} = q^{(l,k)}_{1,0} + q^{(l,k)}_{1,2} \tag{5}$$

$q^{(l,k)}_{0,1} \quad q *^{(l,k)}_{1,0}$

The state transition rate matrix, $Q^{(l,k)}_{i,j}$, for l^{th} MicRA in k^{th} MacRA is given as:

$$Q^{(l,k)}_{i,j} = \begin{bmatrix} q^{(l,k)}_{0,0} & q^{(l,k)}_{0,1} \\ q *^{(l,k)}_{1,0} & q^{(l,k)}_{1,1} \end{bmatrix} \tag{6}$$

The steady state or stationary probability vector, $\pi^{(l,k)}_i$, of a state i, in Markov chain, can be obtained by solving the balance equations:

$$\pi^{(l,k)}_i Q^{(l,k)}_{i,j} = 0 \tag{7}$$

which satisfies $\sum_{i=0}^{1} \pi^{(l,k)}_i = 1 \tag{8}$

MACRO-REGISTRATION MODEL

As shown in Figure 4, the macro-registration model for k^{th} MacRA in the network uses three states namely, 0, 1 and 2. The state 0 is used to represent the GFA in MacRA, which lies at layer 0 of the layered architecture. This also represents the RFA at layer 0, as the GFA can also act as an RFA. The states 1 and 2 symbolize one of the RFAs at layer 1 and layer 2, respectively.

$q^{(k)}_{1,0} \quad q *^{(k)}_{2,0} \quad q^{(k)}_{1,2} \quad q^{(k)}_{0,2} \quad q^{(k)}_{2,1}$

The state transition rate matrix for the kth MacRA is given by:

$$Q^{(k)}_{i,j} = \begin{bmatrix} q^{(k)}_{0,0} & 0 & q^{(k)}_{0,2} \\ q^{(k)}_{1,0} & q^{(k)}_{1,1} & q^{(k)}_{1,2} \\ q *^{(k)}_{2,0} & q^{(k)}_{2,1} & q^{(k)}_{2,2} \end{bmatrix} \tag{9}$$

Here, the transition rate $q^{(k)}_{1,0}$ accounts for the transition of the mobile host from the MicRA having its RFA at layer1, to an FA at layer3. This causes a macro-registration at the visiting FA, and it enters into state 0. Thus,

$$q^{(k)}_{1,0} = q^{(k)}_{1,3} \tag{10}$$

Similarly, the transition rate $q *^{(k)}_{2,0}$ consists of three components as given:

Figure 4. Markov chain for macro-registration in DFHMIP

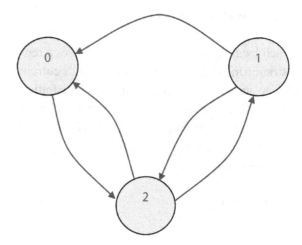

$$q_{2,0}^{*(k)} = q_{2,0}^{(k)} + q_{2,0}^{(k)} + q_{2,3}^{(k)} + q_{2,4}^{(k)} \quad (11)$$

The first term in equation (11) takes account of the transition of a mobile host from MicRA having its RFA at layer 2 to MicRA with its RFA at layer 0. The second and third terms represent the transitions from MicRA with its RFA at layer 2 to an FA either at layer 3 or layer 4, respectively. In either of two cases, the mobile host enters the state 0 of the Markov chain, and performs a macro-registration. The steady state or stationary probability vector, $\pi_i^{(k)}$, of a state i, in Markov chain, can be obtained by solving the balance equations:

$$\pi_i^{(k)} . Q_{i,j}^{(k)} = 0 \quad (12)$$

with requirement that

$$\sum_{i=0}^{2} \pi_i^{(k)} = 1 \quad (13)$$

SIGNALING COSTS

The total signaling cost, in proposed DHMIP, comprises of local registration cost, micro-registration cost, macro-registration cost, and packet delivery cost. All these four components are computed separately as follows:

Local Registration Cost

Each local registration, in MicRA, causes a registration message to travel from the host's current FA to its RFA, resulting in a local registration cost. This registration takes place after each subnet boundary crossing in a MicRA. The average local registration cost per unit time, $Cost_{Local}$, can be expressed as:

$$Cost_{Local} = \lambda_m C_{local} \quad (14)$$

where, C_{local} is the cost of unit local registration.

MICRO-REGISTRATION COST

A user is said to have incurred micro-registration cost on the network, when it moves from one MicRA to the other, within the same MacRA. In this situation, the user requests for a micro-registration with the GFA, via its new RFA.

The average micro-registration cost per unit time, $Cost_{Micro}^{(l,k)}$, in l^{th} MicRA of kth MacRA is given by:

$$Cost_{Micro}^{(l,k)} = \pi_1^{(l,k)} q_{1,2}^{(l,k)} C_{micro}^{(1)} \quad (15)$$

where, $C_{micro}^{(l)}$ is the unit micro-registration cost in l^{th} MicRA, and

$$\pi_1^{(l,k)} = \frac{p_{0,1}^{(l,k)}}{(p_{0,1}^{(l,k)} + p_{1,0}^{*(l,k)})} \quad (16)$$

MACRO-REGISTRATION COST

When a mobile host changes its GFA, it performs registration with its HA, called a macro-registration. The macro-registration cost is due to the macro-registrations taking place in the network. If C_{macro} is the cost of unit macro-registration, then the expression for macro-registration cost per unit time, denoted by $Cost_{Macro}$, can be written as:

$$Cost_{Macro} = \lambda_m \pi_2^{(k)} (p_{2,4}^{(k)} + p_{2,4}^{(k)}) C_{macro} \quad (17)$$

Here,

$$\pi_2^{(k)} = \frac{P_{0,2}^{(k)}(P_{1,3}^{(k)} + P_{1,2}^{(k)})}{[P_{0,2}^{(k)}(P_{1,3}^{(k)} + P_{1,2}^{(k)} + P_{2,1}^{(k)}) + P_{1,3}^{(k)}P_{2,1}^{(k)} \\ + (P_{2,0}^{(k)} + P_{2,3}^{(k)} + P_{2,4}^{(k)})(P_{1,3}^{(k)} + P_{1,2}^{(k)})]}$$

(18)

and, $P_{i,j}^{(k)}$ is the transition probability that the mobile host moves from a MicRA with RFA at layer i to a MicRA whose RFA is at layer j.

Packet Delivery Cost

The packet delivery cost, $Cost_{Packet\_Delivery}$, depends upon the call arrival rate of a mobile host, and can be expressed as:

$$Cost_{Packet\_Delivery} = \lambda_a C_{packet\_delivery} \qquad (19)$$

where, $C_{packet\_delivery}$ is the unit cost for packet delivery.

The total Signaling cost per call arrival can be calculated using following expression:

$$C_T = \frac{1}{\rho}\left[\begin{matrix} Cost_{Local} + Cost_{Micro}^{(k)} \\ + Cost_{Macro} \end{matrix}\right] + C_{Call\_delivery}^{(k)} \quad (20)$$

PERFORMANCE EVALUATION AND COMPARISONS

When a signaling message flows across the network, two types of costs are involved. First, the communication cost between the two entities, and the processing cost at each entity through which the message travels in the network. These costs, for example, can be in terms of delays. To evaluate the performance of proposed DFHMIP, the unit registration costs and unit packet delivery cost need to be defined. Following parameters are used for this purpose:

- $c_{GFA,HA}$: Communication cost of registration messages between a GFA and the HA
- $c_{RFA,GFA}$: Communication cost of registration messages between an RFA and a GFA
- $c_{FA,RFA}$: Communication cost of registration messages between an FA and an RFA
- $c_{MH,FA}$: Communication cost of registration messages between a Host and an FA
- $p_{GFA,HA}$: Processing cost of registration messages between a GFA and the HA
- $p_{RFA,GFA}$: Processing cost of registration messages between an RFA and a GFA
- $p_{FA,RFA}$: Processing cost of registration messages between an RFA and a GFA
- $p_{MH,FA}$: Processing cost of registration messages between a Host and an FA
- $c_{HA,GFA}$: Communication cost of packet delivery between the HA and a GFA
- $c_{GFA,RFA}$: Communication cost of packet delivery between a GFA and an RFA
- $c_{RFA,FA}$: Communication cost of packet delivery between an RFA and an FA
- $c_{FA,MH}$: Communication cost of packet delivery between an FA and a Host
- $p_{HA,GFA}$: Processing cost of packet delivery between the HA and a GFA
- $p_{GFA,RFA}$: Processing cost of packet delivery between a GFA and an RFA
- $p_{RFA,FA}$: Processing cost of packet delivery between an RFA and an FA
- $p_{FA,MH}$: Processing cost of packet delivery between an FA and a Host

Using these parameters, C_{macro}, $C_{micro}^{(l)}$, C_{local}, and $C_{packet\_delivery}^{(l)}$ can be expressed as:

$$C_{macro} = 2c_{MH,GFA} + 2p_{MH,GFA}$$
$$+2dc_{GFA,HA} + p_{GFA,HA}$$
$$C^k_{micro} = 2c_{MH,RFA} + 2p_{MH,RFA}$$
$$+2\alpha\, c_{RFA,GFA} + \log(N^{(k)}_{RFA})p_{RFA,GFA} \qquad (21)$$

$$C_{local} = 2c_{MH,FA} + 2p_{MH,FA}$$
$$+2c_{FA,RFA} + \log(N^{(l)}_{FA})p_{FA,RFA} \qquad (22)$$

$$C^{(k)}_{packet\_delivery} = dc_{HA,GFA} + p_{HA,GFA}$$
$$+\alpha c_{GFA,RFA} + \log(N^{(k)}_{RFA})p_{GFA,RFA}$$
$$+c_{RFA,FA} + \log_{10}(N^{(l)}_{FAs})p_{RFA,FA} + c_{FA,MH} + p_{FA,MH} \qquad (23)$$

The parameter values used for performance evaluation are same as in Ma and Fang (2004), and are listed in Table 1. Unit macro-registration and packet delivery costs depend upon the distance, d, between GFA and HA. The average hop-distance between GFA and RFA, in proposed DFHMIP, is fixed. Therefore, following relationships between the parameters are used in this analysis:

$$c_{HA,GFA} = dc_{RFA,FA}; \qquad c_{GFA,HA} = dc_{FA,RFA} \qquad (24)$$

$$c_{GFA,RFA} = 1.5c_{RFA,FA;} \qquad c_{RFA,GFA} = 1.5c_{FA,RFA} \qquad (25)$$

For analysis purpose, 50 subnets in the network have been considered. The cost analysis uses a MacRA and MicRA with FA1 acting as GFA as well as RFA, as shown in Figure 2. Figure 5 shows that when a mobile host's gateway is near to its home subnet, for example d=2, the macro-registration cost is approximately same as the micro-registration cost. It can be seen from Figures 6 and 7 that the macro-registration cost increases as the distance between HA and GFA of a mobile host becomes longer. This is because the signaling messages from GFA will have to travel long to reach the HA of a mobile host. These figures, also, show how various registration costs, in proposed DFHMIP, change with varying subnet residence time.

It is visible that the three registration costs viz. macro- micro- and local registration costs tend to reduce with increasing subnet residence time. It is obvious because if a mobile host stays either in a MacRA, MicRA, or in a subnet for a longer time, it will generate lesser signaling traffic in the network However, high mobility users, having very small subnet residence time, will cross the registration boundaries quite frequently, and hence, will incur heavy registration costs on the network. It is observed that the local registration

Table 1. Parameter values for analysis

Communication costs of packet delivery from the HA to a Mobile Host	$c_{HA,GFA}$ d*25	$c_{GFA,RFA}$ α*25	$c_{RFA,FA}$ 25	$c_{FA,MH}$ 5
Communication costs of registration messages from a Mobile Host to the HA	$c_{MH,FA}$ 10	$c_{FA,RFA}$ 50	$c_{RFA,GFA}$ α*50	$c_{GFA,HA}$ d*50
Processing costs of packet delivery from the HA to a Mobile Host	$p_{HA,GFA}$ 10	$p_{GFA,RFA}$ 1	$p_{RFA,FA}$ 0.5	$p_{FA,MH}$ 1
Processing costs of registration messages from a Mobile Host to the HA	$p_{MH,FA}$ 2	$p_{FA,RFA}$ 1	$p_{RFA,GFA}$ 0.5	$p_{GFA,HA}$ 20

Figure 5. Variation of registration costs with subnet residence time in proposed DFHMIP; d=2

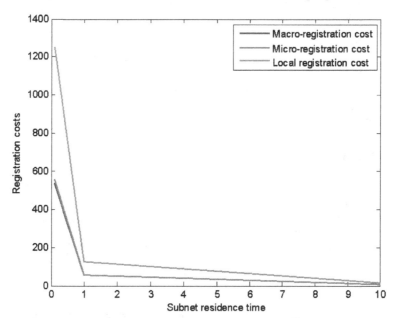

cost is higher than the micro-registration cost. This is due to the fact that the subnet boundary crossings in MicRA are more likely as compared to the MicRA crossings in a MacRA.

The packet delivery cost depends upon two factors; first, packet arrival rate for a mobile host, and second the distance of its GFA from the HA. This is shown in Figure 8. The packet delivery

Figure 6. Variation of registration costs with subnet residence time for proposed DFHMIP; d=4

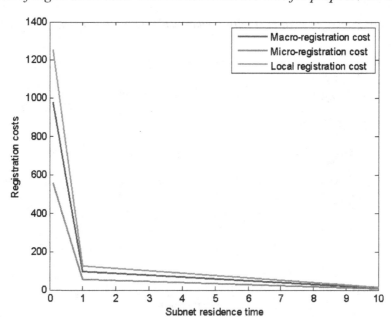

Figure 7. Variation of registration costs with subnet residence time for proposed DFHMIP; d=10

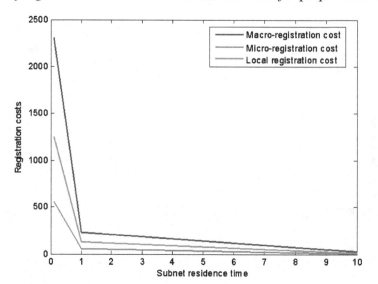

cost increases with increasing packet arrival rate. The reason manifests in frequent delivery of the packets to the intended mobile host. If the GFA of a host is located far away from its HA, the packets will traverse a longer distance to reach their destination. This, further, accounts for the increased packet delivery cost. The effect of varying distance between the host's HA and the GFA on packet delivery cost is, also, shown in figure.

COMPARISON BETWEEN PROPOSED DFHMIP AND DDHMIP

DDHMIP (Xie & Akylidiz, 2002) optimizes the regional network size of a mobile host according to its current CMR value. The regional network of DDHMIP corresponds to the MacRA of the proposed DFHMIP. In Figure 9, a comparison between the two schemes, in terms of total signaling cost,

Figure 8. Variation of Packet arrival cost with packet arrival rate for different hop-distances between HA and GFA

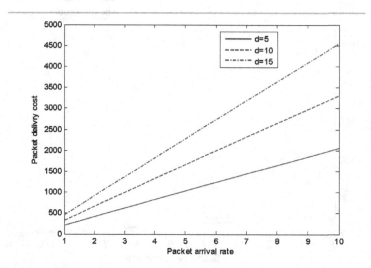

Figure 9. Total signaling cost vs. CMR for DFHMIP and DDHMIP; d=5

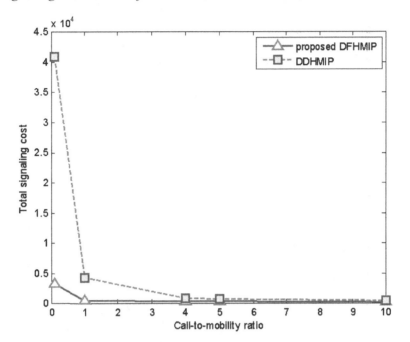

has been made. It can be seen that the proposed DFHMIP performs better than the DDHMIP over the entire range of CMR values of the users. In DDHMIP, the optimum regional network size for low CMR users is very large. These users are characterized as very highly mobile users that cross the subnet boundaries quite frequently, and are expected to stay several hop-distances away from their GFA. Since a mobile host, in DDHMIP, updates its location with its GFA after each subnet crossing, it incurs a heavy registration cost in the regional network. This cost is further fueled by the increasing distance of the host's current FA from its GFA. Also, during packet delivery, the packets intended for low CMR users will have to travel long several hops to finally reach their destinations, resulting in high packet delivery costs. On the contrary, the proposed DFHMIP reduces the update frequency with the GFA by adding a hierarchy in MacRa. The average hop distance between a GFA and a RFA is very small and fixed.

It can, also, be seen from the figure that the proposed DFHMIP performs fairly well over

DDHMIP for high CMR users. This is due to the fact that for high CMRs, DDHMIP behaves like MIP, and the user is frequently required to register with its HA. However, this is not the case with proposed DFHMIP. Figures 10 and 11 show that DFHMIP performs better than DDHMIP irrespective of the distance between GFA and HA.

COMPARISON BETWEEN PROPOSED DFHMIP AND DHMIP

Here, the proposed DFHMIP is compared with DHMIP (Ma & Fang, 2004). Unlike DDHMIP, DHMIP optimizes the signaling cost by establishing an optimum FA hierarchy level in the network according to individual user CMR values. When a mobile host moves in this hierarchy, its new CoA is registered with the previous FA only. As shown in Figure 12, the performance of the proposed DFHMIP, in terms of total signaling cost, is very close to that of DHMIP for higher CMR values (1<CMR<10). However, this gap increases as

Figure 10. Total signaling cost vs. CMR for DFHMIP and DDHMIP; d=10

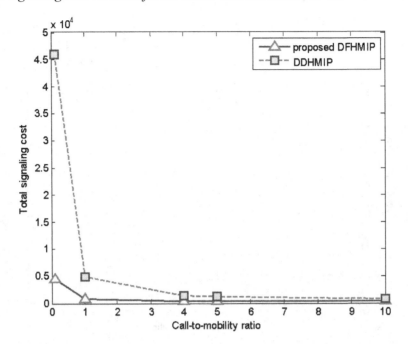

the CMR value reduces. This is because for high CMR values, the length of optimum FA hierarchy levels is very small, and the user is likely to form a new hierarchy level of FAs quite often. This increases the signaling cost of DHMIP as the user is required to update its location to the

Figure 11. Total signaling cost vs. CMR for DFHMIP and DDHMIP; d=15

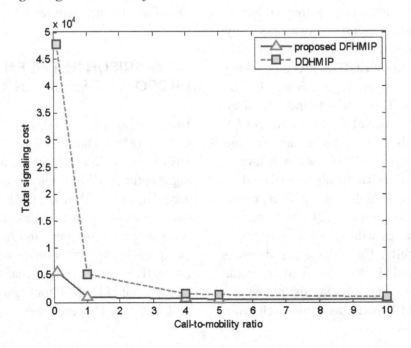

Figure 12. Total signaling cost vs. CMR for DFHMIP and DHMIP; d=2

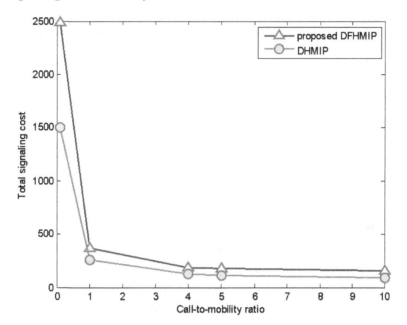

HA after every change in the hierarchy level. Like DDHMIP, DHMIP also behaves similar to MIP for very high CMR users. DHMIP constitutes long hierarchy levels of the FAs for low CMR users. This reduces the update frequency with user's HA. Using similar arguments as for DDHMIP, the signaling cost for low CMR users, in DHMIP, becomes higher. However, this increase is less as compared to that in DDHMIP.

Figure 13. Total signaling cost vs. CMR for DFHMIP and DHMIP; d=5

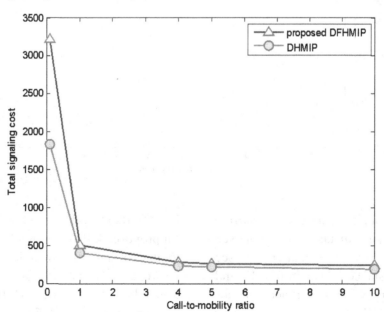

Figure 14. Total signaling cost vs. CMR for DFHMIP and DHMIP; d=10

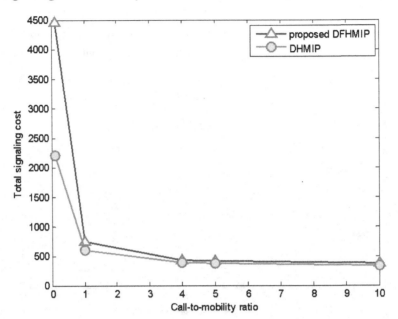

Figure 15. Total signaling cost vs. CMR for DFHMIP and DHMIP; d=15

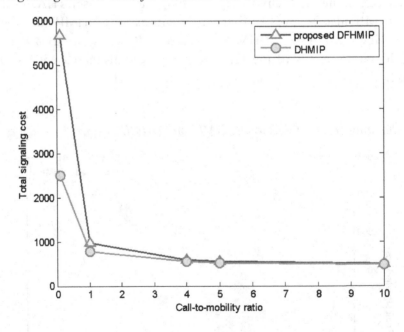

Figures 13 through 15 are used to narrate the effect of varying distances on proposed DFHMIP and DHMIP. It is encouraging to see that for higher distances between the HA and the GFA, the proposed DFHMIP performs almost similar to DHMIP for high CMRs. For low CMRs also, the increase in signaling cost for proposed DHMIP is much less as compared to that of DHMIP with increasing distance between the HA and the GFA of a mobile host. It is to be noted that in DHMIP, the first FA in the hierarchy acts as GFA.

CONCLUSION

In this paper, a new mobility management strategy for IP-based networks has been proposed. This strategy is called as Distributed and Fixed Hierarchical Mobile IP, abbreviated as DFHMIP. An analytical model for signaling cost in DFHMIP has been developed using Markov chains and random-walk mobility model. An extensive performance analysis of proposed DFHMIP in terms of registration costs as the function of subnet residence time, and packet delivery cost as the function of packet arrival rate, has been carried out. The effects of varying distances on these parameters have been examined. The proposed DFHMIP has also been compared with distributed and dynamic schemes like DDHMIP and DHMIP. It has been observed that the proposed DFHMIP performs superior, in terms of reduced signaling cost, than DDHMIP for all CMR values, and for all hop distances between the HA and the GFA. Moreover, its performance is very close to DHMIP over a wide range of high CMRs. It has also been shown that the proposed DFHMIP gives almost similar performance results, as in DHMIP, with the increasing distance between the HA and the GFA of a mobile host. The scheme can be easily implemented, and is free from the ping-pong effect.

REFERENCES

Akyildiz, I. F., & Ho, J. S. M. (1995). Mobile user location update and paging mechanisms under delay constraints. *ACM-Baltzar Journal on Wireless Networks, 1*(4), 413–425. doi:10.1007/BF01985754

Akyildiz, I. F., Xie, J., & Mohanty, S. (2004). A survey of mobility management in next-generation all-IP-based wireless systems. *IEEE Wireless Communications*, 16-28.

Bejerano, Y., & Cidon, I. (2003). An anchor chain scheme for IP mobility management. *Wireless Networks, 9*(5), 409–420. doi:10.1023/A:1024627814601

Campbell, A. T., & Castellanos, J. G. (2001). IP micro-mobility protocols. *Mobile Computing and Communications Review, 4*(4), 45–54. doi:10.1145/380516.380537

Campbell, A. T., Gomez, J., Wan, C.-Y., Kim, S., Turanyi, Z., & Valko, A. (1999). *Cellular IP.* Retrieved from http://tools.ietf.org/html/draft-ietf-mobileip-cellularip-00

Campbell, A. T., Javier, G., Kim, S., Wan, C. Y., Turanyi, Z. R., & Valko, A. G. (2002). Comparison of IP micro-mobility protocols. *IEEE Wireless Communications*, 72-82.

Das, S., McAuley, A., Datta, A., Misra, A., Chakraborty, K., & Das, S. K. (2002). IDMP: An intradomain mobility management protocol for next-generation wireless networks. *IEEE Wireless Communications*, 38-45.

Das, S., Misra, A., & Agrawal, P. (2000). TeleMIP: Telecommunications-enhanced mobile IP architecture for fast intradomain mobility. *IEEE Personal Communications, 7*(4), 50–58. doi:10.1109/98.863996

Gustafsson, E., Jonsson, A., & Perkins, C. E. (2005). *Mobile IPv4 regional registration.* Retrieved from http://tools.ietf.org/

Helmy, A.-G., Jaseemuddin, M., & Bhaskara, G. (2004). Multicast-based mobility: A novel architecture for efficient micromobility. *IEEE Journal on Selected Areas in Communications, 22*(4), 677–690. doi:10.1109/JSAC.2004.826002

Henderson, T. R. (2003). Host mobility for IP networks: A comparison. *IEEE Network, 17*(6), 18–26. doi:10.1109/MNET.2003.1248657

Huang, J., Feng, R., Liu, L., Song, M., & Song, J. (2004, June 27-29). A novel domain re-organizing algorithm for network-layer mobility management in 4G networks. In *Proceedings of the International Conference on Communications, Circuits and Systems* (Vol. 1, pp. 471-474).

Jain, R., & Lin, Y. B. (1995). An auxiliary user location strategy employing forwarding pointers to reduce network impacts of PCS. *Wireless Networks, 1*, 197–210. doi:10.1007/BF01202542

Johnson, A., & Perkins, C. (2002). *Mobile IPv4 regional registration*. Retrieved from http://tools.ietf.org/

Ma, W., & Fang, Y. (2004). Dynamic hierarchical mobility management strategy for mobile IP networks. *IEEE Journal on Selected Areas in Communications, 22*(4), 664–676. doi:10.1109/JSAC.2004.825968

Misra, A., Mcauley, A., Datta, A., & Das, S. K. (2001). *IDMP: An intra-domain mobility management protocol using mobility agents*. Retrieved from http://www.cs.columbia.edu/~dutta/research/idmp.pdf

Otsu, T., Umeda, N., & Yamao, Y. (2001, November 25-29). System architecture for mobile communications systems beyond IMT-2000. In . *Proceedings of the IEEE Global Telecommunications Conference, 1*, 538–542.

Perkins, C. (1997). Mobile IP. *IEEE Communications Magazine*, 84–99. doi:10.1109/35.592101

Perkins, C. (2002). *RFC-3220: IP mobility support for IPv4*. Retrieved from http://tools.ietf.org/html/rfc3220

Pyo, C. W., Li, J., & Kameda, H. (2003a, September 28-October 1). Simulation studies on dynamic and distributed domain-based mobile IPv6 mobility management. In *Proceedings of the 11th IEEE International Conference on Networks* (pp. 239-244).

Pyo, C. W., Li, J., & Kameda, H. (2003b, October 6-9). A dynamic and distributed domain-based mobility management method for mobile IPv6. In *Proceedings of the IEEE 58th Vehicular Technology Conference* (Vol. 3, pp. 1964-1968).

Pyo, C. W., Li, J., & Morikawa, H. (2005). Distance-based localized mobile IP mobility management. In *Proceedings of the 8th International Symposium on Parallel Architectures, Algorithms and Networks*.

Ramjee, R., La Porta, T., Thuel, S., Vardhan, K., & Salgarelli, L. (1999). *IP micro-mobility support using HAWAII*. Retrieved from http://tools.ietf.org/html/draft-ramjee-micro-mobility-hawaii-00

Ramjee, R., Varadhan, K., Salgarelli, L., Thuel, S. R., Wang, S. Y., & La Porta, T. (2002). HAWAII: A domain-based approach for supporting mobility in wide-area wireless networks. *IEEE/ACM Transactions on Networking, 10*(3), 396–410. doi:10.1109/TNET.2002.1012370

Reinbold, P., & Bonaventure, O. (2003). IP micro-mobility protocols. *IEEE Communications Surveys & Tutorials, 5*(1), 40–57. doi:10.1109/COMST.2003.5342229

Saha, D., Mukherjee, A., Misra, I. S., Chakraborty, M., & Subhash, N. (2004). Mobility support in IP: A survey of related protocols. *IEEE Network, 18*(6), 34–40. doi:10.1109/MNET.2004.1355033

Upadhyay, P. C., & Tiwari, S. (2004, June 7-9). A mobile-controlled distributed and dynamic scheme for location management in cellular IP networks. In *Proceedings of the IEEE/IFIP International Conference on Wireless and Optical Networks*, Muscat, Oman (pp. 260-263).

Upadhyay, P. C., & Tiwari, S. (2005, June 19-22). Multiple-gateway environment for tracking mobile hosts in mobile IP networks. In *Proceedings of the 3rd International IEEE Northeast Workshop on Circuits and Systems*, Quebec City, QC, Canada (pp. 304-307).

Upadhyay, P. C., & Tiwari, S. (2006, June 26-29). Location tracking in IP-based networks. In *Proceedings of the International Conference on Communications in Computing, part of The World Congress in Computer Science, Computer Engineering, and Applied Computing*, Las Vegas, NV (pp. 178-184).

Upadhyay, P. C., & Tiwari, S. (2010). Network layer mobility management schemes for IP-based mobile networks. *International Journal of Mobile Computing and Multimedia Communications, 2*(3), 47–60. doi:10.4018/jmcmc.2010070104

Valko, A. G. (1999). Cellular IP: A new approach to internet host mobility. *Computer Communication Review, 29*(1), 50–65.

Wang, Y., Chen, W., & Ho, J. S. M. (1997). *Performance analysis of mobile IP extended with routing agents* (Tech. Rep. No. 97-CSE-13). Dallas, TX: Southern Methodist University.

Xie, H., Tabbane, S., & Goodman, D. J. (1993). Dynamic location area management and performance analysis. In *Proceedings of the IEEE Vehicular Technology Conference* (pp. 536-539).

Xie, J., & Akylidiz, I. F. (2002). A novel distributed dynamic location management scheme for minimizing signaling costs in mobile IP. *IEEE Transactions on Mobile Computing, 1*(3), 163–175. doi:10.1109/TMC.2002.1081753

Xu, Y., Lee, H. C. J., & Thing, V. L. L. (2003, May 11-15). A local mobility agent selection algorithm for mobile networks. In . *Proceedings of the IEEE International Conference on Communications, 2*, 1074–1079.

Yu, L., Yu-mei, W., & Hui-min, Z. (2005, September 22-25). Modeling and analyzing the cost of hierarchical mobile IP. In *Proceedings of the International Conference on Wireless Communications, Networking and Mobile Computing* (Vol. 2, pp. 1056-1059).

Zhang, W., Jaehnert, J., & Dolzer, K. (2003). Design and evaluation of a handover decision strategy for 4th generation mobile networks. In *Proceedings of the 57th IEEE Semiannual Vehicular Technology Conference* (Vol. 3, pp. 1969-1973).

This work was previously published in the International Journal of Interdisciplinary Telecommunications and Networking, Volume 3, Issue 4, edited by Michael R. Bartolacci and Steven R. Powell, pp. 28-48, copyright 2011 by IGI Publishing (an imprint of IGI Global).

Chapter 19

The Media Gatekeeping Model Updated by R and I in ICTs:
The Case of Wireless Communications in Media Coverage of the Olympic Games

Vassiliki Cossiavelou
University of the Aegean, Greece

Evangelia Kavakli
University of the Aegean, Greece

Philemon Bantimaroudis
University of the Aegean, Greece

Laura Illia
IE University, Spain

ABSTRACT

This paper explores the influence of digital technologies on media networks, in particular how they affect the traditional gatekeeping model. Wireless communications are the hot point of all digital technologies, and their application to the transmission of the Olympic Games is a milestone for the global creative industries every two/four years. The authors argue that the research and innovation (R&I) industries' involvement with the media industries needs to be reconsidered within the framework of an updated media gatekeeping model. To investigate this research question, results are reported from a case study examining the gatekeeping processes in the 2008 Olympic Games in Beijing, and the subsequent Olympics up to 2016. Results show the need for a new gatekeeping model that takes into consideration the impact of digital technologies, especially wireless communications. Additionally, new decision models regarding innovation investment in the global media industry are suggested by the impact of R&I on the media gatekeeping model itself.

INTRODUCTION

In the course of the past fifty years, gatekeeping has emerged as one of the most influential communication theories. Kurt Lewin (1947) was the first to use the term, and subsequent scholars

based their work on his assertions in the context of different professional and academic environments. These scholars describe gatekeepers as individuals or organizations/institutions determining the volume and the type of information that can be consumed by an audience. David Manning White (1950) is credited with a seminal work in the field of media studies. White observed

DOI: 10.4018/978-1-4666-2154-1.ch019

that the individual had significant authority in deciding what information should become public. Other researchers have scrutinized the role of individuals in deciding what information should pass the "gates" (Snider, 1967; Peterson, 1981; Singer, 1997; Hollifield et al., 2001; Cohen, 2002; Plaisance & Skewes, 2003; Dimitrova et al., 2003; Wanta & Craft, 2004).

Over the past sixty years, researchers have argued about the complexity of the gatekeeping process. Certainly, gatekeeping should be analyzed beyond the role of individuals/ gatekeepers. According to Dimitrova et al. (2003),

Some practices that reduce uncertainty in making news decisions include: accepting the news definition of opinion leaders within a newsroom or on a particular beat; adopting of a group consensus through daily professional interaction; keying on output of a reference institution, such as the AP or The New York Times; accepting key sources' definition of news; and using attitudes and values of reference groups other than those in the newsroom. (2003, p. 402).

Scholars have also turned their attention toward institutional environments and cultural settings and therefore toward macro analyses, rather than simply micro/individual analyses. For example, while explaining the development of the media sociology field, Reese and Ballinger (2001) argued that European researchers tended to favour macro analysis – institutions, societies, ideologies – in opposition to their American colleagues who routinely scrutinized individuals, and professional practices (2001, p. 641). The authors recommend a holistic research endeavour deriving different data – both qualitative and quantitative – at different levels, from individuals to societies/ ideologies.

In this context, one of the milestone studies, which observed individual/ gatekeepers within their work environments is Breed's (1955) work "Social Control in the Newsroom." Breed did not limit his observations to individual behaviours but rather investigated a process of professional socialization – how a journalist adapts to the organizational environment. Breed's study is deemed significant conceptually as it reached profound conclusions on how a news organization acculturates its employees on what constitutes accepted and unaccepted professional behaviour. The study discovered that reporters were "sensing policy" and conformed to the cultural standards of the organization. These early gatekeeping projects were conducted in newspaper environments, but soon researchers turned their attention toward television (Berkowitz, 1990; Carroll, 1985) and then to the internet (Singer, 1998; Dimitrova et al., 2003).

At the media routines level, Shoemaker and Reese described professional practices and their influence on content. For example, reporters have developed news values that govern news selection and the framing of stories. Also, defensive routines such as "objectivity" allow journalists to protect themselves from accusations of perceived bias. Story structures and narratives are routinized. News stories have the same structure and format. Furthermore, the pack-journalism routine – journalists covering stories together and comparing notes – as well as deadline requirements, force journalists to work within certain boundaries (1996, p. 112). Breed's (1995) work is important for understanding organizations and their influence on media content. Organizations do not always need to define explicitly what constitutes acceptable and unacceptable behaviour. However, there are indirect ways to reward "appropriate" coverage and punish "inappropriate" journalistic behaviour. Every media organization is unique, as they enforce different policies, and promote different priorities, in terms of ownership patterns, pricing policies, business model development strategies, etc.

In terms of influences at the extra-media/ social institution level of analysis, Shoemaker and Vos (2009) refer to interest groups, markets, audiences, advertisers, sources, public relations organizations, government, other media, innovation centres, academia and news consultants. Different types of organizations affect media organizations and their content selection priorities.

Finally, the social system/ideological level of analysis influences media content in various ways. For example, the work of Gans (1979), Gitlin (2007), Herman and Chomsky (2002), Williams (1977), and Entman (2003) present evidence of ideological influences on media content, while ideology is described as a cohesive and consonant environment that exists to support established institutions and their interests.

Currently, scholars focus their interest on digital media, especially wireless communications, which is expected and justified. As traditional media converge on digital platforms, questions of new effects become prominent. Furthermore, digital media arguably challenge old theories. Some of the hypotheses deserve re-examination and re-assessment. Naturally, gatekeeping should be revisited in the context of digital media industries. A useful gatekeeping model which has guided scholarship outlines five dimensions of content filters: individual influences, professional routines, organizational influences, extra-media influences and the ideological environment (Shoemaker & Reese, 1996).

To analyze this complex interplay of influences, scholars need data from different institutions, involving different processes at different stages, from message design to public dissemination. Generating empirical evidence is essential for mapping institutional processes in the modern digitalized environment. In this same environment, the R&I Olympic sponsors need to make decisions more than seven years in advance for the matured wireless technologies to be used for the next Olympic Games, in order to invest in the technologies to be adopted by the global media networks and telecoms industries.

A NEED FOR MODEL UPDATING

The gatekeeping model of Shoemaker and Reese describes a multilevel filtering of media content, without, however, recognizing digital technologies such as a distinct information filter. Shoemaker and Vos (2009) recognize the impact of on-line media on researchers' conceptualization of gatekeeping, but offer little evidence of the impact of digital technologies on every level of media gatekeeping analysis. Why should technology be acknowledged as a distinct gatekeeping filter? And is this an argument for technological determinism? Currently, there is evidence that different information technologies increasingly become information/news filters and thereby acquire a gatekeeping function (Williams & Delli Carpini, 2004; Singer, 1998; Robinson, 2006). The digital media's gatekeeping capabilities are not accounted for in isolation, but in combination with all the other filters described in the Shoemaker/Reese model. According to Reese et al. (2007),

Admittedly, technology has altered the nature of the profession itself, but more broadly journalism has been distributed and interlinked more fluidly with citizen communication. The blogosphere provides an interweaving of these different locations as it pushes users to a network of information, views and perspectives, thus bringing a broader journalistic conversation to life. (p. 238)

The original design of the model took place before the advent of the digital world. Although analogue media technologies, such as electronic news-gathering tools (ENG) had evolved since the early 1960s, transforming the journalism profession, the model does not explicitly identify technological influences on media content. However, the model identifies extra-media pressures on media content produced, as influences of secondary importance. The technological revolution that took place in the 1960s (e.g., analogue video) enhanced the gatekeeping capabilities of news organizations in different ways (increasing the speed of collection, processing and transmission of news content) without, however, providing any gatekeeping authority to audiences. News organizations remained powerful gatekeepers, exercising almost absolute authority in defining and disseminating news content. Virtually all of the available information reaching news organizations was filtered, processed and publicized via mainstream media without significant choices provided to alternative outlets. Thus technology did not receive significant attention in the Shoemaker/Reese model, as the utilization of news-gathering technologies was considered either a professional routine or an extra-media influence on organizational production. Currently at the level of extra-media influences, there is speculation that different industries promote different types of digital technologies and therefore affect how news/content organizations process their cultural products.

As already argued, analogue media technologies have been largely controlled by media organizations. Only in rare instances, individuals not affiliated with media outlets would collect information, which subsequently received significant public attention. There are a few such examples (e.g., in 1991, Los Angeles riots were caused by an amateur video recording of police officers beating Rodney King, an African-American resident of the city). However, even in these cases, individuals still had to deliver their video content to mainstream media for dissemination.

Nowadays, gatekeeping has been challenged by the advent of digital technologies. For example, Williams and Delli Carpini (2004) argue that digital media render the gatekeeping paradigm useless, since media elites no longer have total control over the flow of information toward their publics.

The new media environment creates a multiplicity of gates through which information passes to the public both in terms of the sheer number of *sources of information (i.e. internet, cable television, radio), the speed with which information is transmitted, and the types of genres the public uses for political information (i.e. movies, music, docudramas, talk shows). (p. 1213).*

This multiplicity of choices, both in terms of media sources as well as modes of dissemination, challenges scholars to reconsider old paradigms. Blogs, for example, exercise significant pressures on mainstream news organizations. These pressures are felt in different ways by newspapers, most notably in their declining circulation and the closing of many newspaper outlets worldwide.

According to Singer (2005), since anyone can publicize information on the internet, the role of professional journalists has shifted away from "story selection and toward news judgement" (p. 179). Singer does not diminish the gatekeeping capabilities of 21st century newspapers, but recognizes that the advocacy journalism practised by mainstream media is due to pressures applied by digital non-mainstream media which challenge professional journalists in many ways, but most notably in terms of "accountability and transparency" (p. 179).

Robinson (2006) observes that mainstream media blogs represent an institutional reaction toward independent news blogs. He acknowledges a continuous change in journalists' professional practices, arguing that "blogs are a new form of journalism destined to evolve as time and broadband capabilities advance" (p. 80).

In the current article, we argue that the original gatekeeping model retains its usefulness in the 21st century, but modifications are needed, to account for current digital realities. Although technology is not the independent variable in this study, we present different digital tools as gatekeeping filters that need to be identified and scrutinized. As digital technologies alter core journalistic principles taught in journalism schools prior to the age of the internet, media organizations have a diminished gatekeeping capacity with regards to the form and the quantity of media content that is collected, processed and communicated to different publics. Any individual, group, non-media institution or national entity, which has been traditionally out of the media agenda, can publicize a version of their story/reality without conventional mediators intervening. Certain digital technologies can thereby empower citizens outside the established media organizations to promote their version of reality. This upgrading proposal entails two different components.

First, we do not propose an additional level of analysis. We do propose that technology should be considered an additional gatekeeping filter affecting each of the five previously established levels of analysis. In this revised model, the five levels, described in the literature, should not be scrutinized only as media sources but as nodes of a media gatekeeping network, in which each node is becoming, through every possible direct or indirect interconnection, a source and/or receiver of the others' nodes of filtering, through digital technology.

Second, the upgraded gatekeeping model does not pertain exclusively to journalism but to different types of digital information with different capacities in the modern media industries. Therefore, the concept of the gatekeeper does not pertain only to journalists or journalistic organizations, but different types of organizations and individuals, who under different circumstances provide digital content to different publics.

At the individual level, a multitude of technologies are expected to affect individual content creation and consumption behaviours. Many citizens of the world have editor/gatekeeper capacities as they manage semantic web metadata, become members of social networks, engage in distance learning and master web authoring techniques. Individuals are not necessarily bound by organizational pressures on what to select, unless they are media professionals in established organizations. Even in those cases, their flexibility in

selecting and comparing sources is unprecedented in the history of media. Technology does not minimize individual characteristics addressed in the traditional gatekeeping model. As a matter of fact, gender, cultural background, education, and religious background constitute significant individual influences in current content selections. However, the variety of content available, the ability to verify sources of information and the manner in which individuals interact with content and other individuals throughout the world leads to the formation of a digital techno-culture (Figure 1).

Communication routines have also been affected by technology. Although professionals are still subject to professional norms and widely accepted practices and values in the work environment, digital technologies affect the way things get done. For example, employees' familiarization with ICT constitutes a necessity for all stages of content gathering, packaging and dissemination. Furthermore, digital technologies establish new work practices, professional standards as well as codes of ethics. For example, if virtually anyone can present his/her version of reality, what is the effect of such digital capabilities on the "objectivity" routine, one of the most traditional values in western journalism? Although mainstream media organizations still socialize their personnel, "teaching" them what constitutes appropriate behaviour; one may argue that employees are socialized by technology as well. Furthermore, organizations rely on technology-savvy individuals to carry out their mission. The proliferation of digital media limits the organizations' degree of control of the gates. What an organization chooses to discard, various competing outlets or individuals may choose to publicize. Furthermore, organizational dependence on technology creates a constant urgency in establishing standards and values in order to maintain a strategic advantage.

At the social institution level, influences originating from other individuals, organizations/institutions or national and international authori-

Figure 1. The new media technologies in revisited Shoemaker & Reese model

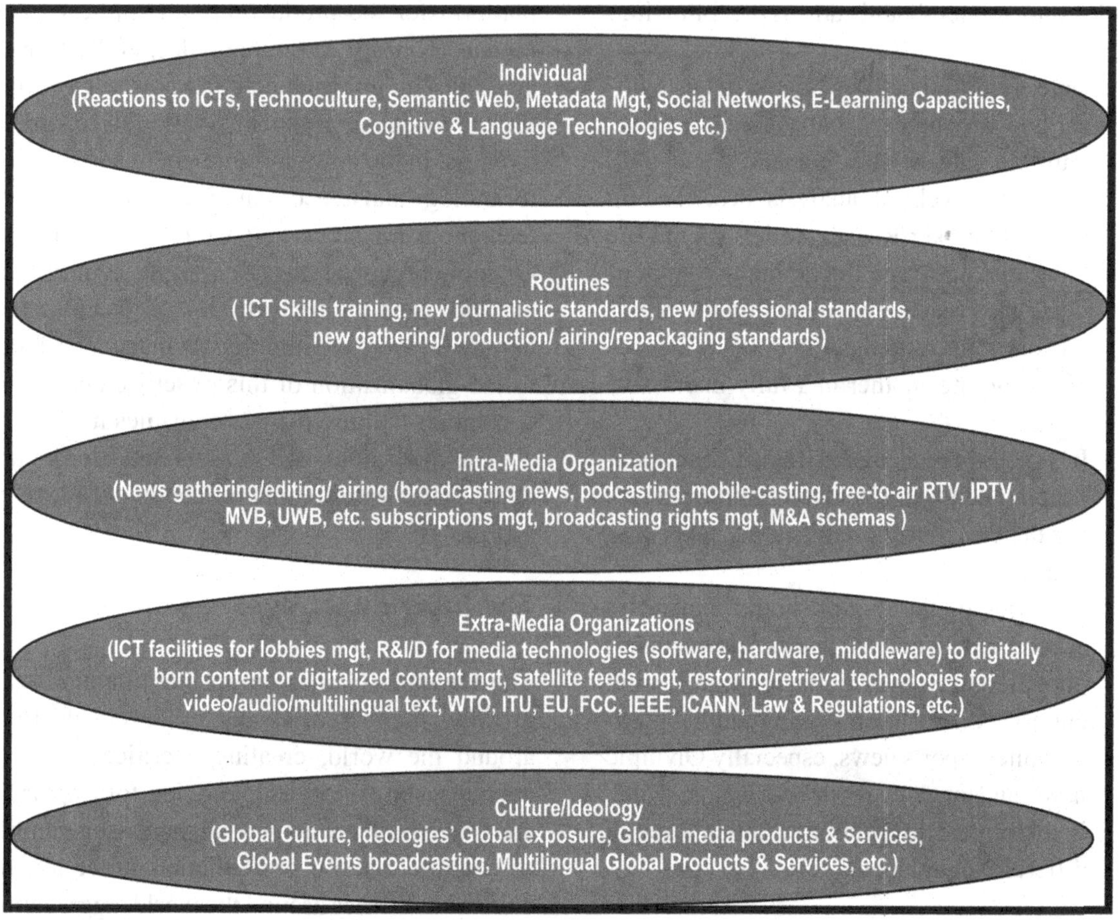

Individual
(Reactions to ICTs, Technoculture, Semantic Web, Metadata Mgt, Social Networks, E-Learning Capacities, Cognitive & Language Technologies etc.)

Routines
(ICT Skills training, new journalistic standards, new professional standards, new gathering/ production/ airing/repackaging standards)

Intra-Media Organization
(News gathering/editing/ airing (broadcasting news, podcasting, mobile-casting, free-to-air RTV, IPTV, MVB, UWB, etc. subscriptions mgt, broadcasting rights mgt, M&A schemas)

Extra-Media Organizations
(ICT facilities for lobbies mgt, R&I/D for media technologies (software, hardware, middleware) to digitally born content or digitalized content mgt, satellite feeds mgt, restoring/retrieval technologies for video/audio/multilingual text, WTO, ITU, EU, FCC, IEEE, ICANN, Law & Regulations, etc.)

Culture/Ideology
(Global Culture, Ideologies' Global exposure, Global media products & Services, Global Events broadcasting, Multilingual Global Products & Services, etc.)

ties have proliferated because of new types of organizations, which emerged after the advent of the internet. Along with advertisers, public relations officers, governments and lobbyists, ICT corporations, software and hardware designers as well as research and development corporations have mounted unprecedented pressures toward news organizations, to examine and/or adopt new digital media and to make choices about the technologies of the future. Not only well-established institutions, but informal networks of internet users put additional pressures on mainstream media, as individual behaviours change in terms of content consumption, most notably diminishing the time previously allocated to mainstream

media consumption. In this context, traditional newspapers face significant challenges, to the point of extinction.

Finally, ideology/social systems influence media content at a macro-societal level. For the first time in history, the average citizen experiences global exposure and interaction. Multiculturalism and multilingualism are widely practised, as citizens easily interact with one another. In this global context, a significant gatekeeping factor remains the diffusion of digital media and the rate of adoption of new technologies. ICT products need to conform to global standards supporting more than one language, and media personnel need to be trained in dealing with culturally diverse consumers. In summary, we propose that a revised gatekeeping model should address the following:

1. Gatekeeping should be treated not as a linear one-way process, but as an interactive, multiple-gate set of influences.
2. The five levels of analysis described in the Shoemaker/Reese model should be redesigned not as a linear, inclusive set of influences (each level including the previous one) but as nodes/filters of influences affecting one another in a fully interactive way.
3. ICTs should be recognized as a distinct gatekeeping catalyst, interacting and influencing the previously established gatekeeping nodes.
4. The uses of the model should expand to accommodate not just "news" (journalism) but different types of digital information and cultural content, such as entertainment programmes, sports news, especially Olympic news, and mega-events broadcasting, which is distributed to different publics through different digital media.

THE OLYMPIC NEWS: WHERE TECHNOLOGY SPONSORS MEET THE GLOBAL MEDIA INDUSTRIES

The relationship between media and sports and more specifically between media industries, sports organizations and cultural evolution is a symbiotic relationship (Rowe, 2005; Miller et al., 1999).

The Olympic Games are an integrated platform to check the media gatekeeping model, within the framework of globalised cultural industries, as they are one of the main cultural products of media culture. Media culture is enjoying increased visibility in global audience reports (Tomlinson, 2005), especially over the last 20 years, following the introduction of digital technologies to Olympic broadcasting.

The Olympic Games are also an integrated platform for the production and application of models (Kavakli & Loukopoulos, 2005; Beis et al., 2006) on information and knowledge within the knowledge society of innovation and competitiveness, in the new environment of information technology and telecommunications. The goal of technology transfer is very well established in the Olympic organizations (Beis et al., 2006; Ma et al., 2007) and the transmission of technological heritage in every Olympic host country contributes to the globalization of this expertise and of the consumers' culture that accompanies it.

The application of Olympic news to the updated media gatekeeping model on each level is as follows:

The Individual Level

The Olympic Games attract the attention of a growing number of media professionals from around the world, creating a critical mass of people testing new technologies for capturing and transmitting content to an increasing number of countries. These professionals move among cultural industries around the world, aggregating and applying all this experience to themselves as media users, and contributing to innovation in the Olympic coverage, and broadcasting the next Olympic Game organization.

The Practice Level

The cultural industry of the Olympics is a global industry and therefore has the characteristics of mobility of people, of services (Atkinson, 2005; Bale & Maguire, 1994) and of renegotiation of the content of their practices themselves, e.g., the profile of the sports reporter in the global media ecosystem (Rowe, 1999; Boyle, 2006, p. 40) and the prominent position of sports coverage within the fusion of news and entertainment over the last 20 years.

Previously, sports reporters were widely criticised for lack of professionalism, questionable business ethics, intentional distortion of meaning (Boyle, 2006, p. 40), and even falsification or denial of truth, lack of objectivity and of the required distance from the athletes and other sources, etc. Over time the impact of digital technology on professional practice won an increased displacement into the "historical, economic, technological and cultural context" (Boyle, 2006; Beck & Bosshart, 2003). Sports journalism and its practices were examined in the light of new possibilities for increased content production, precisely because of easy access to sources. Additionally, as entertainment became predominant in news coverage, the attitude of a journalist covering non-sport reportage did not differ ultimately in content production from sports coverage (Beck & Bosshart, 2003). Moreover, through the new technologies, the practice of coverage of dominant events and the huge global Olympic audience, the Olympic Games became a reliable and tested field of application, serving all television professional practices (close-ups, replays, slow motion, different camera angles from multiple cameras and cameras following the action) and their worldwide transmission.

The Olympic Games rely mostly on the host country, and this practice is consistent with perhaps the most important criterion for news eligibility: the promotion of the prominent position and elitism of the nation-state, at the time of hosting the Olympic Games, over the others (Tai, 2000). This perfect combination of the dramatic and the unexpected gives a ritual sense of belonging and identification with the actors, at a global level (Tomlinson, 1996).

In addition, sports content has a simple and understandable language, minimizing the differentiation of language diversity, producing emotions and tensions, images, material for public consultation (Lamprecht & Stamm, 2002, pp. 140-145) and interpretation. It requires basic journalistic values and practices such as identification and detection bias (Beck & Bosshart, 2003). The final result is media content with the added value of easy repackaging for different audiences with a variety of concepts of social participation, e.g., to the movement of the Olympic Truce (Roche, 2006). Moreover, the hosting of Olympic Games by a country attracts international interest in the culture of the country and in general on issues of human and citizens rights within the country and of the participants as well (Schaffer & Smith, 2000). In all these ways, new technologies are changing practices in sports journalism, particularly during the Olympic coverage. In all these cases, the interdependence of media professionals on the media and telecommunications technologies is obvious, and each time challenged by the local culture of the host country, both at the level of the user and the content producer.

The Intra-Media Level

The new media technologies also challenge ownership patterns of media organizations, public and private, their revenue policies, and their advertising policies. These challenges are identified in a new range of alliances between state, private and international networks. The alliances are differentiated by national and the corresponding international laws and regulations (Chan, 1993, 1994, 1996; Kishore, 1994; Vilanilam, 1996; Hudson, 1998; McChesney, 1998; Price, Verhulst, 1998; Pashupati et al., 2003) within the framework of the implementation of the Olympic Games and world coverage. In addition to this range of alliances, a complex of competing entities is developing (Shoval, 2002). The Olympic Games are a hub of international cooperation in the most advanced areas of world economy and with major market players by sector. Each host country seeks to attract technology sponsors (Cornwell & Maignan, 1998; Townley & Grayson, 1984; Witcher, Craigen, Culligan, & Harvey, 1991) and vice versa. This connection of the sponsors with the Games is a value added promotion of their brand name (Turner & Cusumano, 2000; Burgi, 1997a; Cuneen & Hannan, 1993; Lomax, 1998; McGill & Carroll, 1995; Pope & Voges, 1997; Rathie & Gaspar, 1995; Sandler & Shani, 1993; Turner et al., 1995; Wilber, 1988). The international trend towards convergence and connectivity of new technologies requires new types of sponsorship and in 2012 perhaps the requirement of the Olympic audience, and consequently of Olympic technology sponsors, might be that the content should follow not only the user of a service during the Olympics, but also the potential consumers after their closing ceremony, in terms of smart payments, secure transactions, etc., following the entire value chain of a media product. In addition, each Olympic event serves as a test bed for the convergence of ICTs and furthermore, for the development of social acceptance (social shaping of technology) (MacKenzie & Wajcman, 1985; Flichy, 1995), especially as interactive new media are leading the way for community creation and collective action (Zhao, 2007). This new reality makes the media coverage of Olympics a key event of fixed frequency (2 or 4 years) and a huge showcase of technologies, and of their acceptance, as well as of converging technologies on media production, bringing added value and a longer life cycle, to meet the time-shifted demand from virtual communities.

Particularly after the turning point of September 11, 2001, the Olympic sponsors' technology has always faced issues of national security (Feigenbaum, 2003; Donald et al., 2002) and surveillance (Zinnbauer, 2000; Hughes, 2002) as well as the information flow between intra-organizational and extra-organizational systems of all the actors involved in them.

In the production of the Games, each host country needs the collaboration of the international community, which raises issues of technological standards adoption, related to increasingly complex compatibility issues of human and citizens' rights, in several cultural contexts. Thus "the Olympics are an exemplary case" (Chalip, 1992,

p. 195, as cited in García, 2001, pp. 193-194) to examine multi-layered international marketing.

An example is News Corporation's adoption of sports programmes as a core strategic international marketing tool (Barker, 1997; Gershon, 1997; Herman & McChesney, 1997; Cossiavelou & Bartolacci, 2009) to access new markets, influence organizational structure options, and identify priorities. The selection of this strategy derived from the wide dissemination of new transmission technologies (wired and wireless) through advanced global telecommunications networks; and from the influence of a neo-liberal front that favoured privatization and deregulation (Barker, 1997), as the competition dictates expansion into existing and emerging media markets and leads to multiple global level mergers (Albarran, 2002; Croteau & Hoynes, 2001), especially in sports news production (Law et al., 2002, p. 279). Lastly, the creation of multilevel and multi-platform infotainment media groups (Bagdikian, 1997; Barker, 2000; Albarran & Chan-Olmsted, 1998; Herman & McChesney, 1997), like "communication empires" (Picard, 1996). Media mergers produce fertile territory for a boom in advertising in terms of revenues produced by time and space allocation and also of strategic assertion of broadcasting rights for top sporting events (Law et al., 2002; Miller, 2001), offset by a 'diagonal' extension, i.e., economies of diversity (Albarran & Dimmick, 1996, p. 43), vertical expansion (Holt, 1986) and breadth (Doyle, 2002). These business models are characterized by integration in terms of globalization, which costs less and less as the consumer base widens and produces an increasing investment in new content (Doyle, 2002, p. 62). The added value of the triad of sports, media and advertising has created a surplus value for sports events in the entertainment industries (Rinehart, 1994, p. 25), even for the fragmented audiences of only locally popular sports. Television broadcasting rights used to be the most important source of revenue for the Olympic movement (White, 2001), generating high advertising revenues (Jackson &

Andrews, 2005) and news content. The Olympics are increasingly migrating to new media, and in this ecosystem, the combination of public relations for the Games and advertisers increases gatekeeping (Tambini, 1999) and adds greater complexity, introducing royalties and licence fees to the sport content. The added value of these increased from over $10 billion in the early '80s to over $60 billion at the end of late '90s (Griffith, 2003; The Boston Globe, 06-06-2003). Thus, technology is emerging as a factor in media gatekeeping in terms of final product price policies in conventional, hybrid and new media industries, with global fermentation and consultation among media professionals and their global audience.

Access to media content clarifies the connection between gates and their interdependence with media technology (Zhao, 2007), especially speech processing technologies such as parallel language texts, semi-automatic translations, subtitling technologies, semiautomatic and automatic documentation and retrieval of audiovisual material, linguistic ontologies, cognitive models, voice recognition, text reproduction of voice data, etc.

The Extra-Media Level

In Olympic coverage, the extra-media effect on media gatekeeping is particularly evident. McChesney (1998) and Downey (1998) describe Olympic media coverage as continually concentrating and increasingly controlled by a limited number of corporations with large budgets and international contacts. The Olympic Games require the involvement of a growing number of countries, both participating and watching (Pashupati et al., 2003). The Olympics and all global sporting events have been observed to affect innovation in media production. The Olympics particularly trigger technological developments (Enami, 2005), which serve as gatekeepers in their organization. Technology affects extra-media gatekeeping and periodically creates mature and commercially

proven solutions as well as new and expanding technologically state of the art communities to consume advanced, mostly wireless, services.

The Olympic Games are the test bed for new technologies in different regions, focusing on the user (Rout, 2006) and the culture of the consumer once he has the most positive profile in terms of marketing: maximum consumption, time-focused, with maximized spare time (Yeoman, 2004). The Olympic Games organizers promote both mature and newly tested technologies, with the promise of implementing them in event venues throughout the host country. This is a crucial element for assessing the application files of candidate countries (Makris, 2006).

The Olympic Games provide a focus for technology development and promotion to increasingly demanding users, in terms of universal access to telecommunications products and services; and thus generate policies for each pillar of news gatekeeping (McNulty, 1993, p. 80). In transmission, where the triptych of standards includes "reliability, availability, flexibility" (Papagiannopoulos et al., 2009), an interoperable secure infrastructure is a key requirement for an Olympic event. Thus, the fourth level of gatekeeping analysis is of particular importance and possibly renders the other levels its subsets, even the cultural one. For example, even the hesitant opening of China's telecommunications led to the window of the Olympic Games, from which it became visible throughout the modern world. Norris's (1997) investigation a few years later of researchers, journalists and leaders of media industries highlights the impact of economic and other changes in telecommunications technologies on governments' media policies (Norris, 1997; Nauright, 2004). Strategic alliances between content producers (NBC, Viacom, Time-Warner), manufacturers of content receivers such as computers (Compaq, Apple, IBM, etc.), manufacturers of basic circuits (Intel), software giants (Microsoft), and consumer electronics, such as televisions and mobile phones (General Electric, Philips, Sony, etc.) have experienced increasing mobility. The consumer trend demanding that content follow it, in constant and ever-closer interaction, began a decade ago (Hamelink, 1997). The big media industries such as Turner Broadcasting, ESPN, NBC, MTV Latino, Murdoch's Fox Latin America channel, Bertelsmann etc. are trying to create new trends and consequently new markets, assisted by all these media related industries. The R&D link, recently referred to as global R&I, is evident in every Olympic hosting application (Maguire, 2005). Global cooperation in the sensitive telecommunications sector was not obvious for all WTO countries, e.g., for non-Western political systems, where regulation reforms were needed (Chan, 1994) to increase reliability and transparency. In Western societies, one newly raised issue is the arrangement and use of scarce resources, such as the frequency spectrum and digital dividend (Cossiavelou & Bantimaroudis, 2009a, 2009b).

The Ideology / Culture Level

The final result of technological gatekeeping, imaginary management, has enormous importance. In the case of Olympic Games, the national images of host and participant nations are given the widest possible exposure to the international community through cultural ceremonies (Beck & Bosshart, 2003; Riordan & Krueger, 1999; Rivenburgh, 1993; Rowe et al., 1998; Boyle & Haynes, 2000; Elias & Dunning, 1986; Hargreaves, 1982; Rowe, 1999; Wenner, 1991). The image of the host country is important not only for the promotion of the country as a brand, but also in creating a global consensus (Herman & Chomsky, 1988) for the values of leisure enhancement and management, and for the creation of the image of a global space where national borders have no functional role (Barber, 1997), through a common athletic identity despite the diversity of cultural backgrounds. The Olympic Games are part of the globalization pro-

cess, not producing a homogeneous and culturally synchronized world (Featherstone et al., 1995) but a hybridized, enriched one, contributing to the building of a global culture or world civilization (Pieterse, 1995, p. 62). Technology sponsors generate new structures of production and flow of information and knowledge, promoting ideological and political dynamics within the sponsoring institution (Bourdieu, 1986; Latour, 1987). The symbolic importance of sports events - the social interaction and identification they generate among the audiences - is increasingly recognized (Holt, 1995; Lever, 1983; Melnick, 1993; Slepicka, 1995; Sloan, 1989). The political and sports leaders of all countries, gathered for a month every two years, are treated by the international media as celebrities, (Gitlin, 2003; Andrews & Jackson, 2001), providing new popular and presumed viable media content, internationally recognisable and accepted (Larson & Park, 1993) and very attractive to sponsors and advertising companies, as the language barrier is not that important. The Olympics are periodic major campaigns to create a single market, presenting to the international audience a specific regional culture, not necessarily national, with many different races, cultures and dialects.

In the current article, the authors argue that the original gatekeeping model retains its usefulness in the 21st century, but modifications are needed if the current digital realities are to be accounted for. As digital technologies alter core journalistic principles taught in journalism schools prior the age of wireless communications, media organizations have a diminished capacity in regards to the form and the quantity of media content that is collected, processed and communicated to different publics. Any individual, group, non-media institution or national entity, which has been traditionally out of the media agenda, can publicize a version of their story/ reality without conventional mediators intervening. There is no doubt that digital technologies are primary fac-

The Media Gatekeeping Model Updated by R and I in ICTs

tors that empower citizens outside the established media organizations to promote their version of reality. In the proposed upgrading of the model, the authors do not propose digital technology as an additional level of analysis. They do propose that wireless technology should be considered as an additional gatekeeping catalyst that affects each one of the previously established five levels of analysis. In this revised model, the five levels of analysis, described in the literature, should not be scrutinized only as media sources but as nodes in the networks of the global media industry, becoming sources or receivers or both, and in every possible direct or indirect interconnection.

The interaction between the 3rd and 4th level is becoming especially important, because of the effects of wireless technology and as both content provider and broadcaster rely on technology savvy individuals to carry out the news gathering mission. The proliferation of digital media limits the media organizations' degree of control over the information. What an organization chooses to discard, various competing outlets or individuals may choose to publicize. Furthermore, organizational dependence on technology creates a constant urgency in establishing standards and values in order to maintain a strategic advantage.

In the fourth level, influences originating from other organizations/institutions have proliferated because of new types of organizations, which emerged after the advent of the Internet. Along with advertisers, public relations officers, governments and lobbyists, ICT corporations, software and hardware designers as well as R&D / R&I corporations have mounted unprecedented pressures toward news organizations, to examine and/or adopt new media tools and to make choices about the technologies of the future. Not only well-established institutions, but informal networks of internet users put additional pressures on mainstream media, as individual behaviours change in terms of content consumption, most notably diminishing the time previously allocated for mainstream media consumption.

Media Gatekeeping and ICT Systems

From an IS perspective, the literature has mainly focused on gatekeeping mechanisms for controlling information flow in networked systems. Such mechanisms fall into the following categories (Barzilai-Nahon, 2008):

- Channelling mechanisms, (e.g., search engines, directories, categorizations, hyperlinks)
- Censorship mechanisms (e.g., filtering, blocking, zoning, and deletion of information, users)
- Internationalization mechanisms (localization and translation)
- Security mechanisms (e.g., authentication controls, integrity controls, access controls)
- Cost-effect mechanisms (e.g., cost of joining, cost of usage, and cost of exiting the network)
- Value-adding mechanisms (personalization, contextualization, customization, and integration of information tools)
- Infrastructure mechanisms (e.g., network access, technology channels, and network configuration)
- User interaction mechanisms (e.g., add-on navigation tools)
- Editorial mechanisms (similar to traditional gatekeeping – e.g., technical controls, content controls, and design tools of information content)
- Regulation meta- mechanism (this mechanism is a meta-mechanism that can apply in the area of each of the other mechanisms above - e.g., state regulation of security, self-regulation of categorization of information).

In addition, a number of recent studies (Barzilai-Nahon, 2006; Beard & Olsen, 1999; Detlor, 2001) investigate the role of technological gatekeepers and their practices within the internet context (e.g., web masters, managers of virtual communities, etc). These concentrate either on developing a set of parameters to identify technological gatekeepers, or on using existing parameters to comprehend how gatekeepers affect the flow of information and sub-units' communication in organizations. A limitation of these models is that their focus is solely on individual gatekeepers' roles and professional attitudes and not on the broader context (both social and technical) that may affect gatekeeping (Barzilai-Nahon, 2008).

A gatekeeping model for conceptualizing the complex information ecosystem in technologically determined environments is necessary in order to understand the production, distribution, accessibility, ownership, selection and use of information through the lens of digital technologies. In addition, as an analysis tool, it might reveal issues influencing technology deployment in media networks.

METHODOLOGY

The main research question of the current project is as follows:

- **RQ1:** To what extent does digital technology affect content selection processes as information goes through five different levels of analysis – individual influences, professional routines, organizational structures, extra-media influences and ideology/ social systems?

The second research question is:

- **RQ2:** Is the proposed gatekeeping model affecting the adopted Olympic wireless technologies?

The authors argue that because the Shoemaker/ Reese gatekeeping model was constructed during the era of analogue media, it does not account for the current digital media, as a significant gatekeeping dimension. The authors do not propose a separate technological level of analysis, in addition to the five existing ones. They propose that in the revised model, there should be a distinct technological factor/catalyst influencing each of the five gatekeeping dimensions described by Shoemaker and Reese.

Survey

To support this proposal empirically, the authors conducted an electronic survey of gatekeepers, representing a wide spectrum of organizations, who, in every single case, are related with the previously identified five levels of gatekeeping analysis. The authors focused on certain organizations because of their significant status as global gatekeepers. The questionnaire was sent to different organizations dealing with international news media: media content providers, radio and TV networks, print media, broadcasters, news agencies and journalist unions, international consulting agencies, software, and hardware corporations, Olympic Games authorities, international news correspondents, European Union institutions, telecom operators & CE providers, wireless service operators, technology clusters, R&I centres & institutions, law and regulation authorities and university faculty related to media and ICT studies. The respondents included 49 global and local media organizations, 46 ICT operators &/or providers, 41 law and regulation experts and 78 academics who consult or do research on media and new media. All individuals who participated were included either because of their gatekeeping capacities or because of their expertise on digital media and related industries, R&I organizations. The gatekeeper concept was not treated in a narrow fashion, in the context of news workers (journalists and editors only) but encompassed different institutions, as both media and extra-media workers have gatekeeping capacities. As already established in the literature, virtually anyone can claim gatekeeper capabilities, which represents a challenge for professional gatekeepers.

The questionnaire was distributed electronically by direct email. A database kept detailed records of all returned questionnaires and was updated automatically. Respondents who indicated they did not wish to participate in the survey, were automatically removed from the list. Repeated electronic messages were sent in a time frame of a year and a half, informing the targeted population of the survey and urging their participation. The database built to support the on-line questionnaire kept accurate records of the overall visitation, the organization affiliated with every visitor and the duration of each visit.

The questionnaire was designed to include five sets of questions addressing each of the five levels of analysis, while assessing how technology affects gatekeeping decisions at each level. The questionnaire incorporated the Likert scale as a standard measurement device, utilized in similar descriptive survey projects. A convenience sample of 234 respondents representing 214 organizations participated in the survey with an array of ICT backgrounds. Generalization of the findings to a larger population is thus not intended. The authors attempt to establish whether new gatekeeping principles should be identified from a global media environment.

Analysis

The authors chose a Structural Equation Modelling (SEM) approach to analyze the data. SEM is a useful technique for dealing with multiple variables which might be inter-correlated and, under different circumstances, could lead to ambiguous interpretations. The reason for choosing SEM is that it identifies construct latent variables that are not measured directly, but are estimated in the model from many measured variables. This allows the researcher to capture explicitly the (un)reliability of measurement in a conceptual

model and avoid simplifying the measure in a simple index (either factor or mean). In the present research, the latent variables are the five dimensions of the gatekeeping model (Gefen, Straub, & Bourdeau, 2000; Anderson & Gerbing, 1988; Jöreskog, 1993; Jöreskog & Sörbom, 1993). The measured variables are the items in the questionnaire which refer to digital technologies. SEM is also very useful for testing the relationships among variables, particularly when the direction of the effects/influences is not under scrutiny. This project deals with a broad research question, trying to map a previously uncharted environment. We used SEM purely as an exploratory factor analysis. We refrained from traditional regression analyses, since there is no clear direction of effects /influences and there are no specified relationships among variables. We intend to assess whether a technological filter can be recognized as a distinct functional influence on the existing five-level model. Although certain variables might be correlated, this will not pose a problem for this analysis, if relationships among variables are real and the model has a good fit. Earlier authors have used data with AMOS 17 (Byrne, 2001) by running SEM measurement reflective measurement models (Jarvis, Mackenzie, & Podsakoff, 2003).

Variables

A total of 21 questions/statements were analyzed dealing with all five dimensions of the Shoemaker/Reese model. Respondents were asked to record their agreement or disagreement with statements provided on-line, in a Likert scale from one to five, one meaning strong disagreement and five displaying strong agreement. Choosing zero (not applicable) meant they lacked the relevant professional background to answer the question. The questionnaire was designed in such a way as to include all five dimensions of the model, namely, the individual level, routines, organization, extra-media influences and ideology/social system.

At the individual level, five questions/statements were included as follows:

- The influence of news production and delivery technologies (NPDT) on news ethics (claiming objectivity, giving conflicting evidence, presenting supportive facts, quoting) increases rapidly.
- Behind-the-scenes personnel (visual designers and channel/network technicians or technologically aware staff, media financial counsellors, etc.) have joined the ranks of the news gatekeepers due to changes in technology.
- The role of news professionals becomes more explicative (more graphics, virtual reconstruction, etc.) due to the use of NPDT facilities.
- The dependency on broadcasting NPDT experts increases, especially for media coverage/delivery of sports news.
- During the past decade, the source-journalist relationship has become more independent from the journalist-news industry relationship due to the deployment of new NPDTs (satellite-based videophones, wireless cameras, wireless internet, etc.).

At the routines level, six questions were included as follows:

- Career evolution into media organizations is related to the amount of continuous training on new NPDTs by the personnel working in news-gathering and news production.
- To what extent do you agree with the following statement: the routines of news production (deadlines, decision points and standards) are increasingly affected by NPDTs (broadband internet, WiMAX, mobile TV, etc.).

- Target audiences are formulated according to their familiarity with news-receiving gadgets/equipment (mobile phones, decoders, etc.).
- Advances in broadcasting news technologies have changed the nature of news worthiness (timeliness, proximity, importance, impact, consequence, interest, conflict, controversy, sensationalism, prominence, novelty, oddity, unusual, truthfulness) giving more importance to ease of capture and visual availability.
- The use of telecommunications technologies (web/mobile/satellites) leads to a decline in news production cost.
- Your R&D related departments observed new developments related to the coverage and production of visually attractive and interactive news products for the Beijing Olympics 2008 and successive Olympics.

At the organizational level, two questions were included as follows:

- Corporate policies, and ownership patterns, have changed because of telecommunications NPDT penetration into the nature of news reporting and production.
- The autonomy of remote offices in the news industries has increased due to the globalised telecommunications infrastructure.

At the extra-media level, three questions were included as follows:

- The impact of emerging telecommunications technologies (i.e., GPRS, MIMO, DWDM, etc.) increasingly changes the methods used for monitoring the demographics of news audiences.
- Nowadays, the dependency on paid external sources of information (suppliers) has

become less than prior to the year 2000 due to new technological capabilities available to the public by blogs, photo/video-exchange, online communities, etc.

- An advanced R&D strategy that includes wireless technologies is an important factor for the creation of partnerships in the news industry.

At the ideology level, five questions were included as follows:

- The degree to which new digital technologies are adopted by an audience influences the content of the media.
- The impact of new NPDTs on media content is a crucial factor for understanding the nature and importance of the content's effects on people and society.
- On a given story, the coverage using NPDTs will be less ideologically charged than the coverage by conventional media (i.e., sports covered by virtual on-field cameras, etc.).
- The geographic range of a network (R/TV, satellite, wireless, etc.) is proportional to the standardization applied to its content.
- The greater the cooperation among international R&D players in the wireless broadcasting industry, the greater their influence on news production (see http://www.cossiavelou.gr).

Not all parts of the five-dimensional questionnaire are equally distributed in term of the number of questions. That is because certain items were dropped after the pre-test phase, during which the feedback gathered led to appropriate omissions. Those 21 items were entered into the model. After the initial screenings, several items that did not register as statistically significant were removed from the model. Since the functionality of the model is at question, the authors intended to re-

tain only variables that contributed to the overall explanation of how different levels of analysis interacted with one another

THE FINDINGS AND THE RESULTS OF THE RESEARCH

The Findings of SEM/AMOS Analysis of Quantitative Research

SEM is a useful technique for dealing with multiple variables that might be inter-correlated, which under different circumstances could lead to ambiguous interpretations. For example, multicollineary is often observed in linear regression analysis, rendering regression coefficients unstable and their interpretation problematic. In this project, we are dealing with a broad research question. Therefore, we assess the overall functionality of the five dimensional models when the role of digital technologies is examined in every distinct level of analysis. As we assessed how different variables at different levels of analysis contributed to the overall stability of the model, we analyzed data with AMOS 17 (Byrne, 2001) by running SEM measurement reflective models (Jarvis, Mackenzie, & Podaskoff, 2003). Figure 2 presents graphically the refined model after omitting all problematic variables

Notes to Figure 2

All variances, covariances and factor loadings of latent variables are $p<.001$.

Significant correlations between measurement errors do not appear in the figure. We found a correlation of .27 between e12-e16; a correlation of .20 between e13-e4; a correlation of .20 between e6-e4.

The results show that the variables dealing with digital technologies used for the proposed

Figure 2. SEM measurement model

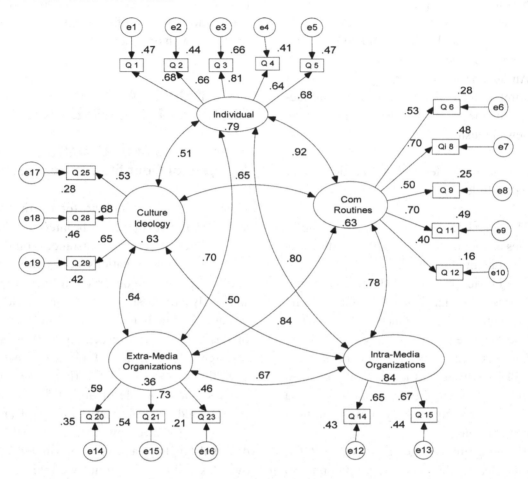

upgrading of the Shoemaker and Reese model represent a reliable additional dimension.

Factor loadings are all significant and always register above .40 (p < 0.001), and the items variance (R-square) always register above .16. There is also a good construct's variance, that is, no R-square of the five dimensional latent variables falls below .36. This confirms the existence of a reflection of the five latent constructs in the measurement items we proposed in the questionnaire. Finally, the Goodness of Fit Indices (GFIs) support our proposed measurement model. GFIs certify the degree of appropriateness and truthfulness of a model, in this case, the updated measurements of the revised gatekeeping model. From a statistical point of view, these indices measure whether,

given the measurement values, the model implies a variance and a covariance suitable to those of the population. In this case the Cmin value is not three times higher than the degrees of freedom (chi-square=169.494 /df=122 that is a proportion of 1.389). Furthermore, the values of the Root Mean Square Error of Approximation (RMSEA), .041, is appropriate because it is lower than .050. The p-value is .003 (thus near to the optimal value of .000) and the pclose value is .848 (thus near to the optimal value of 1.). Finally the Akaike Information Criterion (AIC) and Browne-Cudeck Criterion (BCC) theoretical indices are good, since their values are lower for the default model (that is our tested model) than the saturated one (AICdefault=267.49 ; AICsaturated=342.000;

BCCdefault=276.195; BCCsaturated=372.264). This indicates that altogether the tested model has a high informative value with regard to the theoretical concepts under scrutiny.

The Results of Descriptive Statistical Analysis

On the research question about the driving force of Olympic Games broadcasting into the R&D/I sectors (no 33: The Summer/Winter Olympic Games), 62.60% of all the respondents agreed that the Summer and the Winter Olympic Games serve as a driving force to promote R&D in NPDTs to create ever-more innovative coverage. Adding a modest 23.00% showed that 85.60% agree to some extent that sports mega-events are turning points for the activation of all global innovation capacity in academic and industrial R&D, to maximize the benefits of coverage.

On the research question about the new media Olympic coverage monitoring by organization (no 13: "Your R&D related departments observed new developments related to the coverage and production of visually attractive and interactive news products for the Beijing Olympics 2008 and successive Olympics"), 52.6% of respondents agreed, and only 8.7% disagreed. As the expertise of respondents is split into 26% in media, 10% finance, 18% management, 23% ICT, 18% tele-communications and 5% law, this finding claims that the majority of media related professionals are following up the Olympic Games to predict evolutions in the global media landscape.

On the research question about the Beijing Olympic revenues & broadcasting rights in comparison with the Athens 2004 revenues (no 22: "The Beijing Olympics 2008 is predicted to attract more consumers of news products than Athens Olympics 2004, in term of revenues by sports rights, advertisements and on-demand services"), 46.5% were strongly in favour, but a large percentage of 20.4% did not answer and 25.7% made just a modest assertion that the Olympic

organizational revenue will increase because of broadcasting rights This rate is high, as Olympic broadcasting is traditionally dedicated to specific networks (NBC, ESPN, via EBU, etc.) and there-fore the high cost of these rights are predicted to be refundable. Nevertheless, the new media are more demanding in terms of interactivity, and migration to them seems to be already under consideration by global media professionals.

On the research question about the dominant technology investment from 2008 to 2016, in the Asian, North or South American and European Summer/Winter host candidates, the prominent NPDT and major result of innovation and invest-ment for news services (WiMax, IPTV and/or successors, void, 3G and/or next G, semantic web/ fully interactive (social tagging, etc.), convergence at all levels (devices, technologies, etc.), secure delivery (e/m services, etc.) with full optimiza-tion and heuristics (human-devices interaction, computer science, etc.), the majority of replies, even the limited ones, in all cases not only got approved by the next host cities, having chosen the "right" continent, but even the finally proven technology of transmission. This could mean that the global players are proving that media technol-ogy is leading the way to the next successful bid to host the Olympic Games, that the Olympic Games is a functional tool to consolidate proven media technologies worldwide, and that is a great case study to examine the updated media gatekeeping model, with the catalytic role of technology.

The findings of qualitative research are giving strength to the findings of the quantitative research. The results of the in-depth interviews with 9 global media professionals answering more than 100 questions in total, proved the importance of updating the media gatekeeping model. As all of them had deep knowledge and personal experi-ence of the Beijing Olympic Games 2008, their references are of great value for this project. All of them agree that the gatekeeping process will not remain the same and that a dynamic model must be applied in the global media landscape. Both the findings of the questionnaire and the in-depth interviews with global media industry leaders also illuminate new avenues for future research and identify new questions about specific time periods and about specific media technologies.

CONCLUSION

Initial evidence from a global survey based on a convenience sample of gatekeepers examined the role of different digital technologies at the five levels of analysis, following previous findings described in the Shoemaker and Reese model. This preliminary evidence seems to indicate that a revision of this model, designed during a period of analogue media, is needed to account for technological innovations in the context of different digital information industries.

In the current project, the authors argue that a sixth dimension needs to be included in a revised model which intervenes in the nature of information and affects digital media content at all five levels. Adding a sixth dimension on top of the other five does not constitute an adequate representation of how information gathering, production and dissemination are affected by technology. However, the upgraded model should present two-way relationships, recognizing that information sources and receivers can switch roles randomly, while recognizing new roles for established media institutions. The authors propose a new map of relationships that emerge incrementally in the process. The revised model should provide space for continuous change as technology changes the nature of broadcasting, while initiating new narrowcasting processes. This new design should be supported by different studies accumulating empirical evidence, while leading to a model revision representing a more accurate array of processes which scholars can utilize efficiently in different media markets and cultures.

For example, in terms of the organizational level, technologies seem to be influencing corporate policies as well as the degree of autonomy exercised by remote offices. Digital technologies are described as very important at the extra-media level. Technological dependence on R&D/R&I corporations constitutes a significant extra-media influence added on to previous extra-media pressures, such as lobbies, governments, public relations firms and advertisers. Extra-media pressures have proliferated due to the advent of digital media. Furthermore, there is preliminary evidence demonstrating the importance of international institutions such as ITU, IEEE, and ICANN. These international authorities represent additional extra-media influences.

With the current essay, the authors wish to generate a discussion while arguing that the traditional gatekeeping model is not obsolete. However, an upgrading is essential, recognizing the cataclysmic impact of the digital media on gatekeeping practices worldwide. To achieve those goals, random samples of gatekeepers drawn from different organization/institutional environments are needed to identify influences at all levels. Furthermore, qualitative evidence, utilizing in-depth interviews or testimonies derived from participant observations can complement the existing quantitative evidence providing an accurate picture of processes and practices. Future research utilizing a variety of data should shed more light complementing the current proposal. Additional empirical evidence is needed to assess not only the role of technology as an independent variable but to identify with care gatekeeper differences at the different levels of analysis.

Because of the nature of the data – convenience sample – and the acknowledged difficulty to instigate participation in on-line surveys of this scope, the authors are hesitant to generalize these results to a larger population. Future research utilizing a variety of data should shed more light complementing the current evidence.

On the second research question, the proposed gatekeeping model indicates that there are strong and very strong relations among the intra and extra-media influences (Figure 2) in media industries because of new media technologies. Additionally, the Olympic Games are used by both media organizations and ICT organizations as a test bed for mature wireless and other media technologies in a standard period of time. The R&I organizations - academic and industrial - of both industries - media and ICT- use this time milestone as a driving force to promote R&D into ever-more innovative coverage, investing in it order to maximize the benefits of the coverage. As the expertise of the respondents is split into 26% in media, 10% finance, 18% management, 23% ICT, 18% telecommunications and 5% law, these findings claim that the majority of professionals of media related new technologies are following up the Olympic Games to predict evolutions in the global media landscape.

While Olympic broadcasting rights skyrocket in every new organization, the new media are more demanding in terms of interactivity and increased migration to them seems to be already under consideration by global media professionals.

As the Olympic Games are used as a tool to access global markets, the dominant technology, especially the dominant wireless technology to be promised and/or implemented by the host city and country is another very important point of each candidacy. The R&I of each candidate must therefore have incorporated the global wireless experience, and also promote it into the next 7 years, the period between the nomination of the next Olympic city and the opening ceremony. It seems that the respondents have a clear connection in their minds between the next proven transmission technology and the success of each candidacy.

Thus, the need for an updated media gatekeeping model has not only great theoretical importance, but is a key point for the decision-making procedure in the most dynamic project for the global media industries, the coverage of the Olympic Games. R&I (extra-media influence) lead the way in coverage management by each media industry (intra-media influence), as wireless technologies are focus points for each new Olympic organization worldwide.

ACKNOWLEDGMENT

To Mrs. Yan Mei, Director of News Corp. Asia, Beijing, China and ex CNN Director Asia (CHINA) and her colleagues, for their invaluable support in collecting data from several media and telecoms industries around the world. To Dr. Spilios Makris, Director, Olympic Program Network Reliability & Risk Services, for his generous support in the early stages of this research and his pertinent advice on the Olympic Games technologies space.

REFERENCES

Albarran, A. B. (2002). *Media economics* (2nd ed.). Ames, IA: Iowa State Press.

Albarran, A. B., & Chan-Olmsted, S. M. (1998). *Global media economics – Commercialization, concentration and integration of world media markets*. Ames, IA: Iowa State University Press.

Albarran, A. B., & Dimmick, J. (1996). Economies of multiformity and concentration in the communication industries. *Journal of Media Economics*, *9*, 41–49. doi:10.1207/s15327736me0904_3

Anderson, J. C., & Gerbing, D. W. (1988). Structural equation modelling in practice: A review and recommended two-step approach. *Psychological Bulletin*, *103*, 411–423. doi:10.1037/0033-2909.103.3.411

Andrews, D. L., & Jackson, S. J. (2001). *Sports stars: The cultural politics of sporting celebrity*. New York, NY: Routledge. doi:10.4324/9780203463543

Atkinson, P. (2005). Qualitative research—Unity and diversity, forum qualitative Sozialforschung/Forum. *Qualitative Social Research, 6*(3), 26. Retrieved from http://nbn-resolving.de/urn:nbn:de:0114-fqs0503261

Bagdikian, B. H. (1997). *The media monopoly* (5th ed.). Boston, MA: Beacon.

Bale, J., & Maguire, J. A. (1994). *The global sports arena: Athletic talent migration in an interdependent world.* Staffordshire, UK: University of Keele.

Barber, B. R. (1997). The new telecommunications technology: Endless frontier or the end of democracy? *Constellations, Special Section: Democratizing Technology/Technologizing Democracy, 4*(2), 208-228.

Barker, C. (1997). *Global television: an introduction.* Oxford, UK: Wiley-Blackwell.

Barker, C. (2000). *Cultural studies: Theory and practice.* London, UK: Sage.

Barzilai-Nahon, K. (2006). Gatekeeping in virtual communities: On politics of power in cyberspace. In *Proceedings of the 39th Hawaii International Conference on System Sciences.*

Barzilai-Nahon, K. (2008). Toward a theory of network gatekeeping: A framework for exploring information control. *Journal of the American Society for Information Science and Technology, 59*(9), 1493–1512. doi:10.1002/asi.20857

Beard, F., & Olsen, R. L. (1999). Webmasters as mass media gatekeepers: a qualitative exploratory study. *Internet Research: Electronic Networking Applications and Policy, 9*(3), 200–211. doi:10.1108/10662249910274601

Beck, D., & Bosshart, L. (2003). Sports, media, politics, and national identity. *Communication Research Trends, 22*(4), 1–43.

Beis, D. A., Loucopoulos, P., Pyrgiotis, Y., & Zografos, K. G. (2006). PLATO helps Athens win gold: Olympic Games knowledge modeling for organizational change and resource management. *Interface, 36*(1), 26–42. doi:10.1287/inte.1060.0189

Berkowitz, L. (1990). On the formation and regulation of anger and aggression. A cognitive-neoassociationistic analysis. *The American Psychologist, 45*(4), 494–503. doi:10.1037/0003-066X.45.4.494

Bourdieu, P. (1986). *Distinction: A social critique of the judgement of taste.* London, UK: Routledge.

Boyle, R. (2006). *Sports journalism: Context and issues.* London, UK: Sage.

Boyle, R., & Haynes, R. (2000). *Power play: Sport, the media, and popular culture.* Harlow, UK: Longman.

Breed, W. (1955). Social control in the newsroom: A functional analysis. *Social Forces, 33*(4), 326–335. doi:10.2307/2573002

Burgi, M. (1997). TV exec. sees virtual signs. *Mediaweek, 7*(6), 13.

Byrne, B. M. (2001). *Structural equation modeling with AMOS—Basic concepts, applications, and programming.* Mahwah, NJ: Lawrence Erlbaum.

Carroll, R. L. (1985). Content values in TV news programs in small and large markets. *The Journalism Quarterly, 62*(4), 877–938.

Chalip, L. (1992). The construction and use of polysemic structures: Olympic lessons for sport marketing. *Journal of Sport Management, 4,* 87–98.

Chan, J. M. (1993). Commercialization without independence: Trends and tensions of media development in China. In Cheng, J., & Brosseau, M. (Eds.), *China review.* Hong Kong: Chinese University Press.

Chan, J. M. (1994). Media internationalization in China: Processes and tensions. *The Journal of Communication, 44*(3), 70–88. doi:10.1111/j.1460-2466.1994.tb00689.x

Chan, J. M. (1996). Television development in Greater China: Structure, exports, and market formation. In Sinclair, J., Jacka, E., & Cunningham, S. (Eds.), *New patterns in global television: Peripheral vision* (pp. 126–160). Oxford, UK: Oxford University Press.

Cohen, E. (2002). Online journalism as market-driven journalism. *Journal of Broadcasting & Electronic Media, 46*(4), 532–548. doi:10.1207/s15506878jobem4604_3

Cornwell, T. B., & Maignan, I. (1998). An international review of sponsorship research. *Journal of Advertising, 27*(1), 1–21.

Cossiavelou, V. (2009). Digital dividend aware business models for the creative industries: Challenges and opportunities in EU Market. In Lytras, M. D., De Pablos, P. O., & Damiani, E. (Eds.), *Best practices for the knowledge society - Knowledge, learning, development and technology for all*. New York, NY: Springer. doi:10.1007/978-3-642-04757-2_60

Cossiavelou, V., & Bantimaroudis, Ph. (2009). Mediation of the message in a wireless global environment: revisiting the media gatekeeping model. *International Journal of Interdisciplinary Telecommunications and Networking, 1*(4), 37–53. doi:10.4018/jitn.2009092803

Cossiavelou, V., & Bartolacci, M. (2009). News corporation: Facing the wireless world of the 21st century. In Powell, S., & Shim, J. P. (Eds.), *Wireless technology applications, management and security* (*Vol. 44*, pp. 83–90). Berlin, Germany: Springer-Verlag.

Croteau, D., & Hoynes, W. (2000). *Media society: Industries, images, and audiences*. Thousand Oaks, CA: Pine Forge.

Cuneen, J., & Hannan, M. J. (1993). Intermediate measures and recognition testing of sponsorship advertising at an LPGA tournament. *Sport Marketing Quarterly, 2*(1), 47–56.

Detlor, B. (2001). The influence of information ecology on e-commerce initiatives. *Internet Research: Electronic Networking Applications and Policy, 11*(4), 286–295. doi:10.1108/10662240110402740

Dimitrova, D., Connolly-Ahern, C., Williams, A. P., Kaid, L. L., & Reid, A. (2003). Hyperlinking as gatekeeping: Online newspaper coverage of the execution of an American terrorist. *Journalism Studies, 4*(3), 401–414. doi:10.1080/14616700306488

Downey, J. (1998). Full of Eastern promise? Central and Eastern European media after 1989. In Thussu, D. K. (Ed.), *Electronic empires: Global media and local resistance* (pp. 47–62). London, UK: Arnold.

Doyle, G. (2002). *Understanding media economics*. London, UK: Sage.

Elias, N., & Dunning, E. (1986). *Quest for excitement: Sport and leisure in the civilizing process*. Oxford, UK: Basil Blackwell.

Enami, K. (2005). Future of home media. In *Proceedings of the 13th Annual ACM International Conference on Multimedia*, Hilton, Singapore.

Entman, R. M. (2003). Cascading activation: Contesting the White House's frame after 9/11. *Political Communication, 20*, 415–432. doi:10.1080/10584600390244176

Featherstone, M., Scott, L., & Roland, R. (1995). *Global modernities*. London, UK: Sage.

Feigenbaum, E. A. (2003). Some challenges and grand challenges for computational intelligence. *Journal of the ACM, 50*(1), 32–40. doi:10.1145/602382.602400

Flichy, P. (1995). *L'innovation technique: récents développements en sciences sociales: vers une nouvelle théorie de l'innovation*. Paris, France: Découverte.

Gans, H. J. (1979). *Deciding what's news: A study of CBS Evening News, NBC Nightly News, Newsweek, and Time*. Chicago, IL: Northwestern University Press.

García, B. (2001). Enhancing sport marketing through cultural and arts programs: Lessons from the Sydney 2000 Olympic arts festivals. *Sport Management Review, 4*, 193–219. doi:10.1016/S1441-3523(01)70075-7

Gefen, D., Straub, D., & Bourdeau, M. (2000). Structural equation modelling and regression: Guidelines for research practice. *Communications of AIS, 4*(7), 1–79.

Gershon, R. A. (2000). The transnational media corporation: Environmental scanning and strategy formulation. *Journal of Media Economics, 13*(2), 81–101. doi:10.1207/S15327736ME1302_3

Gitlin, T. (2003). *The whole world is watching: Mass media in the making and unmaking of the new left*. Berkeley, CA: University of California Press.

Gitlin, T. (2007). *Media unlimited: How the torrent of images and sounds overwhelms our lives*. New York, NY: Henry Holt and Co.

Griffith, B. (2003). NBC spends $2.201 billion for broadcast rights for 2010, 2012 Olympic Games. *The Boston Globe*.

Hamelink, C. J. (1997). *New information and communication technologies, social development and cultural change*. Geneva, Switzerland: United Nations Research Institute for Social Development.

Hargreaves, J. (Ed.). (1982). *Sport, culture, and ideology*. London, UK: Routledge & Kegan Paul.

Herman, E., & Chomsky, N. (2002). *Manufacturing consent: The political economy of the mass media*. New York, NY: Pantheon Books.

Herman, E., & McChesney, R. (1997). *The global media: The new missionaries of global capitalism*. London, UK: Cassell.

Hollifield, C., Kosicki, G., & Becker, L. (2001). Organizational vs. professional culture in the newsroom: Television news directors' and newspaper editors' hiring decisions. *Journal of Broadcasting & Electronic Media, 45*(1), 92–117. doi:10.1207/s15506878jobem4501_7

Holt, R. (1989). *Sport and the British – A modern history*. Oxford, UK: Clarendon Press.

Hudson, H. E. (1998). The paradox of ubiquity: Communication satellite policies in Asia. In MacKie-Mason, J. K., & Waterman, D. (Eds.), *Telephony, the Internet and the media, selected papers from the 1997 telecommunications policy research conference*. Mahwah, NJ: Lawrence Erlbaum.

Hughes, C. R. (2002). China and the globalization of ICTs: Implications for international relations. *New Media & Society, 4*(2), 205–224. doi:10.1177/14614440222226343

Jackson, S. J., & Andrews, D. L. (2005). *Sport, culture and advertising: Identities, commodities and the politics of representation*. New York, NY: Routledge. doi:10.4324/9780203462003

Jarvis, C. B., Mackenzie, S. B., & Podsakoff, P. M. (2003). A critical review of construct indicators and measurement model misspecification in marketing and consumer research. *The Journal of Consumer Research, 30*(3), 199–218. doi:10.1086/376806

Jöreskog, K. G. (1993). *Testing structural equation models*. Newbury Park, CA: Sage.

Jöreskog, K. G., & Sörbom, D. (1993). *LISREL 8: Structural equation modelling with the SIMPLIS command language*. Mahwah, NJ: Lawrence Erlbaum.

Kavakli, E., & Loucopoulos, S. (2005). Goal modeling in requirements engineering: Analysis and critique of current methods. In Krogstie, J., Halpin, T., & Siau, K. (Eds.), *Information modeling methods and methodologies* (pp. 102–124). London, UK: Idea Group.

Kishore, K. (1994). The advent of STAR TV in India: Emerging policy issues. *Media Asia, 21*(2), 96–103.

Lamprecht, M., & Stamm, H. (2002). *Sport zwischen Kultur, Kult und Kommerz*. Zürich, Switzerland: Seismo.

Larson, J. F., & Park, H.-S. (1993). *Global television and the politics of the Seoul Olympics*. Boulder, CO: Westview Press.

Latour, B. (1987). *Science in action: How to follow scientists and engineers through society*. Cambridge, MA: Harvard University Press.

Law, A., Harvey, J., & Kemp, S. (2002). The global sport mass media oligopoly. *International Review for the Sociology of Sport, 37*(3-4), 279–302. doi:10.1177/1012690202037004025

Lever, J. (1983). *Soccer madness*. Chicago, IL: University of Chicago Press.

Lewin, K. (1958). Group discussion and social change. In Macoby, E. E., Newcomb, T. M., & Hantley, E. L. (Eds.), *Readings in social psychology* (3rd ed., pp. 197–212). New York, NY: Holt, Rinehart & Winston.

Lomax, H., & Casey, N. (1998). Recording social life: Reflexivity and video methodology. *Sociological Research Online, (3)*: 3–32.

Ma, X., Xiong, J., Gui, L., Liu, C., & Zhang, W. (2007). EU-China roadmap for mobile multimedia industry: the preliminary results. In *Proceedings of the IEEE 2nd International Conference on Communications and Networking in China*.

MacKenzie, D., & Wajcman, J. (Eds.). (1999). *The social shaping of technology*. Buckingham, UK: Open University Press.

Maguire, J. A. (2005). *Power and global sport: Zones of prestige, emulation and resistance*. New York, NY: Routledge.

Makris, S. E. (2006). *Telecommunications for the Athens 2004 Olympic Games: Lessons learned from a consultant's perspective*. Paper presented at the IEEE International Communications Quality and Reliability (CQR) Workshop on Olympic Program Network Reliability & Risk Services.

McChesney, R. (1998). Political economy of communication. In McChesney, R., Wood, E., & Foster, J. B. (Eds.), *Capitalism and the information age*. New York, NY: Monthly Review.

McGill, I., & Carroll, I. (1995). The negotiation and sale of television and other rights associated with a sporting event. In Fewell, M. (Ed.), *Sports law: A practical guide* (pp. 103–165). Sydney, NSW, Australia: LBC Information Services.

McNulty, T. J. (1993). Television's impact on executive decision-making and diplomacy. *Fletcher Foreign World Affaires, (17)*, 67-83.

Melnick, M. (1993). Searching for sociability in the stands: A theory of sports spectating. *Journal of Sport Management, 7*, 44–60.

Miller, T., Lawrence, G., McKay, J., & Rowe, D. (1999). Modifying the sign: sport and globalization. *Journal of Sport and Social Issues, 20*(3), 278–295.

Miller, T., Lawrence, G., McKay, J., & Rowe, D. (2001). *Globalization and sport*. London, UK: Sage.

Nauright, J. (2004). Global games: culture, political economy and sport in the globalised world of the 21st century. *Third World Quarterly, 25*(7), 1325–1336. doi:10.1080/014365904200281302

Norris, P. (1997). *Politics and the press: The news media and their influence.* Boulder, CO: Rienner.

Papagiannopoulos, P., Xenikos, D. G., & Vouddas, P. (2009). Event management and group communications: The case of the 2004 Olympic Games in Athens. *Event Management, 13*(2), 103–116. doi:10.3727/152599509789686281

Pashupati, K., & Lee, J. H. (2003). Web banner ads in online newspaper: A crossnational comparison of India and Korea. *International Journal of Advertising, 22,* 531–564.

Pashupati, K., Sun, H. L., & McDowell, S. D. (2003). Guardians of culture, development communicators, or state capitalists? A comparative analysis of Indian and Chinese policy responses to broadcast, cable and satellite television. *Gazette: The International Journal for Communication Studies, 65*(3), 251–271. doi:10.1177/0016549203065003003

Peterson, S. (1981). International news selection by the elite press: A case study. *Public Opinion Quarterly, 45*(2), 143–163. doi:10.1086/268647

Picard, R. W. (1996). *A society of models for video and image libraries.* Cambridge, MA: MIT Press.

Pieterse, J. N. (1995). Globalization as hybridization. In Featherstone, M., Lash, S., & Robertson, R. (Eds.), *Global modernities* (pp. 45–68). London, UK: Sage.

Plaisance, P., & Skewes, E. (2003). *Personal and professional dimensions of news work: Exploring the link between journalists' values and roles.* Paper presented at the Annual Meeting of the International Communication Association, San Diego, CA.

Pope, N. K. L., & Voges, K. E. (1997). An exploration of sponsorship awareness by product and message location in televised sporting events. *Cyber Journal of Sport Marketing, 1*(1), 16–27.

Price, M. E., & Verhulst, S. G. (Eds.). (1998). *Broadcasting reform in India: Media law from a global perspective.* Oxford, UK: Oxford University Press.

Rathie, K., & Gaspar, T. (1995). Sponsorship agreements. In Fewell, M. (Ed.), *Sports law: A practical guide* (pp. 78–102). Sydney, NSW, Australia: LBC Information Services.

Reese, S., & Ballinger, J. (2001). Roots of a sociology of news: Remembering Mr. Gates and social control in the newsroom. *Journalism & Mass Communication Quarterly, 78*(4), 641–658.

Reese, S. D., Rutigliano, L., Hyun, K., & Jeong, J. (2007). Mapping the blogosphere: Professional and citizen-based media in the global news arena. *Journalism, 8*(3), 235–261. doi:10.1177/1464884907076459

Rinehart, R. (1994). Sport as kitsch: A case study of the American gladiators. *Journal of Popular Culture, 28*(2), 25–35. doi:10.1111/j.0022-3840.1994.2802_25.x

Riordan, J., & Krueger, A. (Eds.). (1999). *The international politics of sport in the 20th century.* London, UK: E & FNSpon and New York. NY: Routledge.

Rivenburgh, N. (1993). Images of nations during the 1992 Barcelona Olympic opening ceremony. In International Olympic Committee (Ed.), *Olympic centennial congress bulletin,* (pp. 32-39). Lausanne, Switzerland: International Olympic Committee.

Robinson, S. (2006). The mission of the j-blog: Recapturing journalistic authority on-Line. *Journalism, 7*(1), 65–83. doi:10.1177/1464884906059428

Roche, M. (2006). Sports mega-events, modernity and capitalist economies: Mega-events and modernity revisited: globalization and the case of the Olympics. *The Sociological Review, 54*(2), 25–40. doi:10.1111/j.1467-954X.2006.00651.x

Rout, P. (2006). London Olympics 2012, a lasting legacy for future generations, British Telecom. In *Proceedings of the IEEE International Communications Quality and Reliability (CQR) Workshop/ World Class Communications for World Class Events.*

Rowe, D. (1999). *Sport, culture and the media: The unruly trinity (Issues in cultural and media studies)*. Buckingham, UK: Open University Press.

Rowe, D. (2005). *Sport, culture and the media. The unruly trinity* (2nd ed.). Philadelphia, PA: Open University Press.

Rowe, D., McKay, J., & Miller, T. (1998). Come together: Sport, nationalism and the media image. In Wenner, L. A. (Ed.), *MediaSport* (pp. 119–133). London, UK: Routledge.

Roy, D. P., & Cornwell, T. B. (2003). Brand equity's influence on responses to event sponsorships. *Journal of Product and Brand Management, 12*(6), 377–393. doi:10.1108/10610420310498803

Samatas, M. (2007). Security and surveillance in the Athens 2004 Olympics: Some lessons from a troubled story. *International Criminal Justice Review, 17*(3), 220–238. doi:10.1177/1057567707306649

Sandler, D. M., & Shani, D. (1993). Sponsorship and the Olympic Games: The consumer perspective. *Sport Marketing Quarterly, 2*(3), 38–43.

Schaffer, K., & Smith, S. (2000). *The Olympics at the millennium: Power, politics and the games*. Rutgers, NJ: Rutgers University Press.

Shoemaker, P. J., & Reese, S. D. (1996). *Mediating the message: Theories of influences on mass media content*. New York, NY: Longman.

Shoemaker, P. J., & Vos, T. P. (2009). *Gatekeeping theory*. Boca Raton, FL: Taylor & Francis.

Shoval, N. (2002). A new phase in the competition for the Olympic gold: The London and New York bids for the 2012 Games. *Journal of Urban Affairs, 24*(5), 583–599. doi:10.1111/1467-9906.00146

Singer, J. (1997). Still guarding the gate? The newspaper journalist's role in an on-line world. *Convergence: The International Journal of Research into New Media Technologies, 3*(1), 72–89. doi:10.1177/135485659700300106

Singer, J. (1998). Online journalists: foundations for research into their changing roles. *Journal of Computer-Mediated Communication, 4*(1). Retrieved September 20, 2011, from http://jcmc.indiana.edu/vol4/issue1/singer.html

Singer, J. B. (2005). The political j-blogger, 'normalizing' a new media form to fit old norms and practices. *Journalism, 6*(2), 173–198. doi:10.1177/1464884905051009

Slepicka, P. (1995). Psychology of the sport spectator. In Stuart Biddle, J. H. (Ed.), *European perspectives on exercise and sport psychology* (pp. 270–289). Champaign, IL: Human Kinetics.

Snider, P. (1967). Mr. Gates revisited: A 1966 version of the 1949 case study. *The Journalism Quarterly, 44*(3), 419–427.

Tai, Z. (2000). Media of the world and world of the media: A cross-national study of the rankings of the 'Top 10 World Events' from 1988 to 1998. *Gazette, 2*(5), 331–353.

Tambini, D. (1999). New media and democracy. *New Media & Society, 1*(3), 305–329. doi:10.1177/14614449922225609

Tomlinson, A. (1996). Olympic spectacle: Opening ceremonies and some paradoxes of globalization. *Media Culture & Society, 18*, 583–602. doi:10.1177/016344396018004005

Townley, S., & Grayson, E. (1984). *Sponsorship of sport, arts and leisure: Law, tax and business relationships*. London, UK: Sweet and Maxwell.

Turner, E. T., Bounds, J., Hauser, D., Motsinger, S., Ozmore, D., & Smith, J. (1995). Television consumer advertising and the sports figure. *Sport Marketing Quarterly, 4*(1), 27–33.

Turner, P., & Cusumano, S. (2000). Virtual advertising: Legal implications for sport. *Sport Management Review*, (3): 47–70. doi:10.1016/S1441-3523(00)70079-9

Vilanilam, J. V. (1996). The socio-cultural dynamics of Indian television: From SITE to insight and privatisation. In French, D., & Richards, M. (Eds.), *Contemporary television, Eastern perspectives* (pp. 61–90). New Delhi, India: Sage.

Wanta, W., & Craft, S. (2004). Women in the newsroom: Influences of female editors and reporters on the news agenda. *Journalism & Mass Communication Quarterly, 81*(1), 124–138.

White, B. (2001). Who are the real competitors in the Olympic Games. *Journal of Contemporary Legal Issues*, (12), 227-239.

White, D. M. (1950). The 'Gate-Keeper': A case study in the selection of news. *The Journalism Quarterly, 27*, 383–390.

Wilber, D. (1988). Linking sports and sponsors. *The Journal of Business Strategy, 9*(4), 8–10. doi:10.1108/eb039234

Williams, B. A., & Delli Carpini, M. X. (2004). Monica and Bill all the time and everywhere: The collapse of gatekeeping and agenda setting in the new media environment. *The American Behavioral Scientist, 47*(9), 1208–1230. doi:10.1177/0002764203262344

Williams, R. (1977). *Marxism and literature*. New York, NY: Oxford University Press.

Witcher, B., Craigen, J. G., Culligan, D., & Harvey, A. (1991). The links between objectives and function in organisational sponsorship. *International Journal of Advertising, 10*(1), 13–33.

Yeoman, I. (2004). *Festival and events management: An international arts & culture perspective*. Oxford, UK: Elsevier Butterworth-Heinemann.

Zhao, Y. (2007). After mobile phones, what? Re-embedding the social in China's 'digital revolution'. *International Journal of Communication, 1*, 92–120.

This work was previously published in the International Journal of Interdisciplinary Telecommunications and Networking, Volume 3, Issue 4, edited by Michael R. Bartolacci and Steven R. Powell, pp. 49-74, copyright 2011 by IGI Publishing (an imprint of IGI Global).

Compilation of References

3rd Generation Partnership Project. (2008). *TS 36.401 V8.4.0: Evolved universal terrestrial radio access network (E-UTRAN); architecture description (Release 8)*. Retrieved from http://www.quintillion.co.jp/3GPP/Specs/36401-840.pdf

3rd Generation Partnership Project. (2008). *UTRA-UTRAN long term evolution (LTE) and 3GPP system architecture evolution (SAE)*. Retrieved from ftp://ftp.3gpp.org/Inbox/2008_web_files/LTA_Paper.pdf

3rd Generation Partnership Project. (2009). *4G SAE specification*. Retrieved from http://www.3gpp.org

3rd Generation Partnership Project. (2009). *LTE-advanced technical specification*. Retrieved from http://www.3gpp.org/LTE-Advanced

3rd Generation Partnership Project. (2010). *TS 32.571, v. 9.0.0: Technical specification group services and system aspects- telecommunication management- home node B (HNB) and home eNode B (HeNB) management- type 2 interface concepts and requirements*. Retrieved from http://www.etsi.org/deliver/etsi_ts/132500_132599/132571/09.00.00_60/ts_132571v090000p.pdf

3rd Generation Partnership Project. (2010). *TS 32.581, v. 10.0.0: Telecommunications management; home node B (HNB) operations, administration, maintenance and provisioning (OAM&P); concepts and requirements for Type 1 interface HNB to HNB management system (HMS)*. Retrieved from http://www.3gpp.org/ftp/specs/html-info/32581.htm

3rd Generation Partnership Project. (2010). *TS 36.300, v. 10.2.0: Evolved universal terrestrial radio access (E-UTRA) and evolved universal terrestrial radio access network (E-UTRAN); overall description; Stage 2*. Retrieved from http://www.3gpp.org/ftp/specs/html-info/36300.htm

3rd Generation Partnership Project. (2010). *TS25.104, v.9.4.0: Base station (BS) radio transmission and reception (FDD)*. Retrieved from http://www.3gpp.org/ftp/specs/html-info/TSG-WG--R4.htm

Abdelhalim, M. B., Salama, A. E., & Habib, S. E.-D. (2006). Hardware Software Partitioning using Particle Swarm Optimization Technique. In *Proceedings of the 6th International Workshop on System on Chip for Real Time Applications*, Cairo, Egypt (pp. 189-194). Los Alamitos, CA: IEEE Press.

AccessData Group. LLC. (2011). *SilentRunner sentinel network forensics software*. Retrieved July 25, 2011, from http://accessdata.com/products/cyber-security-incident-response/ad-silentrunner-sentinel

Achabal, D., Mcintyre, S., Smith, S., & Kalyanam, K. (2000). A decision support system for vendor managed inventory. *Journal of Retailing*, 76(4), 430–454. doi:10.1016/S0022-4359(00)00037-3

AeroSat Corporation. (2008). *About Airborne SatCom*. Retrieved from http://www.aerosat.com/about/about_airborne_satcom.asp

Akunuri, K., Guardiola, I. G., & Phillips, A. (2010). Mobility adds fast-fading impact study on wireless ad hoc networks. *International Journal of Mobile Network Design and Innovation*, 3(3). doi:10.1504/IJMNDI.2010.038097

Akyildiz, I. F., Xie, J., & Mohanty, S. (2004). A survey of mobility management in next-generation all-IP-based wireless systems. *IEEE Wireless Communications*, 16-28.

Akyildiz, I. F., & Ho, J. S. M. (1995). Mobile user location update and paging mechanisms under delay constraints. *ACM-Baltzar Journal on Wireless Networks*, 1(4), 413–425. doi:10.1007/BF01985754

Akyildiz, I. F., Lee, W.-Y., Vuran, M. C., & Mohanty, S. (2010). A survey on spectrum management in cognitive radio networks. *IEEE Communications Magazine, 46*, 40–48. doi:10.1109/MCOM.2008.4481339

Albarran, A. B. (2002). *Media economics* (2nd ed.). Ames, IA: Iowa State Press.

Albarran, A. B., & Chan-Olmsted, S. M. (1998). *Global media economics – Commercialization, concentration and integration of world media markets*. Ames, IA: Iowa State University Press.

Albarran, A. B., & Dimmick, J. (1996). Economies of multiformity and concentration in the communication industries. *Journal of Media Economics, 9*, 41–49. doi:10.1207/s15327736me0904_3

Alcatel. (2005). *Open IMS solutions for innovative applications*. Paris, France: Author.

American Medical Network. (2005). *Hands-free phones just as dangerous in cars*. Retrieved from http://www.health.am/ab/more/hands_free_phones_just_as_dangerous_in_cars

Amrhein, D., & Quint, S. (2009, April 8). *Cloud computing for the enterprise: Part 1: Capturing the cloud*. Retrieved July 18, 2011, from IBM: http://www.ibm.com/developerworks/websphere/techjournal/0904_amrhein/0904_amrhein.html

Anbar, M., & Vidyarthi, D. P. (2009). Buffer Management in Cellular IP Network using GA. In *Proceedings of the 2nd International Conference on Advanced Computer Theory and Engineering (ICACTE 2009),* Egypt, Cairo (Vol. 2, pp. 1163-1173). New York, NY: ASME Press.

Anbar, M., & Vidyarthi, D. P. (2010). Comparative study of two CPU router time management algorithms in cellular IP networks. *International Journal of Network Management*.

Anbar, M., & Vidyarthi, D. P. (2009). On Demand Bandwidth Reservation for Real-Time Traffic in Cellular IP Network using Particle Swarm Optimization. *International Journal of Business Data Communications and Networking, 5*(3), 53–66. doi:10.4018/jbdcn.2009070104

Anbar, M., & Vidyarthi, D. P. (2009). Router CPU Time Management using Particle Swarm Optimization in Cellular IP Networks. *International Journal of Advancements in Computing Technology, 1*(2), 48–55.

Anderson, J. C., & Gerbing, D. W. (1988). Structural equation modelling in practice: A review and recommended two-step approach. *Psychological Bulletin, 103*, 411–423. doi:10.1037/0033-2909.103.3.411

Andrews, D. L., & Jackson, S. J. (2001). *Sports stars: The cultural politics of sporting celebrity*. New York, NY: Routledge. doi:10.4324/9780203463543

Angulo, A., Nachtmann, H., & Waller, M. A. (2004). Supply chain information sharing in a vendor managed inventory partnership. *Journal of Business Logistics, 25*(1), 101–120. doi:10.1002/j.2158-1592.2004.tb00171.x

Athanasios, B. (2006). A mobility sensitive approach for efficient routing in ad hoc mobile networks. In *Proceedings of the 9th ACM International Symposium on Modeling Analysis and Simulation of Wireless and Mobile Systems,* Terromolinos, Spain.

Atkinson, P. (2005). Qualitative research—Unity and diversity, forum qualitative Sozialforschung/Forum. *Qualitative Social Research, 6*(3), 26. Retrieved from http://nbn-resolving.de/urn:nbn:de:0114-fqs0503261

Aziz, S. R. A., Endut, N. A., Abdullah, S., & Doud, M. N. M. (2009, March). Performance evaluation of AODV, DSR and DYMO routing protocol in MANET. In *Proceedings of the Conference on Scientific and Social Research*.

Badis, H., Munaretto, A., Agha, K. A., & Pujolle, G. (2003). QoS for Ad hoc Networking Based on Multiple Metrics: Bandwidth and Delay. In *Proceedings of the 5th IFIP-TC6 International Conference on Mobile and Wireless Communication Networks* (pp. 15-18).

Bagdikian, B. H. (1997). *The media monopoly* (5th ed.). Boston, MA: Beacon.

Bai, F., & Helmy, A. (2007). A survey of mobility models. In *Wireless ad hoc networks*.

Bai, F., Sadagopan, N., & Helmy, A. (2003). IMPORTANT: A framework to systematically analyze the impact of mobility on performance of routing protocols for ad hoc networks. In *Proceedings of the 22nd Annual IEEE INFOCOM Joint Conference* (pp. 825-835).

Bale, J., & Maguire, J. A. (1994). *The global sports arena: Athletic talent migration in an interdependent world*. Staffordshire, UK: University of Keele.

Balsamo, S., De Nitto, V., & Onvural, R. (2001). *Analysis of queueing networks with blocking*. Boston, MA: Kluwer Academic.

Barabasi, A. L., & Albert, R. (1999). Emergence of scaling in random networks. *Science, 286*, 509–512. doi:10.1126/science.286.5439.509

Barber, B. R. (1997). The new telecommunications technology: Endless frontier or the end of democracy? *Constellations, Special Section: Democratizing Technology/Technologizing Democracy, 4*(2), 208-228.

Barker, C. (1997). *Global television: an introduction*. Oxford, UK: Wiley-Blackwell.

Barker, C. (2000). *Cultural studies: Theory and practice*. London, UK: Sage.

Barratt, M. (2004). Understanding the meaning of collaboration in the supply chain. *Supply Chain Management: An International Journal, 9*(1), 30–42. doi:10.1108/13598540410517566

Barzilai-Nahon, K. (2006). Gatekeeping in virtual communities: On politics of power in cyberspace. In *Proceedings of the 39th Hawaii International Conference on System Sciences*.

Barzilai-Nahon, K. (2008). Toward a theory of network gatekeeping: A framework for exploring information control. *Journal of the American Society for Information Science and Technology, 59*(9), 1493–1512. doi:10.1002/asi.20857

Bayer, T. J., Cooney, L. A., Delp, C. L., Dutenhoffer, C. A., Gostelow, R. D., Ingham, M. D., et al. (2010). An operations concept for Integrated Model-Centric Engineering at JPL. In *Proceedings of the 2010 IEEE Aerospace Conference*, Big Sky, MT (pp. 1-14).

Beard, F., & Olsen, R. L. (1999). Webmasters as mass media gatekeepers: a qualitative exploratory study. *Internet Research: Electronic Networking Applications and Policy, 9*(3), 200–211. doi:10.1108/10662249910274601

Beck, D., & Bosshart, L. (2003). Sports, media, politics, and national identity. *Communication Research Trends, 22*(4), 1–43.

Beis, D. A., Loucopoulos, P., Pyrgiotis, Y., & Zografos, K. G. (2006). PLATO helps Athens win gold: Olympic Games knowledge modeling for organizational change and resource management. *Interface, 36*(1), 26–42. doi:10.1287/inte.1060.0189

Bejerano, Y., & Cidon, I. (2003). An anchor chain scheme for IP mobility management. *Wireless Networks, 9*(5), 409–420. doi:10.1023/A:1024627814601

Bello, P. A. (1973). Aeronautical channel characterization. *IEEE Transactions on Communications, 21*(5), 548–563. doi:10.1109/TCOM.1973.1091707

Berkowitz, L. (1990). On the formation and regulation of anger and aggression. A cognitive-neoassociationistic analysis. *The American Psychologist, 45*(4), 494–503. doi:10.1037/0003-066X.45.4.494

Bettstetter, C. (2006). Smooth is better than sharp: A random mobility model for simulation of wireless networks. In *Proceedings of the 4th ACM International Workshop on Modeling, Analysis and Simulation of Wireless and Mobile Systems*, Rome, Italy.

Blumenstein, J. (2007). *Aircell: Inflight Wi-Fi Built for the Airline Business*. Retrieved from http://blog.aircell.com

Boci, E. (2009). RF Coverage analysis methodology as applied to ADS-B design. In *Proceedings of the 2009 IEEE Aerospace Conference*, Big Sky, MT (pp. 17).

Boci, E., Sarkani, S., & Mazzuchi, T. A. (2009). *Optimizing ADS-B RF coverage*. Paper presented at the Integrated Communications, Navigation and Surveillance Conference, Arlington, VA.

Boeglen, H., & Chatellier, C. (2006). On the robustness of a joint source-channel coding scheme for image transmission over non frequency selective Rayleigh fading channels. *In Proceedings of the 2nd Conference on Information and Communication Technologies* (pp. 2320-2324).

Bourdieu, P. (1986). *Distinction: A social critique of the judgement of taste*. London, UK: Routledge.

Boyle, R. (2006). *Sports journalism: Context and issues*. London, UK: Sage.

Boyle, R., & Haynes, R. (2000). *Power play: Sport, the media, and popular culture*. Harlow, UK: Longman.

Breed, W. (1955). Social control in the newsroom: A functional analysis. *Social Forces*, *33*(4), 326–335. doi:10.2307/2573002

Brink, S. (2000). Design of serially concatenated codes based on iterative decoding convergence. In *Proceedings of the Second International Symposium on Turbo Codes and Related Topics*, Brest, France (pp. 319-322).

Brink, S. (2001). Convergence behavior of iteratively decoded parallel concatenated codes. *IEEE Transactions on Communications*, *49*(10), 1727–1737. doi:10.1109/26.957394

Broch, J., Maltz, D. A., Johnson, D. B., Hu, Y. C., & Jetcheva, J. (1998). A performance comparison of multihop wireless ad hoc network routing protocols. In *Proceedings of the ACM/IEEE Conference on Mobile Communications* (pp. 85-97).

Broderick, S. (2008). *Airports and NextGen*. Retrieved from http://www.itt.com/adsb/pdf/ITT-AirportsandNext-Gen.pdf

Brodkin, J. (2009, May 18). *10 cloud computing companies to watch*. Retrieved July 18, 2011, from Network World, Inc.: http://www.networkworld.com/supp/2009/ndc3/051809-cloud-companies-to-watch.html

Brown, T. X., & Mohan, S. (1997). Mobility management for personal communications systems. *IEEE Transactions on Vehicular Technology*, *46*(2), 269–278. doi:10.1109/25.580765

Brule, J. F. (1985). *Fuzzy Systems – A Tutorial*. Retrieved from http://www.ortech-engr.com/fuzzy/tutor.txt

Bruno, R., & Dyer, G. (2008). *Engineering a US national Automatic Dependent Surveillance - Broadcast (ADS-B) radio frequency solution*. Paper presented at the Tyrrhenian International Workshop on Digital Communications - Enhanced Surveillance of Aircraft and Vehicles, Capri, Italy.

Buenz, A. (2007, April 11). *Reliable repositories: Using Microsoft Forefront security for SharePoint to defend collaboration* (Microsoft, Producer). Retrieved July 24, 2011, from http://technet.microsoft.com/en-us/library/cc512661.aspx

Buhmann, J. (2002). *Data clustering and learning: Handbook of brain theory and neural networks* (2nd ed.). Cambridge, MA: MIT Press.

Bunting, S., & Anson, S. (2007). *Mastering Windows network forensics and investigation*. Indianapolis, IN: Wiley.

Bureau of Transportation Statistics. (n.d.). *Airlines and Airports*. Retrieved from http://www.bts.gov/programs/airline_information

Burgi, M. (1997). TV exec. sees virtual signs. *Mediaweek*, *7*(6), 13.

Burr, A. (2001). *Modulation and coding for wireless communications*. Upper Saddle River, NJ: Pearson/Prentice Hall. Engels, M. (2002). *Wireless OFDM systems: How to make them work?* Boston, MA: Kluwer Academic.

Bush, G. W. (2001). *Executive order establishing office of homeland security*. Retrieved from http://www.whitehouse.gov/news/releases/2001/10/20011008-2.html

Bush, G. W. (2002). Executive order on critical infrastructure protection. *Communications of the ACM*.

Bychkovskiy, V., Megerian, S., Estrin, D., & Potkanjak, M. (2003). A collaborative approach to in-place sensor calibration. In F. Zhao & L. Guibas (Eds.), *Proceedings of the 2nd International Conference on Information Processing in Sensor Networks* (LNCS 2634, p. 556).

Byrne, B. M. (2001). *Structural equation modeling with AMOS—Basic concepts, applications, and programming*. Mahwah, NJ: Lawrence Erlbaum.

Campbell, A. T., Gomez, J., Wan, C.-Y., Kim, S., Turanyi, Z., & Valko, A. (1999). *Cellular IP*. Retrieved from http://tools.ietf.org/html/draft-ietf-mobileip-cellularip-00

Campbell, A. T., Javier, G., Kim, S., Wan, C. Y., Turanyi, Z. R., & Valko, A. G. (2002). Comparison of IP micro-mobility protocols. *IEEE Wireless Communications*, 72-82.

Campbell, A. T., & Castellanos, J. G. (2001). IP micro-mobility protocols. *Mobile Computing and Communications Review*, *4*(4), 45–54. doi:10.1145/380516.380537

Campbell, A. T., Gomez, J., Kim, S., Valko, A. G., Wan, C.-Y., & Turanyi, Z. R. (2000). Design, implementation, and evaluation of cellular IP. *IEEE Personal Communications*, *7*(4), 42–49. doi:10.1109/98.863995

Carlton, G. H. (2007). *A protocol for the forensic data acquisition of personal computer workstations* (Doctoral dissertation). Available from ProQuest Dissertations and Theses database (UMI No. 3251043).

Carlton, G. H. (2008). An evaluation of Windows-based computer forensics application software running on a Macintosh. *Journal of Digital Forensics. Security and Law, 3*(3), 43–60.

Carroll, R. L. (1985). Content values in TV news programs in small and large markets. *The Journalism Quarterly, 62*(4), 877–938.

Castelliccia, C., Hartenstein, H., Paar, C., & Westhoff, D. (Eds.). (2005). *Security in ad hoc and sensor networks.* New York, NY: Springer.

Catbird Networks, Inc. (2008, September 27). *Virtualization security: The Catbird Primer.* Retrieved July 25, 2011, from http://www2.catbird.com/news/analyst_reports.php

Celidonio, M., Di Zenobio, D., & Pulcini, L. (2010, May). A broadband integrated radio LAN. In *Proceedings of the 17th IEEE Workshop on Local and Metropolitan Area Networks.*

Celidonio, M., Di Zenobio, D., Pulcini, L., & Rufini, A. (2011, June). Femtocell technology combined to a condominium cabled infrastructure. In *Proceedings of the IEEE International Symposium on Broadband Multimedia Systems and Broadcasting*, Nuremberg, Germany.

Center for Disease Control (CDC). (2002). *Protecting the nation's health in an era of globalization: CDC's global infectious disease strategy.* Retrieved from http://www.cdc.gov/globalidplan/

Center for Disease Control (CDC). (2011). *About the tracking program.* Retrieved from http://ephtracking.cdc.gov/showAbout.action

Center for Disease Control (CDC). (2011). *National public tracking program.* Retrieved from http://www.cdc.gov/nceh/tracking/

Chalip, L. (1992). The construction and use of polysemic structures: Olympic lessons for sport marketing. *Journal of Sport Management, 4*, 87–98.

Chandrasekhar, V., & Andrews, J. (2008). Femtocells networks: A survey. *IEEE Communications Magazine, 46*(9), 59–67. doi:10.1109/MCOM.2008.4623708

Chan, J. M. (1993). Commercialization without independence: Trends and tensions of media development in China. In Cheng, J., & Brosseau, M. (Eds.), *China review.* Hong Kong: Chinese University Press.

Chan, J. M. (1994). Media internationalization in China: Processes and tensions. *The Journal of Communication, 44*(3), 70–88. doi:10.1111/j.1460-2466.1994.tb00689.x

Chan, J. M. (1996). Television development in Greater China: Structure, exports, and market formation. In Sinclair, J., Jacka, E., & Cunningham, S. (Eds.), *New patterns in global television: Peripheral vision* (pp. 126–160). Oxford, UK: Oxford University Press.

Chaomei, C., Song, I.-Y., & Zhu, W. (2007, June 25-27). *Trends in Conceptual Modeling: Citation Analysis of the ER Conference Papers (1979-2005).* Paper presented at the Conference on the International Society for Scientometrics and Informatrics, Madrid, Spain.

Chaouchi, H. (2009). *Wireless and mobile network security.* New York, NY: John Wiley & Sons. doi:10.1002/9780470611883

Charlton, S. G. (2009). Driving while conversing: Cell phones that distract and passengers who react. *Accident; Analysis and Prevention, 41*, 160–173. doi:10.1016/j.aap.2008.10.006

Chen, J., Rauber, P., Singh, D., Sundarraman, C., Tinnakornsrisuphap, P., & Yavuz, M. (2010, Jan) *Femtocells-architecture & network aspects.* Retrieved from http://www.qualcomm.com/common/documents/white_papers/Femto_Overview_Rev_C.pdf

Chen, Y., & Yang, Y. (2009). Cellular based machine to machine communication with un-peer2peer protocol stack. In *Proceedings of the 70th IEEE Conference on Vehicular Technology* (pp. 1-5).

Chen, P. P.-S. (1976). The Entity-Relationship Model - Toward a Unified View of Data. *ACM Transactions on Database Systems, 1*(1), 9–36. doi:10.1145/320434.320440

Chen, P. P.-S. (1983). English sentence structure and entity-relationship diagrams. *Information Sciences, 29*(2-3), 127–149. doi:10.1016/0020-0255(83)90014-2

Chin, W. W. (1998). The partial least squares approach to structural equation modeling. In Marcoulides, G. A. (Ed.), *Modern methods for business research* (pp. 295–336). Mahwah, NJ: Lawrence Erlbaum.

Chlamtac, I., Conti, M., & Liu, J. (2003). Mobile ad hoc networking: Imperatives and challenges. *Ad Hoc Networks, 1*(1), 13–64. doi:10.1016/S1570-8705(03)00013-1

Cisco Systems Inc. (2011). *Cisco IOS Netflow*. Retrieved July 25, 2011, from http://www.cisco.com/en/US/products/ps6601/products_ios_protocol_group_home.html

Clausen, T., & Banerjee, P. J. (2003). *Optimized Link State Routing Protocol*. Retrieved from http://www.ietf.org/rfc/rfc3626.txt

Clevorn, T., Godtmann, S., & Vary, P. (2006). BER prediction using EXIT charts for BICM with iterative decoding. *IEEE Communications Letters, 10*(1), 49–51. doi:10.1109/LCOMM.2006.1576566

Cochennec, J. Y. (2002). Activities on next generation networks under global information infrastructure in ITU-T. *IEEE Communications Magazine*, 98–101. doi:10.1109/MCOM.2002.1018013

Cockburn, A. (2001). *Writing effective use cases*. Reading, MA: Addison-Wesley.

Cohen, E. (2002). Online journalism as market-driven journalism. *Journal of Broadcasting & Electronic Media, 46*(4), 532–548. doi:10.1207/s15506878jobem4604_3

Cole, R. G., & Rosenbluth, J. H. (2001). Voice over IP performance monitoring. *ACM SIGCOMM Computer Communication Review, 31*(2), 9–24. doi:10.1145/505666.505669

Cooper, G. F., Dash, D. H., Levander, J. D., Wong, W. K., Hogan, W. R., & Wagner, M. M. (2004). Bayesian biosurveillance of disease outbreaks. In *Proceedings of the 20th Conference on Uncertainty in Artificial Intelligence* (pp. 94-103).

Cornwell, T. B., & Maignan, I. (1998). An international review of sponsorship research. *Journal of Advertising, 27*(1), 1–21.

Corporation, E. M. C. (2011). *NetWitness Investigator*. Retrieved July 24, 2011, from Netwitness Corporation: http://netwitness.com/products-services/investigator

Corson, S., & Macker, J. (1999). *Mobile ad hoc networking (MANET): Routing protocol performance issues and evaluation considerations*. Retrieved from http://www.ietf.org/rfc/rfc2501.txt

Cossiavelou, V. (2009). Digital dividend aware business models for the creative industries: Challenges and opportunities in EU Market. In Lytras, M. D., De Pablos, P. O., & Damiani, E. (Eds.), *Best practices for the knowledge society - Knowledge, learning, development and technology for all*. New York, NY: Springer. doi:10.1007/978-3-642-04757-2_60

Cossiavelou, V., & Bantimaroudis, Ph. (2009). Mediation of the message in a wireless global environment: revisiting the media gatekeeping model. *International Journal of Interdisciplinary Telecommunications and Networking, 1*(4), 37–53. doi:10.4018/jitn.2009092803

Cossiavelou, V., & Bartolacci, M. (2009). News corporation: Facing the wireless world of the 21st century. In Powell, S., & Shim, J. P. (Eds.), *Wireless technology applications, management and security* (Vol. 44, pp. 83–90). Berlin, Germany: Springer-Verlag.

Cristaldi, L., Faifer, M., Grande, F., & Ottoboni, R. (2005, February 8-10). An improved M2M platform for multisensors agent application. In *Proceedings of the Sensors for Industry Conference* (pp. 79-83).

Croteau, D., & Hoynes, W. (2000). *Media society: Industries, images, and audiences*. Thousand Oaks, CA: Pine Forge.

Cummings, M., & Haruyama, S. (1999). FPGA in the Software Radio. *IEEE Communications Magazine, 37*(2), 108–112. doi:10.1109/35.747258

Cuneen, J., & Hannan, M. J. (1993). Intermediate measures and recognition testing of sponsorship advertising at an LPGA tournament. *Sport Marketing Quarterly, 2*(1), 47–56.

Curran, I., & Pluta, S. (2008, April 22-23). Overview of machine to machine and telematics. In *Proceedings of the IEEE 6th Institution of Engineering and Technology Water Event* (pp. 1-33).

Dahlberg, T., & Jung, J. (2001). Survivable Load Sharing Protocols: A Simulation Study. *ACM Wireless Networks Journal, 7*(3), 283–296. doi:10.1023/A:1016630206995

Das, S. R., Castaneda, R., Yan, J., & Sengupta, R. (1998). Comparative performance evaluation of routing protocols for mobile, ad hoc networks. In *Proceedings of the IEEE 7th International Conference on Computer Communications and Networks* (pp. 153-161).

Das, S. R., Perkins, C. E., & Royer, E. M. (2000, March). Performance comparison of two on-demand routing protocols for ad hoc networks. In *Proceedings of the IEEE Conference on Computer Communications.*

Das, S., McAuley, A., Datta, A., Misra, A., Chakraborty, K., & Das, S. K. (2002). IDMP: An intradomain mobility management protocol for next-generation wireless networks. *IEEE Wireless Communications*, 38-45.

Dash, D. K. (2009). *India leads world in road deaths: WHO*. Retrieved from http://articles.timesofindia.indiatimes.com/2009-08-17/india/ 28181973_1_road-accidents-road-fatalities-global-road-safety

Das, S., Misra, A., & Agrawal, P. (2000). TeleMIP: Telecommunications-enhanced mobile IP architecture for fast intradomain mobility. *IEEE Personal Communications*, *7*(4), 50–58. doi:10.1109/98.863996

De Toni, A. F., & Zamolo, E. (2005). From a traditional replenishment system to vendor-managed inventory: A case study from the household electrical appliances sector. *International Journal of Production Economics*, *96*(1), 63–79. doi:10.1016/j.ijpe.2004.03.003

Detlor, B. (2001). The influence of information ecology on e-commerce initiatives. *Internet Research: Electronic Networking Applications and Policy*, *11*(4), 286–295. doi:10.1108/10662240110402740

Dimitrova, D., Connolly-Ahern, C., Williams, A. P., Kaid, L. L., & Reid, A. (2003). Hyperlinking as gatekeeping: Online newspaper coverage of the execution of an American terrorist. *Journalism Studies*, *4*(3), 401–414. doi:10.1080/14616700306488

Disney, S. M., & Towill, D. R. (2003). The effect of vendor managed inventory (VMI) dynamics on the bullwhip effect in supply chains. *International Journal of Production Economics*, *85*(2), 199–215. doi:10.1016/S0925-5273(03)00110-5

Divsalar, D., Dolinar, S., & Pollara, F. (2000). Serial turbo trellis coded modulation with rate-1 inner code. In *Proceedings of the International Symposium on Information Theory*, San Francisco, CA (pp. 194-200).

Dong, Y., Xu, K., & Dresner, M. (2007). Environmental determinants of VMI adoption: An exploratory analysis. *Transportation Research Part E, Logistics and Transportation Review*, *43*(4), 355–369. doi:10.1016/j.tre.2006.01.004

Downey, J. (1998). Full of Eastern promise? Central and Eastern European media after 1989. In Thussu, D. K. (Ed.), *Electronic empires: Global media and local resistance* (pp. 47–62). London, UK: Arnold.

Doyle, G. (2002). *Understanding media economics*. London, UK: Sage.

Elias, N., & Dunning, E. (1986). *Quest for excitement: Sport and leisure in the civilizing process*. Oxford, UK: Basil Blackwell.

Elnoubi, S. M. (1992, May 10-13). A simplified stochastic model for the aeronautical mobile radio channel. In *Proceedings of the Vehicular Technology Conference (VTC 1999)*, Denver, CO (Vol. 2, pp. 960-963).

El-Rabbanny, A. (2002). *Introduction to GPS: The global positioning system*. Boston, MA: Artech House.

El-Rabbany, A. (1994). *The effect of physical correlations on the ambiguity and accuracy estimation in GPS differential positioning* (Tech. Rep. No. 170). Fredericton, NB, Canada: Department of Geodesy and Geomatics Engineering.

Emery, D., & Hilliard, R. (2009, September 14-17). *Every architecture description needs a framework: Expressing architecture frameworks using ISO/IEC 42010*. Paper presented at the IEEE/IFIP European Conference on Software Architecture (WICSA/ECSA).

Emigh, J. (1999). Vendor-managed inventory. *Computerworld*, *33*(34), 52.

Enami, K. (2005). Future of home media. In *Proceedings of the 13th Annual ACM International Conference on Multimedia*, Hilton, Singapore.

Enterprise, J. T. R. S. (n.d.). *Connecting the tactical edge.* Retrieved from http://jpeojtrs.mil

Entman, R. M. (2003). Cascading activation: Contesting the White House's frame after 9/11. *Political Communication, 20,* 415–432. doi:10.1080/10584600390244176

Erturk, M. C., Haque, J., & Arslan, H. (2010, March 6-13). Challenges in Aeronautical Data Networks. In *Proceedings of the IEEE Aerospace Conference,* Big Sky, MT (pp. 1-7).

Estefan, J. A. (2007). *Survey of Model-Based Systems Engineering (MBSE) Methodologies.* Retrieved from http://syseng.omg.org/MBSE_Methodology_Survey_RevA.pdf

Ethiopian Review. (2010). *India has the highest number of road accidents in the world.* Retrieved from http://www.ethiopianreview.com/news/90301

European Organization for the Safety of Air Navigation (Eurocontrol). (2008). *STATFOR - Air Traffic Statistics and Forecasts.* Retrieved from http://www.eurocontrol.int/statfor/

Eveleigh, T. J., Mazzuchi, T. A., & Sarkani, S. (2007). Spatially-aware systems engineering design modeling applied to natural hazard vulnerability assessment. *Systems Engineering, 10*(3), 187–202. doi:10.1002/sys.20073

FAA. (2010). *2010 NextGen Implementation Plan.* Washington, DC: Author.

Fall, K., & Varadhan, K. (Eds.). (2009). *The ns manual.* Retrieved from http://www.isi.edu/nsnam/ns/doc/ns_doc.pdf

Fan, Y.-F., & Chiu, I.-G. (2007). *U.S. Patent No. 7248835: Method for automatically switching a profile of a mobile phone.* Washington, DC: United States Patent and Trademark Office.

Fang, Y., Chlamtac, I., & Fei, H. (2000). Analytical results for optimal choice of location update interval for mobility database failure restoration in PCS networks. *IEEE Transactions on Parallel and Distributed Systems, 11*(6), 615–624. doi:10.1109/71.862211

Fang, Y., Chlamtac, I., & Lin, Y. (2000). Portable movement modeling for PCS networks. *IEEE Transactions on Vehicular Technology, 49*(4), 1356–1363. doi:10.1109/25.875258

Fazekas, P., Imre, S., & Telek, M. (2002). Modeling and analysis of broadband cellular networks with multimedia connections. *Telecommunication Systems, 19*(3-4), 263–288. doi:10.1023/A:1013821901357

Featherstone, M., Scott, L., & Roland, R. (1995). *Global modernities.* London, UK: Sage.

Feigenbaum, E. A. (2003). Some challenges and grand challenges for computational intelligence. *Journal of the ACM, 50*(1), 32–40. doi:10.1145/602382.602400

Femto Forum. (2009). *Femtocells - natural solution for offload.* Retrieved from http://www.femtoforum.org

Femto Forum. (2009). *Interference management and performance analysis of UMTS/HSPA + femtocells.* Retrieved from http://www.femtoforum.org

Femto Forum. (2009). *Interference management in OFDMA femtocells.* Retrieved from http://www.femtoforum.org

Femto Forum. (2009). *The best that LTE can be: Why LTE needs femtocells.* Retrieved from http://www.femtoforum.org

Fernandez-Duran, A., Perez Leal, R., & Alonso, J. I. (2008). Dimensioning method for conversational video applications in wireless convergent networks. *EURASIP Journal on Wireless Communications and Networking.*

Fisher, G. H. (1998). Model-based systems engineering of automotive systems. In *Proceedings of the 1998 AIAA/IEEE/SAE Digital Avionics Systems Conference,* Bellevue, WA (pp. B15/1-B15/7).

Fisher, J. (1998). *Model-Based Proposal Development.* Retrieved from http://www.vitechcorp.com/whitepapers/files/200701031635570.fisher98.pdf

Flichy, P. (1995). *L'innovation technique: récents développements en sciences sociales: vers une nouvelle théorie de l'innovation.* Paris, France: Découverte.

Fondazione Ugo Bordoni. (2002). *Italian patent IT 14114241: Cabled- Wireless END USER ACCESS (C-W EUA).* Brussels, Belgium: European Patent Office.

Fornell, C., & Larcker, D. F. (1981). Evaluating structural equation models with observable variables and measurement error. *JMR, Journal of Marketing Research, 18*(1), 39–59. doi:10.2307/3151312

Fowler, M. (1997). *UML distilled: Applying the standard object modeling language*. Reading, MA: Addison-Wesley.

Franken, N., & Engelbrecht, A. P. (2005). Particle swarm optimization approaches to coevolve strategies for the iterated prisoner's dilemma. *IEEE Transactions on Evolutionary Computation, 9*(6), 562–579. doi:10.1109/TEVC.2005.856202

Gans, H. J. (1979). *Deciding what's news: A study of CBS Evening News, NBC Nightly News, Newsweek, and Time*. Chicago, IL: Northwestern University Press.

Gantenbein, R. E. (2004, September 1-5). Establishing a rural telehealth project: The Wyoming network for telehealth. In *Proceedings of the IEEE 26th Annual International Conference on Engineering in Medicine and Biology* (pp. 3089-3092).

García, B. (2001). Enhancing sport marketing through cultural and arts programs: Lessons from the Sydney 2000 Olympic arts festivals. *Sport Management Review, 4*, 193–219. doi:10.1016/S1441-3523(01)70075-7

Garrison, C., Lillard, T. V., Schiller, C. A., & Steele, J. (2010). *Digital forensics for network, internet, and cloud computing*. Amsterdam, The Netherlands: Elsevier Science.

Gattorna, J. L., & Walters, D. W. (1996). *Managing the supply chain: A strategic perspective*. London, UK: Palgrave Macmillan.

Gefen, D., Straub, D., & Bourdeau, M. (2000). Structural equation modelling and regression: Guidelines for research practice. *Communications of AIS, 4*(7), 1–79.

Gerhardt, W., & Medcalf, R. (2010, March). *Femtocells: Implementing a better business model to increase SP profitability*. Retrieved from http://www.cisco.com/en/US/solutions/ns341/ns973/ns941/femtocell_point_of_view.pdf

Gerla, M., & Raychaudhari, D. (2007). *Mobility in wireless network*. Rutgers, NJ: Rutgers University.

Gershon, R. A. (2000). The transnational media corporation: Environmental scanning and strategy formulation. *Journal of Media Economics, 13*(2), 81–101. doi:10.1207/S15327736ME1302_3

Ghosh, C., Cordeiro, C., Agrawal, D. P., & Rao, M. B. (2009). Markov chain existence and hidden Markov models in spectrum sensing. In *Proceedings of the IEEE International Conference on Pervasive Computing and Communications* (pp. 1-6).

Gibson, J. D. (1996). *The mobile communications handbook*. Boca Raton, FL: CRC Press.

Gilbert, T., & Bruno, R. (2009). *Surveillance and Broadcast Services - An effective nationwide solution*. Paper presented at the Integrated Communications, Navigation and Surveillance Conference, Arlington, VA.

Gilbert, T., Jin, J., Berger, J., & Henriksen, S. (2008). *Future Aeronautical Communication Infrastructure Technology Investigation*. Huntsville, AL: NASA.

Gitlin, T. (2003). *The whole world is watching: Mass media in the making and unmaking of the new left*. Berkeley, CA: University of California Press.

Gitlin, T. (2007). *Media unlimited: How the torrent of images and sounds overwhelms our lives*. New York, NY: Henry Holt and Co.

Gomez, E. A. (2008). Connecting communities of need with public health: Can SMS text-messaging improve outreach communication? In *Proceedings of the 41st Hawaii International Conference on Systems Science* (p. 128).

Gomez, E. A. (2010, December). Towards sensor networks: Improved ICT usage behavior for business continuity. In *Proceedings of the SIGGreen Pre-ICIS Workshop*.

Gomez, E. A., & Bartolacci, M. (2007). Beyond email: The need for immediacy in a wireless world. In *Proceedings of the Networking and Electronic Commerce Research Conference*, Riva Del Garda, Italy.

Gomez, E. A., & Bartolacci, M. (2011, May). Crisis management and mobile devices: Extending usage of sensor networks within an integrated framework. In *Proceedings of the 8th International Information Systems for Crisis Response and Management Conference*, Lisbon, Portugal.

Griffith, B. (2003). NBC spends $2.201 billion for broadcast rights for 2010, 2012 Olympic Games. *The Boston Globe*.

Gross, D., & Harris, C. (1998). *Fundamentals of queueing theory*. New York, NY: John Wiley & Sons.

Gruber, I., & Li, H. (2002). Link expiration times in mobile ad hoc networks. In *Proceedings of the 27th Annual IEEE International Conference on Local Computer Networks* (pp. 743-750).

Guardiola, I. G. (2007). Mitigating the stochastic effects of fading in mobile wireless ad-hoc networks. In *Proceedings of the IIE Annual Conference.*

Guardiola, I. G., & Matis, T. I. (2007). Fast-fading, an additional mistaken axiom of wireless-network research. *International Journal of Mobile Network Design and Innovation, 2,* 153–158. doi:10.1504/IJMNDI.2007.017319

Guidance Software, Inc. (2011). *EnCase Enterprise.* Retrieved July 25, 2011, from http://www.guidancesoftware.com/computer-forensics-fraud-investigation-software.htm

Gulati, R., Nohria, N., & Zaheer, A. (2000). Strategic networks. *Strategic Management Journal, 21*(3), 203–215. doi:10.1002/(SICI)1097-0266(200003)21:3<203::AID-SMJ102>3.0.CO;2-K

Guo, S., Yang, O., & Shu, Y. (2005). Improving source routing reliability in mobile ad hoc networks. *IEEE Transactions on Parallel and Distributed Systems, 16*(4).

Gustafsson, E., Jonsson, A., & Perkins, C. E. (2005). *Mobile IPv4 regional registration.* Retrieved from http://tools.ietf.org/

Haas, E. (2002). Aeronautical channel modeling. *IEEE Transactions on Vehicular Technology, 51*(2), 254–264. doi:10.1109/25.994803

Halperin, W., Baker, E. L., & Monson, R. R. (1992). *Public health surveillance.* New York, NY: Van Nostrand Reinhold.

Hamelink, C. J. (1997). *New information and communication technologies, social development and cultural change.* Geneva, Switzerland: United Nations Research Institute for Social Development.

Haque, J., Erturk, M. C., & Arslan, H. (2010, April). Doppler Estimation for OFDM based Aeronautical Data Communication. In *Proceedings of the IEEE Wireless Telecommunications Symposium (WTS),* Tampa, FL (pp. 1-6).

Hargreaves, J. (Ed.). (1982). *Sport, culture, and ideology.* London, UK: Routledge & Kegan Paul.

Harris, S. (2008). *CISSP all-in-one exam guide* (4th ed.). New York, NY: McGraw-Hill.

Hassan, Y. K., El-Aziz, M. H. A., & El-Rad, A. S. A. (2010). Performance evaluation of mobility speed over MANET routing protocols. *International Journal of Network Security, 11*(3), 128–138.

Health Indicators. (n. d.). *Overview of the HIW web services.* Retrieved from http://healthindicators.gov/Developers/Overview

Healthy People. (2010). *Health related quality of life and well-being.* Retrieved from http://www.healthypeople.gov/2020/about/QoLWBabout.aspx

Healthy People. (2010). *Initiative.* Retrieved from http://www.healthypeople.gov/

Heimlicher, S., & Karaliopoulos, M. (2007). End-to-end vs. hop-by-hop transport under intermittent connectivity. In *Proceedings of the 1st International Conference on Autonomic Computing and Communication Systems* (Article No. 20). Brussels, Belgium: ICST Press.

Helmy, A.-G., Jaseemuddin, M., & Bhaskara, G. (2004). Multicast-based mobility: A novel architecture for efficient micromobility. *IEEE Journal on Selected Areas in Communications, 22*(4), 677–690. doi:10.1109/JSAC.2004.826002

Henderson, T. R. (2003). Host mobility for IP networks: A comparison. *IEEE Network, 17*(6), 18–26. doi:10.1109/MNET.2003.1248657

Herman, E., & Chomsky, N. (2002). *Manufacturing consent: The political economy of the mass media.* New York, NY: Pantheon Books.

Herman, E., & McChesney, R. (1997). *The global media: The new missionaries of global capitalism.* London, UK: Cassell.

Herzog, E., & Torne, A. (2000). Support for representation of functional behaviour specifications in AP-233. In *Proceedings of the IEEE International Conference and Workshop on the Engineering of Computer Based Systems (ECBS 2000),* Edinburgh, UK (pp. 351-358).

Hodges, C. (2011, April 8). *Cloud computing rains cost savings, productivity benefits.* Retrieved July 2, 2011, from Industry Week Social Media: http://www.industryweek.com/articles/cloud_computing_rains_cost_savings_productivity_benefits_24337.aspx?SectionID=3

Hoeher, P., & Haas, E. (1999, September). Aeronautical channel modeling at VHF-band. In *Proceedings of the Vehicular Technology Conference (VTC 1999),* Amsterdam, The Netherlands (Vol. 4, pp. 1961-1966).

Ho, J., & Akyildiz, I. (1997). Dynamic hierarchical database architecture for location management in PCS networks. *IEEE/ACM Transactions on Networking, 5*(5), 646–660. doi:10.1109/90.649566

Hollifield, C., Kosicki, G., & Becker, L. (2001). Organizational vs. professional culture in the newsroom: Television news directors' and newspaper editors' hiring decisions. *Journal of Broadcasting & Electronic Media, 45*(1), 92–117. doi:10.1207/s15506878jobem4501_7

Holt, R. (1989). *Sport and the British – A modern history.* Oxford, UK: Clarendon Press.

Hong, X., Kwon, T., Gerla, M., Gu, D., & Pei, G. (2007, January). A mobility framework for ad hoc wireless networks. In *Proceedings of the ACM 2nd International Conference on Mobile Data Management.*

Horstkotte, E. (2000). *Fuzzy Logic Overview.* Retrieved from http://www.austinlinks.com/Fuzzy/overview.html

Huang, J., Feng, R., Liu, L., Song, M., & Song, J. (2004, June 27-29). A novel domain re-organizing algorithm for network-layer mobility management in 4G networks. In *Proceedings of the International Conference on Communications, Circuits and Systems* (Vol. 1, pp. 471-474).

Hudson, H. E. (1998). The paradox of ubiquity: Communication satellite policies in Asia. In MacKie-Mason, J. K., & Waterman, D. (Eds.), *Telephony, the Internet and the media, selected papers from the 1997 telecommunications policy research conference.* Mahwah, NJ: Lawrence Erlbaum.

Hughes, C. R. (2002). China and the globalization of ICTs: Implications for international relations. *New Media & Society, 4*(2), 205–224. doi:10.1177/14614440222226343

Hulland, J. (1999). Use of partial least squares (PLS) in strategic management research: A review of four recent studies. *Strategic Management Journal, 20*(2), 195–204. doi:10.1002/(SICI)1097-0266(199902)20:2<195::AID-SMJ13>3.0.CO;2-7

IEEE Standards Association. (1993). *American national standard, Canadian standard graphic symbols for electrical and electronics diagrams (including reference designation letters).* Retrieved from http://ieeexplore.ieee.org/xpl/freeabs_all.jsp?isnumber=21239&arnumber=985670&count=1

IEEE Standards Association. (2004). *Standard for local and metropolitan area networks. part 16: Air interface for fixed broadband wireless access systems.* Retrieved from http://ieeexplore.ieee.org/xpl/freeabs_all.jsp?tp=&isnumber=33683&arnumber=1603394

IETF. (n.d.). Mobile. *Ad Hoc Networks.* Retrieved from http://www.ietf.org/html.charters/manet-charter.html.

Iliev, T. (2007). Analysis and design of combined interleaver for turbo codes. In *Proceedings of the Romanian Technical Sciences Academy*, Bacau, Romania (Vol. pp. 148-153).

Iliev, T., & Radev, D. (2007). Turbo codes in the CCSDS standard for wireless data. In *Proceedings of the Papers ICEST*, Ohrid, Macedonia (pp. 31-35).

Iliev, T., Lokshina, I., & Radev, D. (2009). Use of extrinsic information transfer chart to predict behavior of turbo codes. In *Proceedings of the IEEE Wireless Telecommunications Symposium*, Prague, Czech Republic (pp. 1-4).

INCOSE. (2007). *Systems Engineering Vision 2020.* San Diego, CA: International Council of Systems Engineering.

Information Sciences Institute. (2010). *Ns Manual.* Retrieved from http://www.isi.edu/nsnam/ns/ns/documentation.html

Inhyok Cha Shah, Y., Schmidt, A. U., Leicher, A., & Meyerstein, M. V. (2009). Trust in M2M communication. *IEEE Vehicular Technology Magazine, 4*(3), 69–75. doi:10.1109/MVT.2009.933478

Institute of Medicine (IOM). (2003). *The future of the public's health in the 21ˢᵗ century.* Washington, DC: The National Academies Press.

International Air Transport Association (IATA). (n.d.). *Schedule Reference Service (SRS)*. Retrieved from http://www.iata.org/ps/publications/srs/

International Organization for Standardization. (1994). *Information technology - open systems interconnection - basic reference model - conventions for the definition of OSI services*. Retrieved from http://www.iso.org/iso/iso_catalogue/catalogue_tc/catalogue_detail.htm?csnumber=18824

International Organization for Standardization. (1994). *JPEG image coding system*. Retrieved from http://www.iso.org/iso/home.html

International Organization for Standardization. (2000). *JPEG 2000 image coding system*. Retrieved from http://www.iso.org/iso/home.html

International Telecommunications Union. (2001). *Propagation data and prediction models for the planning of indoor radiocommunication systems and radio local area networks in the frequency range 900 MHz to 100 GHz* (Tech. Rep. No. ITU-R P.1238). Geneva, Switzerland: International Telecommunications Union.

ISO/IEC. (1993). *Coding of moving pictures and associated audio for digital storage media at up to about 1.5 Mbit/s – Part 2: Video (ISO/IEC 11172)*. Geneva, Switzerland: Author.

ISO/IEC. (1999). *Coding of audio-visual objects – Part 2: Visual (ISO/IEC 14496-2)*. Geneva, Switzerland: Author.

ITU-T and ISO/IEC. (1994). *Generic coding of moving pictures and associated audio information – Part 2: Video (ISO/IEC 13818-2)*. Geneva, Switzerland: International Telecommunications Union.

ITU-T. (1993). *Video codec for audiovisual services at px64 kbits/s*. Geneva, Switzerland: International Telecommunications Union.

ITU-T. (2000). *Video coding for low bit rate communication*. Geneva, Switzerland: International Telecommunications Union.

Jackson, S. J., & Andrews, D. L. (2005). *Sport, culture and advertising: Identities, commodities and the politics of representation*. New York, NY: Routledge. doi:10.4324/9780203462003

Jain, R., & Lin, Y. B. (1995). An auxiliary user location strategy employing forwarding pointers to reduce network impacts of PCS. *Wireless Networks*, *1*, 197–210. doi:10.1007/BF01202542

Jain, R., Lin, Y., & Mohan, S. (1994). A caching strategy to reduce network impacts on PCS. *IEEE Journal on Selected Areas in Communications*, *12*(8), 1434–1445. doi:10.1109/49.329333

Jakobowicz, E., & Derquenne, C. (2007). A modified PLS path modeling algorithm handling reflective categorical variables and a new model building strategy. *Computational Statistics & Data Analysis*, *51*(8), 3666–3678. doi:10.1016/j.csda.2006.12.004

Jarvis, C. B., Mackenzie, S. B., & Podsakoff, P. M. (2003). A critical review of construct indicators and measurement model misspecification in marketing and consumer research. *The Journal of Consumer Research*, *30*(3), 199–218. doi:10.1086/376806

Jayaram, R., Sen, S. K., Kakani, N. K., & Das, S. K. (2000). Call Admission and Control for Quality-of-Service (QoS) Provisioning in Next Generation Wireless Networks. *ACM Wireless Networks Journal*, *6*(1), 17–30. doi:10.1023/A:1019160708424

Johansson, P., Larsson, T., Hedman, N., Mielczarek, B., & Degermark, M. (1999). Scenario-based performance analysis of routing protocols for mobile ad hoc networks. In *Proceedings of the IEEE/ACM Conference on Mobile Communications* (pp. 195-206).

Johnson, A., & Perkins, C. (2002). *Mobile IPv4 regional registration*. Retrieved from http://tools.ietf.org/

Johnson, D., & Maltz, D. A. (1996). Dynamic Source Routing in Ad Hoc Wireless Networks. In *Mobile Computing* (pp. 153–181). Dordrecht, The Netherlands: Kluwer Academic Publishers. doi:10.1007/978-0-585-29603-6_5

Joint Commission on Accreditation of Healthcare Organizations. (2002). *Sentinel events and alerts*. Retrieved from http://www.premierinc.com/safety/topics/patient_safety/index_3.jsp

Jöreskog, K. G. (1993). *Testing structural equation models*. Newbury Park, CA: Sage.

Jöreskog, K. G., & Sörbom, D. (1993). *LISREL 8: Structural equation modelling with the SIMPLIS command language*. Mahwah, NJ: Lawrence Erlbaum.

Jung, S., Chang, T. W., Sim, E., & Park, J. (2004). Vendor managed inventory and its effect in the supply chain. In D. K. Baik (Ed.), *Proceedings of the Third Asian Simulation Conference on Systems Modeling and Simulation: Theory and Applications* (LNCS 3398, pp. 545-552).

Kaehler, S. D. (1998). *Fuzzy Logic – An Introduction*. Retrieved from http://www.seattlerobotics.org/encoder/mar98/fuz/fl_part1.html

Kaipia, R., Holmström, J., & Tanskanen, K. (2002). VMI: What are you losing if you let your customer place orders? *Production Planning and Control*, *13*(1), 17–25. doi:10.1080/09537280110061539

Kaipia, R., & Tanskanen, K. (2003). Vendor managed category management—An outsourcing solution in retailing. *Journal of Purchasing and Supply Management*, *9*(4), 165–175. doi:10.1016/S1478-4092(03)00009-8

Kass-Hout, T., & Zhuang, X. (Eds.). (2010). *Biosurveillance: Methods and case studies*. Boca Raton, FL: Taylor & Francis. doi:10.1201/b10315

Kavakli, E., & Loucopoulos, S. (2005). Goal modeling in requirements engineering: Analysis and critique of current methods. In Krogstie, J., Halpin, T., & Siau, K. (Eds.), *Information modeling methods and methodologies* (pp. 102–124). London, UK: Idea Group.

Kehoe, D., & Boughton, N. (2001). Internet based supply chain management: A classification of approaches to manufacturing planning and control. *International Journal of Operations & Production Management*, *21*(4), 516–525. doi:10.1108/01443570110381417

Kent, P. (2008). *Cell phone and automobile accidents on the rise*. Retrieved from http://www.articlesbase.com/automotive-articles/ cell-phone-and-automobile-accidents-on-the-rise-390037.html#ixzz10HgLel8M

Khanbary, L. M. O., & Vidyarthi, D. P. (2009). Reliability Based Channel Allocation using Genetic Algorithm in Mobile Computing. *IEEE Transactions on Vehicular Technology*, *58*(8), 4248–4256. doi:10.1109/TVT.2009.2019666

Kilgore, S., & Orlov, L. M. (2002). *Balancing supply and demand*. Cambridge, UK: TechStrategy Research/Forrester Research.

Kishore, K. (1994). The advent of STAR TV in India: Emerging policy issues. *Media Asia*, *21*(2), 96–103.

Kjeldsen, E., Dill, J., & Lindsey, A. (2003). Exploiting the synergies of circular simplex turbo block coding and wavelet packet modulation. In. *Proceedings of the Military Communications Conference*, *2*, 1202–1207.

Klamargias, A. D., Parsopoulos, K. E., Alevizos, P. D., & Vrahatis, M. N. (2008). Particle filtering with particle swarm optimization in systems with multiplicative noise. In *Proceedings of 10th Annual Conference on Genetic and Evolutionary Computation*, Atlanta, GA (pp. 57-62). New York, NY: ACM Press.

Klis, G. J., & Yuan, B. O. (2001). *Fuzzy sets and Fuzzy Logic Theory and applications*. Delhi, India: Prentice Hall India.

Kohonen, T. (1997). *Self-organizing maps* (2nd ed.). Berlin, Germany: Springer-Verlag.

Kordon, M., Wall, S., Stone, H., Blume, W., Skipper, J., Ingham, M., et al. (2007). Model-Based Engineering Design Pilots at JPL. In *Proceedings of the 2007 IEEE Aerospace Conference*, Big Sky, MT (pp. 1-20)

Kuk, G. (2004). Effectiveness of vendor-managed inventory in the electronics industry: Determinants and outcomes. *Information & Management*, *41*, 645–654. doi:10.1016/j.im.2003.08.002

Kumar, B. P. V., & Venkataram, P. (2003). Reliable multicast routing in mobile networks: a neural-network approach. *IEEE Proceedings Communications*, *150*(5), 377–384. doi:10.1049/ip-com:20030649

Lackey, R. J., & Upmal, D. W. (1995). Speakeasy: The Military Software Radio. *IEEE Communications Magazine*, *33*(5), 56–61. doi:10.1109/35.392998

Lai, J. (1998). *Broadband Wireless Communication Systems Provided by Commercial Airplanes*.

Lamprecht, M., & Stamm, H. (2002). *Sport zwischen Kultur, Kult und Kommerz*. Zürich, Switzerland: Seismo.

Larman, C. (1998). *Applying UML and patterns: An introduction to object-oriented analysis and design.* Upper Saddle River, NJ: Prentice Hall.

Larmo, A., Lindstrom, M., Meyer, M., Pelletier, G., Torsner, J., & Weimann, H. (2009). The LTE link-layer design. *IEEE Communications Magazine, 47*(4), 52–59. doi:10.1109/MCOM.2009.4907407

Larson, J. F., & Park, H.-S. (1993). *Global television and the politics of the Seoul Olympics.* Boulder, CO: Westview Press.

Latour, B. (1987). *Science in action: How to follow scientists and engineers through society.* Cambridge, MA: Harvard University Press.

Law, A., Harvey, J., & Kemp, S. (2002). The global sport mass media oligopoly. *International Review for the Sociology of Sport, 37*(3-4), 279–302. doi:10.1177/1012690202037004025

Lee, J. D., McGehee, D. V., Brown, T. L., & Reyes, M. L. (2002). Collision warning timing, driver distraction, and driver response to imminent rear-end collisions in a high-fidelity driving simulator. *Human Factors, 44*, 314–334. doi:10.1518/0018720024497844

Lee, W.-Y., & Akyildiz, I. F. (2008). Optimal spectrum sensing framework for cognitive radio networks. *IEEE Transactions on Wireless Communications, 7*(10).

Lemme, P. W., Glenister, S. M., & Miller, A. W. (1999). Iridium(R) aeronautical satellite communications. *IEEE AES Systems Magazine, 14*(11), 11–16. doi:10.1109/62.809197

Lenders, V., Wagner, J., & May, M. (2006). Analyzing the impact of mobility in ad hoc networks. In *Proceedings of the ACM/SIGMOBILE Workshop on Multi-hop Ad Hoc Networks: From Theory to Reality*, Florence, Italy.

Lever, J. (1983). *Soccer madness.* Chicago, IL: University of Chicago Press.

Levy, B. S. (1996). Toward a holistic approach to public health surveillance. *American Journal of Public Health, 86*(5), 624–625. doi:10.2105/AJPH.86.5.624

Levy, M., & Grewal, D. (2000). Supply chain management in a networked economy. *Journal of Retailing, 76*(4), 415–429. doi:10.1016/S0022-4359(00)00043-9

Lewin, K. (1958). Group discussion and social change. In Macoby, E. E., Newcomb, T. M., & Hantley, E. L. (Eds.), *Readings in social psychology* (3rd ed., pp. 197–212). New York, NY: Holt, Rinehart & Winston.

Li, J., Pan, Y., & Xiao, Y. (2004). A dynamic HLR location management scheme for PCS networks. In *Proceedings of the IEEE Annual Joint Conference INFOCOM* (Vol. 1).

Liang, Y.-C., Zeng, Y., Peh, E. C. Y., & Anh, T. H. (2008). Sensing-throughput tradeoff for cognitive radio networks. *IEEE Transactions on Wireless Communications, 7*(4).

Liao, W., Ke, C.-A., & Lai, J.-R. (2000). Reliable Multicast with Host Mobility. In *Proceedings of the IEEE Global Telecommunication Conference (GLOBECOM '00)*, San Francisco, CA (Vol. 3, pp. 1692-1696). Los Alamitos, CA: IEEE Press.

Li, J., Kameda, H., & Li, K. (2000). Optimal dynamic mobility management for PCS networks. *IEEE/ACM Transactions on Networking, 8*(3), 319–327. doi:10.1109/90.851978

Lindsey, A. (1996). Improved spread-spectrum communication with a wavelet packet based transceiver. In *Proceedings of the IEEE-SP International Symposium on Time-Frequency and Time-Scale Analysis* (pp. 417-420).

Lindsey, A., & Dill, J. (1995). Wavelet packet modulation: A generalized method for orthogonally multiplexed communications. In *Proceedings of the Twenty-Seventh Southeastern Symposium on System Theory* (pp. 392-396).

Lindsey, A. (1997). Wavelet packet modulation for orthogonally multiplexed communication. *IEEE Transactions on Signal Processing, 45*(5), 1336–1339. doi:10.1109/78.575704

Lin, Y. (1997). Reducing location update cost in a PCS network. *IEEE/ACM Transactions on Networking, 5*(1), 25–33. doi:10.1109/90.554719

Lissy, K. S., Cohen, J. T., Park, M. Y., & Graham, J. D. (2000). *Cellular phone use while driving: Risks and benefits.* Retrieved from http://www.cellphonefreedriving.ca/media/harvard.pdf

Lo, C., Wolff, R., & Bernhardt, R. (1992). Expected network database transaction volume to support personal communications services. In *Proceedings of the 1st International Conference on Universal Personal Communications* (pp. 1-6).

Loeb, P. D., & Clarke, W. A. (2009). The cell phone effect on pedestrian fatalities. *Transportation Research Part E: Logistics, 45*, 284–290. doi:10.1016/j.tre.2008.08.001

Lokshina, I. (2009). Analysis of Reed-Solomon codes and their application to digital video broadcasting systems. In *Proceedings of NAEC*, Riva del Garda, Italy (pp. 31-36).

Lokshina, I., & Bartolacci, M. R. (2008). Effective assessment of mobile communication networks performance with clustering and neural modeling. In *Proceedings of the 7th Annual Wireless Telecommunications Symposium*, Pomona, CA (pp. 9-16).

Lomax, H., & Casey, N. (1998). Recording social life: Reflexivity and video methodology. *Sociological Research Online*, (3): 3–32.

Lopez-Perez, D., Valcarce, A., de la Roche, G., & Zhang, J. (2009, June). OFDMA femtocells: A roadmap on interference avoidance. *IEEE Communications Magazine, 47*(9), 41–48. doi:10.1109/MCOM.2009.5277454

Ma, X., Xiong, J., Gui, L., Liu, C., & Zhang, W. (2007). EU-China roadmap for mobile multimedia industry: the preliminary results. In *Proceedings of the IEEE 2nd International Conference on Communications and Networking in China.*

MacKenzie, D., & Wajcman, J. (Eds.). (1999). *The social shaping of technology.* Buckingham, UK: Open University Press.

Maguire, J. A. (2005). *Power and global sport: Zones of prestige, emulation and resistance.* New York, NY: Routledge.

Makris, S. E. (2006). *Telecommunications for the Athens 2004 Olympic Games: Lessons learned from a consultant's perspective.* Paper presented at the IEEE International Communications Quality and Reliability (CQR) Workshop on Olympic Program Network Reliability & Risk Services.

Malek, J. (2002). *Trace graph program.* Retrieved from http://www.angelfire.com/al4/esorkor/

Manvi, S. S., Kakkasageri, M. S., Paliwal, S., & Patil, R. (2010). ZLERP: Zone and link expiry based routing protocol for MANETs. *International Journal on Advanced Networking and Applications, 2*(3), 650–655.

Marasli, R., Amer, P. D., & Conrad, P. T. (1996). Retransmission-Based Partially Reliable Transport Service: An Analytic Model. In *Proceedings of the 15th Annual Joint Conference: IEEE Computer Societies: Networking the next generation,* San Francisco, CA (Vol. 2, pp. 24-28). Los Alamitos, CA: IEEE Press.

Masuda, M., & Ori, K. (2001). Delay Variation Metrics for Speech Quality Estimation of VoIP. *IEIC Technical Report, 101*(11), 101–106.

Ma, W., & Fang, Y. (2002). Two-level pointer forwarding strategy for location management in PCS networks. *IEEE Transactions on Mobile Computing, 1*(1), 32–45. doi:10.1109/TMC.2002.1011057

Ma, W., & Fang, Y. (2004). Dynamic hierarchical mobility management strategy for mobile IP networks. *IEEE Journal on Selected Areas in Communications, 22*(4), 664–676. doi:10.1109/JSAC.2004.825968

McAdams, J. (2006). *SMS for SOS: Short message service earns valued role as a link of last resort for crisis communications.* Retrieved from http://www.fcw.com/article92790-04-03-06-Print

McCartt, A. T., Hellinga, L. A., & Braitman, K. A. (2006). Cell phones and driving: Review of research. *Traffic Injury Prevention, 7*, 89–106. doi:10.1080/15389580600651103

McChesney, R. (1998). Political economy of communication. In McChesney, R., Wood, E., & Foster, J. B. (Eds.), *Capitalism and the information age.* New York, NY: Monthly Review.

McDonald, M. D. (2002). Key participants in combating terrorism: The role of American citizens and their communities in homeland security. *IEEE Engineering in Medicine and Biology Magazine, 21*(5), 34–37. doi:10.1109/MEMB.2002.1044158

McEvoy, S. P., Stevenson, M. R., McCartt, A. T., Woodward, M., Haworth, C., Palamara, P., & Cercarelli, R. (2005). Role of mobile phones in motor vehicle crashes resulting in hospital attendance: A case-crossover study. *BMJ (Clinical Research Ed.), 331*, 20–27. doi:10.1136/bmj.38537.397512.55

McGill, I., & Carroll, I. (1995). The negotiation and sale of television and other rights associated with a sporting event. In Fewell, M. (Ed.), *Sports law: A practical guide* (pp. 103–165). Sydney, NSW, Australia: LBC Information Services.

McNulty, T. J. (1993). Television's impact on executive decision-making and diplomacy. *Fletcher Foreign World Affaires*, (17), 67-83.

Medina, D., Hoffman, F., Ayaz, S., & Rokitansky, C. H. (2008, June 16-20). Feasibilty of an Aeronautical Mobile Ad-Hoc Network over the North Atlantic Corridor. In *Proceedings of the 5th IEEE Conference on Sensor, Mesh, and Ad hoc Communications and Networks,* San Francisco, CA (pp. 109-116).

Meier-Hellstern, K., & Alonso, E. (1992). The use of SS7 and GSM to support high density personal communications. In *Proceedings of the International Conference on Discovering a New World of Communications* (p. 1698).

Mell, P., & Grance, T. (2009). *The NIST definition of cloud computing*. Washington, DC: U.S. Department of Commerce, National Institute of Standards and Technology.

Melnick, M. (1993). Searching for sociability in the stands: A theory of sports spectating. *Journal of Sport Management*, *7*, 44–60.

Merida, S. N., & Saha, R. A. (2005). An operations based systems engineering approach for large-scale systems. In *Proceedings of the 2005 IEEE Aerospace Conference,* Big Sky, MT (pp. 4239-4250).

Microsoft Corporation. (2009). *Microsoft forefront threat management gateway 2010*. Retrieved July 25, 2011, from http://www.microsoft.com/forefront/threat-management-gateway/en/us/default.aspx

Microsoft Corporation. (2011). *Computer Online Forensic Evidence Extractor (COFEE)*. Retrieved July 25, 2011, from http://www.microsoft.com/industry/government/solutions/cofee/default.aspx

Miller, T., Lawrence, G., McKay, J., & Rowe, D. (1999). Modifying the sign: sport and globalization. *Journal of Sport and Social Issues*, *20*(3), 278–295.

Miller, T., Lawrence, G., McKay, J., & Rowe, D. (2001). *Globalization and sport*. London, UK: Sage.

Misra, A., Mcauley, A., Datta, A., & Das, S. K. (2001). *IDMP: An intra-domain mobility management protocol using mobility agents*. Retrieved from http://www.cs.columbia.edu/~dutta/research/idmp.pdf

Mitola, J. (1995). The software Radio architecture. *IEEE Communications Magazine*, *33*(5), 26–38. doi:10.1109/35.393001

Mitola, J., & Maguire, G. Q. (1999). Cognitive Radio:bMaking Software Radios More Personal. *IEEE Personal Communications*, *6*(4), 13–18. doi:10.1109/98.788210

Mohebbi, B., Filho, E. C., Maestre, R., Davies, M., & Kurdahi, F. J. (2003, October). A case study of mapping a software-defined radio (SDR) application on a reconfigurable DSP core. In *Proceedings of the 1st IEEE/ACM/IFIP International Conference on Hardware/Software Codesign and System Synthesis,* Newport Beach, CA (pp. 103-108).

Mohr, J., & Spekman, R. (1994). Perfecting partnerships. *Marketing Management*, *4*(4), 52–60.

Morris, D., & Aghvami, A. H. (2008). A novel location management scheme for cellular overlay networks. *IEEE Transactions on Broadcasting*, *52*(1).

Moura, N. T., Vianna, B. A., Albuquerque, C. V. N., Rebello, V. E. F., & Boeres, C. (2007). MOS-Based Rate Adaption for VoIP Sources. In *Proceedings of the IEEE International Conference on Communication* (pp. 628-633).

Mueller, S., & Ghosal, D. (2005, April). Analysis of a distributed algorithm to determine multiple routes with path diversity in ad hoc networks. In *Proceedings of the Conference on Modeling and Optimization in Mobile, Ad Hoc, and Wireless Networks* (pp. 277-285).

Mullen, J., & Hong, H. (2005, October 10-13). Impact of multipath fading in wireless ad hoc networks. In *Proceedings of the 2nd ACM Workshop on Performance Evaluation of Wireless Ad Hoc, Sensor, and Ubiquitous Networks*.

Narbutt, M., & Davis, M. (2006). Gauging VoIP Call Quality from 802.11 WLAN. In *Proceedings of the 2006 International Symposium on World of Wireless, Mobile and Multimedia Networks* (pp. 315-324).

NASA. (2009). *Advanced CNS Architectures and System Technologies*. Retrieved from http://acast.grc.nasa.gov/main/projects/

Nasar, J., Hecht, P., & Wener, R. (2008). Mobile telephones, distracted attention, and pedestrian safety. *Accident; Analysis and Prevention*, 40, 69–75. doi:10.1016/j.aap.2007.04.005

National Center for Statistics and Analysis. (2009). *Driver electronic device use in 2008* (Research Report No. []). Washington, DC: National Highway Traffic Safety Administration.]. *DOT HS*, *811*, 184.

National Safety Council. (2010). *Attribute risk estimate model*. Washington, DC: National Safety Council.

Nauright, J. (2004). Global games: culture, political economy and sport in the globalised world of the 21st century. *Third World Quarterly*, *25*(7), 1325–1336. doi:10.1080/014365904200281302

Newlin, H. (1998). Developments in the use of wavelets in communication systems. In. *Proceedings of the Military Communications Conference*, *1*, 343–349.

Newsky. (n.d.). NEWSKY – Networking the Sky for Aeronautical Communications. Retrieved from http://www.newsky-fp6.eu/

Neyensa, D. M., & Boyle, L. N. (2007). The effect of distractions on the crash types of teenage drivers. *Accident; Analysis and Prevention*, *39*, 206–212. doi:10.1016/j.aap.2006.07.004

Ng, C. K., & Chan, H. W. (2005). Enhanced distance-based location management of mobile communication systems using a cell coordinates approach. *IEEE Transactions on Mobile Computing*, *4*(1). doi:10.1109/TMC.2005.12

Nicholas, D. (2009, November 6). *Siren.gif: Microsoft COFEE law enforcement tool leaks all over the Internet~!* Retrieved July 14, 2011, from http://techcrunch.com/2009/11/06/siren-gif-microsoft-cofee-law-enforcement-tool-leaks-all-over-the-internet/

Nikkei Electronics Asia. (2009). *Mobile phone sales fall; Smartphone sales up, Q1*. Retrieved from http://techon.nikkeibp.co.jp/article/HONSHI/20090629/172377

Norris, P. (1997). *Politics and the press: The news media and their influence*. Boulder, CO: Rienner.

O'Shaughnessy, D. (1999). *Speech communications: Human and machine* (2nd ed.). New York, NY: John Wiley & Sons.

Obama, B. (2011). *President Obama details plan to win the future through expanded wireless access*. Retrieved from http://www.whitehouse.gov/the-press-office/2011/02/10/president-obama-details-plan-win-future-through-expanded-wireless-access

Ogg, E. (2006). *Cell phones as dangerous as drunk driving*. Retrieved from http://news.cnet.com/8301-10784_3-6090342-7.html#ixzz10HaYIDvx

Oliveira, R., Luis, M., Bernardo, L., Dinis, R., Pinto, R., & Pinto, P. (2010). Impact of node's mobility on link-detection based on routing hello messages. In *Proceedings of the Wireless Communications and Networking Conference*.

Olivena, C., Kim, J. B., & Suda, T. (1998). An Adaptive Bandwidth Reservation Scheme for High-Speed Multimedia Wireless Networks. *IEEE Journal on Selected Areas in Communications*, *16*(6), 858–874. doi:10.1109/49.709449

Otsu, T., Umeda, N., & Yamao, Y. (2001, November 25-29). System architecture for mobile communications systems beyond IMT-2000. In *Proceedings of the IEEE Global Telecommunications Conference*, *1*, 538–542.

P, M. J. (2009, November 9). *More COFEE please, on second thought. ..* Retrieved July 14, 2011, from Praetorian Perfect: http://praetorianprefect.com/archives/2009/11/more-cofee-please-on-second-thought/

Papagiannopoulos, P., Xenikos, D. G., & Vouddas, P. (2009). Event management and group communications: The case of the 2004 Olympic Games in Athens. *Event Management*, *13*(2), 103–116. doi:10.3727/152599509789686281

Pashupati, K., & Lee, J. H. (2003). Web banner ads in online newspaper: A crossnational comparison of India and Korea. *International Journal of Advertising*, *22*, 531–564.

Pashupati, K., Sun, H. L., & McDowell, S. D. (2003). Guardians of culture, development communicators, or state capitalists? A comparative analysis of Indian and Chinese policy responses to broadcast, cable and satellite television. *Gazette: The International Journal for Communication Studies*, *65*(3), 251–271. doi:10.1177/0016549203065003003

Paudel, B., & Guardiola, I. G. (2009). *On the effects of small-scale fading and mobility in mobile wireless communication network*. Rolla, MO: Graduate School of Missouri S&T.

Pérez, L., & Fernández, P. C. (2006). Aplicaciones innovadorasen el entorno IMS/TISPAN. In *Proceedings of the Telecom I+D Conference*, Madrid, Spain (p. 9).

Perkins, C. (2002). *RFC-3220: IP mobility support for IPv4*. Retrieved from http://tools.ietf.org/html/rfc3220

Perkins, C. E., & Royer, E. M. (1999). Ad hoc on-demand distance vector routing. In *Proceedings of the 2nd IEEE Workshop on Mobile Computing Systems and Applications*, New Orleans, LA (pp. 90-100).

Perkins, C., Royer, E. M., & Das, S. R. (2002). *Ad Hoc On - Demand Distance Vector routing*. Retrieved from http://www.ietf.org/rfc/rfc3561.txt

Perkins, C. (1997). Mobile IP. *IEEE Communications Magazine*, 84–99. doi:10.1109/35.592101

Peterson, S. (1981). International news selection by the elite press: A case study. *Public Opinion Quarterly, 45*(2), 143–163. doi:10.1086/268647

Picard, R. W. (1996). *A society of models for video and image libraries*. Cambridge, MA: MIT Press.

Pieterse, J. N. (1995). Globalization as hybridization. In Featherstone, M., Lash, S., & Robertson, R. (Eds.), *Global modernities* (pp. 45–68). London, UK: Sage.

Pilgrim, D. (2008). *Simplifying RF front-end design in multiband handsets*. Retrieved from http://rfdesign.com/microwave_millimeter_tech/rf_front_end_mmic/radio_simplifying_rf_frontend/index1.html

Plaisance, P., & Skewes, E. (2003). *Personal and professional dimensions of news work: Exploring the link between journalists' values and roles*. Paper presented at the Annual Meeting of the International Communication Association, San Diego, CA.

Pope, N. K. L., & Voges, K. E. (1997). An exploration of sponsorship awareness by product and message location in televised sporting events. *Cyber Journal of Sport Marketing, 1*(1), 16–27.

Prakash, R., Shivaratri, N. G., & Singhal, M. (1999). Distributed Dynamic Fault-Tolerant Channel Allocation for Cellular Networks. *IEEE Transactions on Vehicular Technology, 48*(6), 1874–1888. doi:10.1109/25.806780

Pratt, W. K. (1978). *Digital image processing*. New York, NY: John Wiley & Sons.

Price, M. E., & Verhulst, S. G. (Eds.). (1998). *Broadcasting reform in India: Media law from a global perspective*. Oxford, UK: Oxford University Press.

Pullin, A. (2009, November 11). *Microsoft's not bothered about COFEE leak*. Retrieved July 14, 2009, from The Inquirer: http://www.theinquirer.net/inquirer/news/1561911/microsoft-bothered-cofee-leak

Pyo, C. W., Li, J., & Kameda, H. (2003a, September 28-October 1). Simulation studies on dynamic and distributed domain-based mobile IPv6 mobility management. In *Proceedings of the 11th IEEE International Conference on Networks* (pp. 239-244).

Pyo, C. W., Li, J., & Kameda, H. (2003b, October 6-9). A dynamic and distributed domain-based mobility management method for mobile IPv6. In *Proceedings of the IEEE 58th Vehicular Technology Conference* (Vol. 3, pp. 1964-1968).

Pyo, C. W., Li, J., & Morikawa, H. (2005). Distance-based localized mobile IP mobility management. In *Proceedings of the 8th International Symposium on Parallel Architectures, Algorithms and Networks*.

Rácz, S., Tari, A., & Telek, M. (2003). A distribution estimation method for bounding the reward measures of large MRMs. In *Proceedings of the International Conference on Numerical Solution of Markov Chains. Urbana (Caracas, Venezuela), IL*, 341–342.

Radev, D., & Lokshina, I. (2007). Modeling of media gateway nodes for next generation networks. In *Proceedings of the 6th Annual Wireless Telecommunications Symposium*, Pomona, CA (pp. 1-8).

Radev, D., & Lokshina, I. (2008). Modeling and simulation of traffic with compression at media gateways for next generation networks based on Markov rewards models. In *Proceedings of the Industrial Simulation Conference*, Lyon, France (pp. 199-209).

Radev, D., Lokshina, I., & Radeva, S. (2006). Evaluation of the queuing network equilibrium based on clustering analysis and self-organizing map. In *Proceedings of the 13th Annual European Concurrent Engineering Conference*, Athens, Greece (pp. 59-63).

Radev, D., & Lokshina, I. (2007). Clustering and neural modelling for performance evaluation of mobile communication networks. *Journal of Electrical Engineering, 58*(3), 152–160.

Rajaee, A., Saedy, M., & Sahebalam, A. (2010). Competitive spectrum sharing for cognitive radio on scale-free wireless networks. In *Proceedings of the International Wireless Communications and Mobile Computing Conference*, Istanbul, Turkey.

Ramjee, R., La Porta, T., Thuel, S., Vardhan, K., & Salgarelli, L. (1999). *IP micro-mobility support using HAWAII.* Retrieved from http://tools.ietf.org/html/draft-ramjee-micro-mobility-hawaii-00

Ramjee, R., Varadhan, K., Salgarelli, L., Thuel, S. R., Wang, S. Y., & La Porta, T. (2002). HAWAII: A domain-based approach for supporting mobility in wide-area wireless networks. *IEEE/ACM Transactions on Networking, 10*(3), 396–410. doi:10.1109/TNET.2002.1012370

Ramzan, N., & Izquierdo, E. (2006). Scalable video transmission using double binary turbo code. In *Proceedings of the IEEE International Conference on Image Processing* (pp. 1309-1312).

Rappaport, T. S. (2002). *Wireless communications: Principles and practice.* Upper Saddle River, NJ: Prentice Hall.

Rathie, K., & Gaspar, T. (1995). Sponsorship agreements. In Fewell, M. (Ed.), *Sports law: A practical guide* (pp. 78–102). Sydney, NSW, Australia: LBC Information Services.

Raybeck, M. R. (2010). *Information about cell phones & driving.* Retrieved from http://www.ehow.com/about_6402532_information-cell-phones-driving.html#ixzz10HfO3dT3

Reese, S. D., Rutigliano, L., Hyun, K., & Jeong, J. (2007). Mapping the blogosphere: Professional and citizen-based media in the global news arena. *Journalism, 8*(3), 235–261. doi:10.1177/1464884907076459

Reese, S., & Ballinger, J. (2001). Roots of a sociology of news: Remembering Mr. Gates and social control in the newsroom. *Journalism & Mass Communication Quarterly, 78*(4), 641–658.

Reinbold, P., & Bonaventure, O. (2003). IP micro-mobility protocols. *IEEE Communications Surveys & Tutorials, 5*(1), 40–57. doi:10.1109/COMST.2003.5342229

Rhim, A., & Dziong, Z. (2009). Routing based on link expiration time for MANET performance improvement. In *Proceedings of the IEEE 9th Malaysia International Conference on Communications* (pp. 555-560).

Rice, R., & Katz, J. (2001). *The Internet and health communication: Experiences and expectations.* Thousand Oaks, CA: Sage.

Richardson, T. (2000). The geometry of turbo decoding dynamics. *IEEE Transactions on Information Theory, 46*, 9–23. doi:10.1109/18.817505

Rinehart, R. (1994). Sport as kitsch: A case study of the American gladiators. *Journal of Popular Culture, 28*(2), 25–35. doi:10.1111/j.0022-3840.1994.2802_25.x

Riordan, J., & Krueger, A. (Eds.). (1999). *The international politics of sport in the 20th century. London, UK: E & FNSpon and New York.* NY: Routledge.

Rittinghouse, J. W., & Ransome, J. F. (2010). *Cloud computing: Implementation, management, and security.* Boca Raton, FL: CRC Press/Taylor & Francis Group.

Rivenburgh, N. (1993). Images of nations during the 1992 Barcelona Olympic opening ceremony. In International Olympic Committee (Ed.), *Olympic centennial congress bulletin,* (pp. 32-39). Lausanne, Switzerland: International Olympic Committee.

Robertazzi, T. G. (2000). *Computer Networks and Systems, Queuing theory and performance evaluation.* New York, NY: Springer-Verlag.

Robinson, S. (2006). The mission of the j-blog: Recapturing journalistic authority on-Line. *Journalism, 7*(1), 65–83. doi:10.1177/1464884906059428

Roche, M. (2006). Sports mega-events, modernity and capitalist economies: Mega-events and modernity revisited: globalization and the case of the Olympics. *The Sociological Review, 54*(2), 25–40. doi:10.1111/j.1467-954X.2006.00651.x

Ross, S. M. (2007). *Introduction to Probability Models*. Amsterdam, The Netherlands: Elsevier.

Rout, P. (2006). London Olympics 2012, a lasting legacy for future generations, British Telecom. In *Proceedings of the IEEE International Communications Quality and Reliability (CQR) Workshop/World Class Communications for World Class Events.*

Rowe, D. (1999). *Sport, culture and the media: The unruly trinity (Issues in cultural and media studies)*. Buckingham, UK: Open University Press.

Rowe, D. (2005). *Sport, culture and the media. The unruly trinity* (2nd ed.). Philadelphia, PA: Open University Press.

Rowe, D., McKay, J., & Miller, T. (1998). Come together: Sport, nationalism and the media image. In Wenner, L. A. (Ed.), *MediaSport* (pp. 119–133). London, UK: Routledge.

Royal Society for the Prevention of Accidents. (2001). *The risk of using a mobile phone while driving.* Retrieved from http://www.rospa.com/roadsafety/info/mobile_phone_report.pdf

Roy, D. P., & Cornwell, T. B. (2003). Brand equity's influence on responses to event sponsorships. *Journal of Product and Brand Management*, *12*(6), 377–393. doi:10.1108/10610420310498803

RTCA. (2005). *DO-278, Guidelines for Communication, Navigation, Surveillance, and Air Traffic Management (CNS/ATM) Systems Software Integrity Assurance.* Washington, DC: Author.

Saha, D., Mukherjee, A., Misra, I. S., Chakraborty, M., & Subhash, N. (2004). Mobility support in IP: A survey of related protocols. *IEEE Network*, *18*(6), 34–40. doi:10.1109/MNET.2004.1355033

Sakhaee, E., & Jamalipour, A. (2006). The Global In-Flight Internet. *IEEE Journal on Selected Areas in Communications*, *24*(9), 1748–1757. doi:10.1109/JSAC.2006.875122

Salkintzis, A. K., Hong, N., & Mathiopoulos, P. T. (1999). ADC and DSP challenges in the development of software radio base stations. *IEEE Personal Communications*, *6*(4), 47–55. doi:10.1109/98.788215

Salkintzis, A. K., Hong, N., & Mathiopoulos, P. T. (1999). ADC and DSP Challenges in the Development of Software Radio Base Stations. *IEEE Personal Communications*, *6*, 47–55. doi:10.1109/98.788215

Samarajiva, R. (2001). The ITU consider problems of fixed-mobile interconnection. *Telecommunications Policy*, 155–160. doi:10.1016/S0308-5961(00)00085-9

Samatas, M. (2007). Security and surveillance in the Athens 2004 Olympics: Some lessons from a troubled story. *International Criminal Justice Review*, *17*(3), 220–238. doi:10.1177/1057567707306649

Sandler, D. M., & Shani, D. (1993). Sponsorship and the Olympic Games: The consumer perspective. *Sport Marketing Quarterly*, *2*(3), 38–43.

Saxe, J. G. (n. d.). *Blind men and the elephant.* Retrieved July 18, 2011, from http://wordinfo.info/unit/1?letter=B&spage=3

Schaffer, K., & Smith, S. (2000). *The Olympics at the millennium: Power, politics and the games*. Rutgers, NJ: Rutgers University Press.

Schneier, B. (2009, June 4). *Schneier on security: Cloud computing.* Retrieved July 18, 2011, from http://www.schneier.com/blog/archives/2009/06/

Shavers, B. (2008). *Virtual forensics: A discussion of virtual machines related to forensics analysis.* Retrieved July 20, 2011, from http://www.forensicfocus.com/downloads/virtual-machines-forensics-analysis.pdf

Sheikh, H., & Bovik, A. (2006). Image information and visual quality. *IEEE Transactions on Image Processing*, *15*(2), 430–444. doi:10.1109/TIP.2005.859378

Shivakumar, N., & Widom, J. (1995). User profile replication for faster location lookup in mobile environments. In *Proceedings of the 1st Annual International Conference on Mobile Computing and Networking* (pp.161-169).

Shoemaker, P. J., & Reese, S. D. (1996). *Mediating the message: Theories of influences on mass media content.* New York, NY: Longman.

Shoemaker, P. J., & Vos, T. P. (2009). *Gatekeeping theory.* Boca Raton, FL: Taylor & Francis.

Shoval, N. (2002). A new phase in the competition for the Olympic gold: The London and New York bids for the 2012 Games. *Journal of Urban Affairs*, *24*(5), 583–599. doi:10.1111/1467-9906.00146

Si, P., Sun, E., Yang, R., & Zhang, Y. (2010). Cooperative and distributed spectrum sharing in dynamic spectrum pooling networks. In *Proceedings of the Wireless and Optical Communications Conference* (pp. 1-5).

Signal Processing Design Line. (n.d.). *EE times design*. Retrieved from http://www.dspdesignline.com

Simchi-Levi, D., Kaminsky, P., & Simchi-Levi, E. (2008). *Designing and managing the supply chain: concepts, strategies & case studies*. New York, NY: McGraw-Hill.

Singer, J. (1998). Online journalists: foundations for research into their changing roles. *Journal of Computer-Mediated Communication*, *4*(1). Retrieved September 20, 2011, from http://jcmc.indiana.edu/vol4/issue1/singer.html

Singer, J. (1997). Still guarding the gate? The newspaper journalist's role in an on-line world. *Convergence: The International Journal of Research into New Media Technologies*, *3*(1), 72–89. doi:10.1177/135485659700300106

Singer, J. B. (2005). The political j-blogger, 'normalizing' a new media form to fit old norms and practices. *Journalism*, *6*(2), 173–198. doi:10.1177/1464884905051009

Si, P., Ji, H., Yu, F. R., & Leung, V. C. M. (2010). Optimal cooperative internetwork spectrum sharing for cognitive radio systems with spectrum pooling. *IEEE Transactions on Vehicular Technology*, *59*, 1760–1768. doi:10.1109/TVT.2010.2041941

Slepicka, P. (1995). Psychology of the sport spectator. In Stuart Biddle, J. H. (Ed.), *European perspectives on exercise and sport psychology* (pp. 270–289). Champaign, IL: Human Kinetics.

Snider, P. (1967). Mr. Gates revisited: A 1966 version of the 1949 case study. *The Journalism Quarterly*, *44*(3), 419–427.

Sobrinho, J. L. (2002). Algebra and algorithms for QoS path computation and hop-by-hop routing in the internet. *IEEE/ACM Transactions on Networking*, *10*(4), 541–550. doi:10.1109/TNET.2002.801397

Southard, P. B., & Swenseth, S. R. (2008). Evaluating vendor-managed inventory (VMI) in non-traditional environments using simulation. *International Journal of Production Economics*, *116*(2), 275–287. doi:10.1016/j.ijpe.2008.09.007

Soyjaudah, K., & Fowdur, T. (2006). An integrated unequal error protection scheme for the transmission of compressed images with ARQ. In *Proceedings of the International Conference on Networking, International Conference on Systems and International Conference on Mobile Communications and Learning Technologies* (p. 103).

Srikanteswara, R., Chembil Palat, R., Reed, J. H., & Athanas, P. (2003). An Overview of Configurable Computing Machines for Software Radio Handsets. *IEEE Communications Magazine*, *41*(7), 134–141. doi:10.1109/MCOM.2003.1215650

Stank, T. P., Keller, S. B., & Daugherty, P. J. (2001). Supply chain collaboration and logistical service performance. *Journal of Business Logistics*, *22*(1), 29–48. doi:10.1002/j.2158-1592.2001.tb00158.x

Storey, V. C. (1991). Relational database design based on the entity-relationship model. *Data & Knowledge Engineering*, *7*(1), 47–83. doi:10.1016/0169-023X(91)90033-T

Strain, R. (2007). *Surveillance and Broadcast Services Coverage*. Retrieved from http://www.faa.gov/about/office_org/headquarters_offices/ato/service_units/enroute/surveillance_broadcast/program_office_news/ind_day/media/MITRE.pdf

Strayer, D. L., & Drews, F. A. (2007). Cell-phone-induced driver distraction. *Current Directions in Psychological Science*, *16*, 128–131. doi:10.1111/j.1467-8721.2007.00489.x

Stuedi, P., & Alonso, G. (2007). Wireless ad hoc VoIP. In *Proceedings of the 2007 Workshop on Middleware for Next-Generation Converged Networks and Applications*, Newport Beach, CA (p. 8).

Sullivan, G. J., Topiwala, P., & Luthra, A. (2004, August). The H.264/AVC Advanced Video Coding Standard: Overview and Introduction to the Fidelity Range Extensions. In *Proceedings of the SPIE Conference on Applications of Digital Image Processing XXVII Special Session on Advances in the New Emerging Standard: H.264/AVC* (pp. 1-24).

Sun, S., Krzymien, W. A., & Jalai, A. (1998). Optimal forward link power allocation for data transmission in CDMA systems. In *Proceedings of the IEEE International Symposium on Personal, Indoor and Mobile Radio Communication,* Boston, MA (Vol. 2, pp.848-852). Los Alamitos, CA: IEEE Press.

Sundaresan, K., & Rangarajan, S. (2009). Efficient resource management in OFDMA femto cells. In *Proceedings of the Tenth ACM International Symposium on Mobile Ad Hoc Networking and Computing* (pp. 33-42).

Susaki, H. (2002). A Fast Algorithm for High-Accuracy Frequency Measurement: Application to Ultrasonic Doppler Sonar. *IEEE Journal of Oceanic Engineering, 27*(1). doi:10.1109/48.989878

Su, W., Lee, S.-J., & Gerla, M. (2001). Mobility prediction and routing in ad hoc wireless networks. *International Journal of Network Management,* 3–30. doi:10.1002/nem.386

Suzuki, N., Fujimoto, M., Shibata, T., Itoh, N., & Nishikawa, K. (1999). Maximum likelihood decoding for wavelet packet modulation. In *Proceedings of the 50th IEEE Vehicular Technology Conference* (Vol. 5, pp. 2895-2898).

SWOV. (2010). *Fact sheet: Use of the mobile phone while driving.* Retrieved from http://www.swov.nl/rapport/Factsheets/UK/FS_Mobile_phones.pdf

Symantec Corporation. (2011). *Network security software solutions: Symantic.* Retrieved July 25, 2011, from http://www.symantec.com/business/network-access-control

Tai, Z. (2000). Media of the world and world of the media: A cross-national study of the rankings of the 'Top 10 World Events' from 1988 to 1998. *Gazette, 2*(5), 331–353.

Tambini, D. (1999). New media and democracy. *New Media & Society, 1*(3), 305–329. doi:10.1177/14614449922225609

Tan, C. K. (2001). *The Use of Fuzzy Metric in QoS Based OSPF Network.* Retrieved from http://www.ee.ucl.ac.uk/lcs/previous/LCS2001/LCS054.pdf

Tanenbaum, A. S. (2004). *Computer Networks.* New Delhi, India: Pearson Education.

Teo, H. H., Chan, H. C., & Wei, K. K. (2006). Performance effects of formal modeling language differences: a combined abstraction level and construct complexity analysis. *IEEE Transactions on Professional Communication, 49*(2), 160–175. doi:10.1109/TPC.2006.875079

Texas Instruments. (2010). *TMS320C6472.* http://focus.ti.com/docs/prod/folders/print/tms320c6472.html

Thaindian News. (2008). *Strong link between mobile use and road accidents found.* Retrieved from http://www.thaindian.com/newsportal/world-news/strong-%20link-between-mobile-use-and-road-accidents-%20found_10096231.html#ixzz10Hublokz

Thomos, N., Boulgouris, N., & Strintzis, M. (2005). Wireless image transmission using turbo codes and optimal unequal error protection. *IEEE Transactions on Image Processing, 14*(11), 1890–1901. doi:10.1109/TIP.2005.854482

Tipper, D., & Dahlberg, T. (2002). Providing fault tolerance in wireless access networks. *IEEE Communications Magazine, 40*(1), 17–30. doi:10.1109/35.978050

Tomlinson, A. (1996). Olympic spectacle: Opening ceremonies and some paradoxes of globalization. *Media Culture & Society, 18,* 583–602. doi:10.1177/016344396018004005

Tonguz, O. K., & Ferrari, G. (2006). *Ad hoc wireless networks: A communication perspective.* New York, NY: John Wiley & Sons. doi:10.1002/0470091126

Toroujeni, S. M. M., Sadough, S. M.-S., & Ghorashi, S. A. (2010). Time-frequency spectrum leasing for OFDM-based dynamic spectrum sharing systems. In *Proceedings of the 6th Conference on Wireless Advanced* (pp. 1-5).

Townley, S., & Grayson, E. (1984). *Sponsorship of sport, arts and leisure: Law, tax and business relationships.* London, UK: Sweet and Maxwell.

Tsao, C.-L., Wu, Y.-T., Liao, W., & Kuo, J.-C. (2006). Link duration of the random way point model in mobile ad hoc networks. In *Proceedings of the IEEE Wireless Communications and Networking Conference* (pp. 367-371).

Turn Off the Cell Phone. (2010). *PhonEnforcer: Turn off cell phones while driving.* Retrieved from http://turnoffthecellphone.com/

Turner, E. T., Bounds, J., Hauser, D., Motsinger, S., Ozmore, D., & Smith, J. (1995). Television consumer advertising and the sports figure. *Sport Marketing Quarterly, 4*(1), 27–33.

Turner, P., & Cusumano, S. (2000). Virtual advertising: Legal implications for sport. *Sport Management Review*, (3): 47–70. doi:10.1016/S1441-3523(00)70079-9

University of Southern California. (1995). *Image database: Lenna test image.* Retrieved from http://sipi.usc.edu/database/database.php

Upadhyay, P. C., & Tiwari, S. (2004, June 7-9). A mobile-controlled distributed and dynamic scheme for location management in cellular IP networks. In *Proceedings of the IEEE/IFIP International Conference on Wireless and Optical Networks*, Muscat, Oman (pp. 260-263).

Upadhyay, P. C., & Tiwari, S. (2005, June 19-22). Multiple-gateway environment for tracking mobile-hosts in mobile IP networks. In *Proceedings of the 3rd International IEEE Northeast Workshop on Circuits and Systems*, Quebec City, QC, Canada (pp. 304-307).

Upadhyay, P. C., & Tiwari, S. (2006, June 26-29). Location tracking in IP-based networks. In *Proceedings of the International Conference on Communications in Computing, part of The World Congress in Computer Science, Computer Engineering, and Applied Computing*, Las Vegas, NV (pp. 178-184).

Upadhyay, P. C., & Tiwari, S. (2010). Network layer mobility management schemes for IP-based mobile networks. *International Journal of Mobile Computing and Multimedia Communications, 2*(3), 47–60. doi:10.4018/jmcmc.2010070104

Valko, A. G. (1999). Cellular IP: A new approach to internet host mobility. *Computer Communication Review, 29*(1), 50–65.

Van der Vlist, P., Kuik, R., & Verheijen, B. (2007). Note on supply chain integration in vendor-managed inventory. *Decision Support Systems, 44*(1), 360–365. doi:10.1016/j.dss.2007.03.003

Varshney, U., & Vetter, R. (2000). Emerging mobile and wireless networks. *Communications of the ACM*, 73–81. doi:10.1145/336460.336478

VendorManagedInventory.com. (2009). *Definition.* Retrieved May 5, 2011, from http://www.vendormanagedinventory.com

Vigtil, A., & Dreyer, H. C. (2006). Critical aspects of information and communication technology in vendor managed inventory. In *Proceedings of the APMS Conference* (pp. 443-451).

Vilanilam, J. V. (1996). The socio-cultural dynamics of Indian television: From SITE to insight and privatisation. In French, D., & Richards, M. (Eds.), *Contemporary television, Eastern perspectives* (pp. 61–90). New Delhi, India: Sage.

Vitech. (2010). *CORE Software.* Retrieved from http://www.vitechcorp.com/products/index.html

Waller, M. A., Johnson, M. E., & Davis, T. (1999). Vendor-managed inventory in the retail supply chain. *Journal of Business Logistics, 20*(1), 183–203.

Walsh, S. P., White, K. M., Hyde, M. K., & Watson, B. (2008). Dialling and driving: Factors influencing intentions to use a mobile phone while driving. *Accident; Analysis and Prevention, 40*, 1893–1900. doi:10.1016/j.aap.2008.07.005

Wang, J., & Wu, Q. (2008). Porting VoIP applications to DCCP. In *Proceedings of the International Conference on Mobile Technology, Applications, And Systems*, Yilan, Taiwan (p. 8).

Wang, S., Dai, J., Hou, C., & Liu, X. (2006). Progressive image transmission over wavelet packet based OFDM. In *Proceedings of the Canadian Conference on Electrical and Computer Engineering* (pp. 950-953).

Wang, X. Y., Wong, A., & Ho, P. (2010). Dynamic Markov-chain Monte Carlo channel negotiation for cognitive radio. In *Proceedings of the IEEE INFOCOM Conference on Computer Communications* (pp. 1-5).

Wang, Y., Chen, W., & Ho, J. S. M. (1997). *Performance analysis of mobile IP extended with routing agents* (Tech. Rep. No. 97-CSE-13). Dallas, TX: Southern Methodist University.

Wanta, W., & Craft, S. (2004). Women in the newsroom: Influences of female editors and reporters on the news agenda. *Journalism & Mass Communication Quarterly, 81*(1), 124–138.

Webb, A. (1999). *Statistical pattern recognition*. London, UK: Arnold.

Wertz, J. R., & Larson, W. J. (1999). *Space Mission Analysis and Design* (3rd ed.). Bloomington, IN: Microcosm Press.

White, B. (2001). Who are the real competitors in the Olympic Games. *Journal of Contemporary Legal Issues*, (12), 227-239.

White, D. M. (1950). The 'Gate-Keeper': A case study in the selection of news. *The Journalism Quarterly*, *27*, 383–390.

Wilber, D. (1988). Linking sports and sponsors. *The Journal of Business Strategy*, *9*(4), 8–10. doi:10.1108/eb039234

Williams, B. A., & Delli Carpini, M. X. (2004). Monica and Bill all the time and everywhere: The collapse of gatekeeping and agenda setting in the new media environment. *The American Behavioral Scientist*, *47*(9), 1208–1230. doi:10.1177/0002764203262344

Williams, R. (1977). *Marxism and literature*. New York, NY: Oxford University Press.

Witcher, B., Craigen, J. G., Culligan, D., & Harvey, A. (1991). The links between objectives and function in organisational sponsorship. *International Journal of Advertising*, *10*(1), 13–33.

Wittfeld, K., Helferich, A., & Herzwurm, G. (2008). Vendor-managed inventory as a knowledge-intensive service in procurement logistics in the automotive industry. *VIMation Journal, Knowledge, Service & Production. IT as an Enabler*, *1*, 17–25.

Wong, W., & Leung, V. (2000). Location management for next generation personal communication networks. *IEEE Network*, 8-14.

Wu, Z., Tao, J.-S., & Qian, Y. (2010). An information model for process operation system integration based on XML and STEP standard. In *Proceedings of the 2nd International Conference on Computer Engineering and Technology (ICCET)*, Chengdu, China (pp. V4-407-V4-410)

Xiaojiang, D., & Dapeng, W. (2006). Adaptive cell relay routing protocol for mobile ad hoc networks. *IEEE Transactions on Vehicular Technology*, 278–285.

Xiao, Y. (2003). A dynamic anchor-cell assisted paging with an optimal timer for PCS networks. *IEEE Communications Letters*, *7*(6).

Xie, H., Tabbane, S., & Goodman, D. J. (1993). Dynamic location area management and performance analysis. In *Proceedings of the IEEE Vehicular Technology Conference* (pp. 536-539).

Xie, J., & Akylidiz, I. F. (2002). A novel distributed dynamic location management scheme for minimizing signaling costs in mobile IP. *IEEE Transactions on Mobile Computing*, *1*(3), 163–175. doi:10.1109/TMC.2002.1081753

Xilinx. (n.d.). *Xilinx DSP Platform*. Retrieved from http://www.xilinx.com/technology/dsp.htm

Xiuchao, W. (2004). *Simulate 802.11b channel within ns-2*. Singapore: National University of Singapore.

Xu, Y., Lee, H. C. J., & Thing, V. L. L. (2003, May 11-15). A local mobility agent selection algorithm for mobile networks. In. *Proceedings of the IEEE International Conference on Communications*, *2*, 1074–1079.

Yang, W., Bi, G., & Yum, T.-S. (1997). A multirate wireless transmission system using wavelet packet modulation. In *Proceedings of the 47th IEEE Vehicular Technology Conference* (Vol. 1, pp. 368-372).

Yao, Y., Feng, Z., & Miao, D. (2010). Markov-based optimal access probability for dynamic spectrum access in cognitive radio networks. In *Proceedings of the 71ˢᵗ IEEE Conference on Vehicular Technology* (pp. 1-5).

Yao, Y., Evers, P. T., & Dresner, M. E. (2007). Supply chain integration in vendor-managed inventory. *Decision Support Systems*, *43*(2), 663–674. doi:10.1016/j.dss.2005.05.021

Yeoman, I. (2004). *Festival and events management: An international arts & culture perspective*. Oxford, UK: Elsevier Butterworth-Heinemann.

Ying, T., Yang, Y.-P., & Zeng, J.-C. (2006). An Enhanced Hybrid Quadratic Particle Swarm Optimization. In *Proceedings of the Sixth International Conference on Intelligent Systems Design and Applications*, Jinan, China (Vol. 2, pp. 980-985). Los Alamitos, CA: IEEE Press.

Yonghui, F., & Rajesh, P. (2004). Supply-side collaboration and its value in supply chains. *European Journal of Operational Research*, *152*(1), 281–288. doi:10.1016/S0377-2217(02)00670-7

Yu, L., Yu-mei, W., & Hui-min, Z. (2005, September 22-25). Modeling and analyzing the cost of hierarchical mobile IP. In *Proceedings of the International Conference on Wireless Communications, Networking and Mobile Computing* (Vol. 2, pp. 1056-1059).

Yu, Z., Yan, H., & Cheng, T. C. E. (2001). Benefits of information sharing with supply chain partnerships. *Industrial Management & Data Systems*, *101*(3), 114–121. doi:10.1108/02635570110386625

Zanikopoulos, A., Hegt, H., & van Roermund, A. (2006). Programmable/Reconfigurable ADCs for Multi-standard Wireless Terminals. In *Proceedings of the IEEE Conference on Communications, Circuits and Systems* (Vol. 2, pp. 1337-1341).

Zerfos, P., Meng, X., & Wong, S. (2006, October 25-27). A study of the short message service of a nationwide cellular network. In *Proceedings of the 6th ACM SIGCOMM Conference on Internet Measurement* (pp. 263-268).

Zetter, K. (2009, December 14). *Hackers brew self-destruct code to counter police forensics*. Retrieved July 14, 2011, from Wired Magazine: http://www.wired.com/threatlevel/2009/12/decaf-cofee/

Zhang, D., Ray, S., Kannan, R., & Iyengar, S. S. (2003). A Recovery Algorithm for Reliable Multicasting in Reliable Networks. In *Proceedings of the IEEE 32nd International Conference on Parallel Processing*, Kaohsiung, Taiwan (pp. 493-500). Los Alamitos, CA: IEEE Press.

Zhang, W., Jaehnert, J., & Dolzer, K. (2003). Design and evaluation of a handover decision strategy for 4th generation mobile networks. In *Proceedings of the 57th IEEE Semiannual Vehicular Technology Conference* (Vol. 3, pp. 1969-1973).

Zhang, J. (2002). Location management in cellular networks. In Stojmenovic, I. (Ed.), *Handbook of wireless networks and mobile computing* (pp. 27–49). New York, NY: John Wiley & Sons. doi:10.1002/0471224561.ch2

Zhang, Y., & Ansari, N. (2010). Wireless Telemedicine service over integrated IEEE802.11/Wlan and 802.16/Wimax networks. *IEEE Wireless Communications*, *17*(1), 30–36. doi:10.1109/MWC.2010.5416347

Zhao, D., Shen, X., & Mark, J. W. (2006). Soft Handoff and Connection Reliability in Cellular CDMA Downlinks. *IEEE Transactions on Wireless Communications*, *5*(2), 354–365. doi:10.1109/TWC.2006.1611059

Zhao, X., Xie, F., & Zhang, W. F. (2002). The impact of information sharing and ordering co-ordination on supply chain performance. *Supply Chain Management*, *7*(1), 24–40. doi:10.1108/13598540210414364

Zhao, Y. (2007). After mobile phones, what? Re-embedding the social in China's 'digital revolution'. *International Journal of Communication*, *1*, 92–120.

Zheng, Y., He, D., Xu, L., & Tang, X. (2005). Security scheme for 4G wireless systems. In *Proceedings of the International Conference on Communications, Circuits and Systems* (Vol. 1, pp. 397-401).

Zheng, Y., He, D., Yu, W., & Tang, X. (2005). Trusted computing-based security architecture for 4G mobile networks. In *Proceedings of the Sixth International Conference on Parallel and Distributed Computing, Applications and Technologies* (p. 251).

Zhu, Y., & Leung, C. M. (2008). Optimization of distance-based location management for PCS networks. *IEEE Transactions on Wireless Communications*, *7*(9).

About the Contributors

Michael R. Bartolacci is an Associate Professor of Information Sciences and Technology at the Pennsylvania State University's Berks Campus. He teaches courses in the areas of telecommunications and computer networking, systems integration, information security, and business information systems. He has a PhD in Industrial Engineering with a concentration in information systems from Lehigh University and a MBA from Lehigh University. He was formerly an elected member of the INFORMS Telecommunications Section's governing board and sits on the editorial boards of several telecommunications journals. His research interests include wireless telecommunications modeling, and cultural aspects of information technology.

Steven R. Powell is Professor Emeritus at California State Polytechnic University (Pomona, USA), where he led the university's efforts in interdisciplinary telecommunications education and research. He is the founder and General Chair of the Wireless Telecommunications Symposium, a leading interdisciplinary mobile communications and wireless networking conference supported by the IEEE Communications Society. Dr. Powell has taught telecommunications, engineering management, electrical engineering, IS, and computer science courses at the California State Polytechnic University - Pomona, Claremont Graduate University, Long Island University – C.W. Post Campus, and the University of Southern California. His industrial experience includes positions as Director of Corporate Planning at American Express Company and Member of the Technical Staff at Bell Laboratories. He organized and chairs the IEEE Communications Society Foothill Chapter. Powell's research interests are primarily in the areas of telecommunications strategy, telecommunications management, telecommunications policy, international telecommunications, and wireless communications. He holds an MBA from Columbia University's Graduate School of Business, a PhD and MS in Electrical Engineering from the University of Southern California, and a BS in Electrical Engineering from the Massachusetts Institute of Technology.

* * *

Hesham A. Ali is a Prof in Computer Engineering and systems and an assoc. Prof. in Information systems. He received a BSc in electrical Eng. (electronics), and MSc and PhD in computer Eng. and automatic control from the Fac. of Eng., Mansoura Univ., in 1987,1991 and 1997, respectively. From January 2000 up to September 2001, he was joined as Visiting Professor to the Department of Computer Science, University of Connecticut, Storrs. He was awarded with the Highly Commended Award From Emerald Literati Club 2002. He is a founder member of the IEEE SMC Society Technical Committee on Enterprise Information Systems (EIS). He has many book chapters published by international press and about 150 published papers in international (conf. and journal). He has served as a reviewer for many high quality journals,. His interests are in the areas of network security, mobile agent, Network management, Search engine, pattern recognition, distributed databases, and performance analysis.

Mohammad Anbar received his B.Tech in Electronics Engineering from Tishreen University, Lattakia, Syria, in the year 2003, and M.Tech in Computer Science from Jawaharlal Nehru University, New Delhi, India in the year 2007. Currently, Mohammad Anbar is Ph.D student at the school of Computer and Systems Sciences, Jawaharlal Nehru University, New Delhi, India. His research interest includes Wireless Communication, Mobile Computing, Particle Swarm Optimization, and Genetic Algorithms.

Hüseyin Arslan, received his PhD. Degree in 1998 from Southern Methodist University (SMU), Dallas, Tx. From January 1998 to August 2002, he was with the research group of Ericsson Inc., NC, USA, where he was involved with several project related to 2G and 3G wireless cellular communication systems. Since August 2002, he has been with the Electrical Engineering Dept. of University of South Florida. He has worked for Anritsu Company, Morgan Hill, CA as a visiting professor during the summers of 2005 and 2006 and as a part-time consultant between August 2005 and August 2007. In addition, he has worked for The Scientific and Technological Research Council of Turkey- TUBITAK as a visiting professor over the summer 2008. His involvement with Tubitak has been continuing as a part time consultant since August 2008. Dr. Arslan's research interests are related to advanced signal processing techniques at the physical layer, with cross-layer design for networking adaptively and Quality of Service (QoS) control. He is interested in many forms of wireless technologies including cellular, wireless PAN/LAN/MANs, fixed wireless access, and specialized wireless data networks like wireless sensors networks and wireless telemetry. The current research interests are on UWB, OFDM based wireless technologies with emphasis on WIMAX and IMT-Advanced, and cognitive and software defined radio. He has served as technical program committee chair, technical program committee member, session and symposium organizer, and workshop chair in several IEEE conferences. He is a member of the editorial board for wireless communication and mobile computing journal and research letters in communications. Dr. Arslan is a senior member of IEEE

Philemon Bantimaroudis, (PhD), is associate professor and department chair in the department of cultural technology and communication at the University of the Aegean (2002-present). He is holding a PhD from the University of Texas at Austin, U.S.A (1999). He was a visiting assistant professor in the department of communication and performance studies at Northern Michigan University, USA (1999-2000). Recently, he was an academic visitor at the Judge Business School of the University of Cambridge, U.K (Spring, 2008). His research interests include media and culture, political and international communication. He has been published in different journals such as: The Harvard International Journal of Press/Politics, The Journal of Communication, Communication Research, Media War & Conflict, The Journal of Media and Religion, Mass Communication & Society, Mediterranean Quarterly, etc. He is a contributor and/or editor to several peer-reviewed volumes. At present, he is on the editorial board of The Open Communication Journal.

Erton S. Boci received his M.S. in Electrical Engineering in 2001 and his M.S. in Systems Engineering in 2004 from The George Washington University (GWU), Washington, D.C. He joined ITT Corporation in 1998. Currently, he is leading the Service Volume Engineering design effort for the FAA's ADS-B program. He is working toward his Ph.D. degree in Engineering Management and Systems Engineering at GWU.

Vassiliki Cossiavelou, (PhD), is currently the Communications Secretary A' for the Greek Embassy – Press Office in Brussels, Belgium (2009-now), after her diplomatic post in Beijing, China (2003-2005). Dr. Cossiavelou is an expert on the global media and telecommunications industry. She has extensive industry experience in the area of ICTs in Europe, North Africa and Middle East, and she has worked on ICT related projects with international organizations such as UNESCO, NATO and the European Union. She holds a PhD from Aegean University - Department of Cultural Informatics (Greece) on news media industries. She was an inviting lecturer in Pennsylvania State University (USA), and at the joined Master program of the Donau - Universität Krems/ "Athena" RIC (Austria/ Greece, 2005-2009), while she is awarded with invited keynote speeches in international conferences of high impact factor. Her research interests include media gatekeeping models, business models in global media industries, emerging asian media wireless markets, digital dividend management, EU policies on communications. Her papers have appeared in journals and special issues, such as the International Journal of Mobile Network Design and Innovation (IJMNDI), International Journal of Interdisciplinary Telecommunications and Networking (IJITN) and in conferences such as IEEE (Institute of Electrical and Electronics Engineers) Communications Society, Springer Lecture Notes in Electrical Engineering - Wireless Technology Applications, Management and Security, Association for Computing Machinery (ACM), Institute for Operations Research and the Management Sciences (INFORMS), Wireless Telecommunications Symposium (WTS), Video and Broadband Telecommunications Symposium (VBTS), International Conference on Telecommunication Systems - Modelling and Analysis (ICTSM). She is on editorial board of the IJITN and VBTS.

M. Cenk Erturk, received his B. S. and M. S. degrees in Electrical Engineering from Nigde and Bilkent Universities, Turkey in 2005 and 2008 respectively. From 2006 to 2009, he was with the Scientific and Technological Research Council of Turkey, where he was involved in several projects. He is currently working toward his Ph.D. degree in Electrical Engineering at the University of South Florida, FL, USA. His research interests include OFDM/OFDMA systems, software defined and cognitive radio systems, aeronautical networks, adaptive cross-layer design for network and physical layers and quality of service control.

Jamal Haque, received his B.S. & M.S. degree in Electrical Engineering from University of South Florida. He is currently pursuing his Ph.D. in Electrical Engineering at the University of South Florida, Tampa, FL. His research interests include wireless systems, OFDM-based systems in high mobile platform, synchronization, channel estimation, cognitive, software defined radio and channel coding.

Laura Illia is academic director of the master in corporate communication (IE University) and professor in corporate communication in the same university (ES). Before joining IE, she was researcher in UK at the Judge Business School (University of Cambridge) y London School of Economics (University of London). From 2001 to 2006, she worked in Switzerland (CH), at the Institute of Marketing and Communication Management (University of Lugano), were she got her PhD in communication. Dr. Illia centers her research on how problems of identity, brand, image and reputation are linked with organizational management and change. Recently she published a book and various academic articles in international journals on these themes (for example Journal of Business Research, International Journal of wine business research, Museum management and curatorship, Journal of Applied Behavioural Science, Corporate Reputation Review, Corporate Communication: an International Journal). She is member of the editorial board of the Corporate Reputation Review (Palgrave) and Corporate Communication: an International Journal (Emerald), and works as ad-hoc reviewer for international journals as European Journal of Marketing and Journal of Business Ethics.

Mohd Nazri Ismail. Ph.D, is lecturing at University of Kuala Lumpur. He is completed his PhD (November 2010) in Computer Industry Department (FTSM faculty) and specializing in Computer Network, at UKM. He has many years experience in R&D Computer Networking. Before joining University of Kuala Lumpur, he was System Analyst at Bank Simpanan Nasional (2 years) and Network Engineer at MIMOS Berhad (4 years). In MIMOS, he has been exposed in R&D for Wireless, Infrared, WAP, Remote Access Server (RAS), VoIP and IP telephony technology. He holds a Master's Degree by Research in Engineering Science at Multimedia University (MMU). He has researched under "Prepaid and Postpaid VoIP Architecture in Rural Areas" during his Master Degree at MMU. He received a B.S. degree in Computer Science from UKM, Malaysia. He has experience in Network Simulation tool such as Matlab, NS-2, OPNET and OMNET++.

Evangelia Kavakli, PhD is an assistant professor at the department of cultural technology and communication, University of the Aegean. She has received the BS degree in computer science from the University of Crete, and the MS and PhD degrees in computation from the University of Manchester, Institute of Science and Technology (UMIST). She is head of the Cultural Informatics Laboratory (CILab) of the University of the Aegean. Her research work has been financially supported by the U.K. Engineering and Physical Sciences and Research Council, the Greek General Secretariat for Research and Technology, and the European Commission. She is the co-editor of the book: Aspects of representation – Studies on art and technology. She has published over 40 papers in international conference proceedings, edited books and academic journals, including the Requirements Engineering Journal, Information Systems Journal, IEEE Transactions on Systems, Man, and Cybernetics and Springer Lecture Notes on Computer Science. She is a member of the CIDOC Conceptual Reference Model SIG. Her current research focuses on information systems ecologies. She may be 'virtually' reached at www.ct.aegean.gr/people/vkavakli.

Izabella Lokshina, PhD is Associate Professor of MIS and chair of Management, Marketing and Information Systems Department at SUNY Oneonta, USA. Her main research interests are Intelligent Information Systems and Communication Networks, Queuing Systems and Performance Analysis, and Complex System Modeling and Simulation; lokshiiv@oneonta.

Thomas A. Mazzuchi received his M.S. in 1979 and his D.Sc. in 1982, both in Operations Research from The George Washington University (GWU), Washington D.C. Currently, he is a Professor of Engineering Management and Systems Engineering in the School of Engineering and Applied Science at GWU. He is also the Chair of the Department of Engineering Management and Systems Engineering at GWU where he has served as the Chair of the Operations Research Department and as Interim Dean of the School of Engineering and Applied Science. Dr. Mazzuchi has been engaged in consulting and research in the area of reliability, risk analysis, and quality control for over twenty years.

Ahmed Ibrahim Saleh is a Prof in Computer engineering and systems Department, Faculty of Engineering, Mansoura University, Mansoura, Egypt. His interests are in Mobile computing, Grid Scheduling, Web caching and Mobile Agents.

Shahram Sarkani holds the Ph.D. in Civil Engineering from Rice University, and B.S. and M.S. degrees in Civil Engineering from Louisiana State University. Professor Sarkani joined The George Washington University (GWU) faculty in 1986, where his administrative appointments include chair of the Civil, Mechanical, and Environmental Engineering Department (1994-1997); School of Engineering and Applied Science Interim Associate Dean for Research and Development (1997-2001); and Faculty Adviser, Academic Director and Head of EMSE Off-Campus Programs (since 2001). In his current role, Professor Sarkani designs and administers off-campus programs on behalf of the Department of EMSE at GWU. He is author of over 150 technical publications and presentations.

Mohammed Helmy Ali Sharara is an IT Engineer in Sharm El-Sheik International Airport, South Sinai, Egypt. His interests are in Mobile computing, Networking, Security, Parallel Processing, Database and Mobile Agents.

Sudarshan Tiwari did his B.Tech. in electronics engineering, and M. Tech. in communication engineering from Institute of Technology, BHU, Varanasi, India, in year 1976 and 1978, respectively. He received his PhD in electronics and compter engineering from University of Roorkee (now, Indian Institute of Technology) Roorkee, India in year 1993. He is professor in department of electronics and communication engineering at Motilal Nehru National Institute of Technology, Allahabad, India, since year 1999, where he undertook the responsibilities of head of department and dean (research and consultancy). He was visiting professor at J. M. Liverpool, U.K., during year 1998-99 under Indo-UK joint research project. He has been reviewer of several international/national journals and conferences. He has published more than 50 research papers in reputed International journals and conferences, and has guided 4 PhD students in the area of Communications & Networking. He has successfully completed research projects sponsored by AICTE/MHRD/DST, Govt. of India, and Govt. of U. K. His current research areas are communication engineering, wireless communications and networks, WDM optical networks, ATM networks, and broadband ISDN.

Paramesh C. Upadhyay received his BE degree in electronics and communication engineering from Regional Engineering College Srinagar (now, National Institute of Technology, Srinagar), J&K, India, in year 1989, and ME degree in electronics engineering from Punjab University, Chandigarh, India, in year 2001. He earned his PhD in electronics and communication Eengineering from Motilal Nehru National

Institute of Technology, Allahabad, India, in year 2007. He is serving as associate professor in department of electronics and communication engineering, Sant Longowal Institute of Engineering and Technology, Longowal. His current research interests are in the areas of mobility management, performance evaluation, digital signal processing, and mobile-IP networks. He has been reviewer of Elsevier's Computers and Electrical Engineering Journal and IEEE ICNSC'06, and TPC member of IEEE GLOBECOM'06.

Deo Prakash Vidyarthi received master's degree in computer application from MMM Engineering College and Ph.D in Computer Science from Jabalpur University (work done at Banaras Hindu University, Varanasi). Taught UG and PG students in the Department of Computer Science of Banaras Hindu University, Varanasi for more than 12 years. Joined JNU in 2004 and currently working as associate professor in the school of Computer and Systems Sciences, Jawaharlal Nehru University, New Delhi. Dr. Vidyarthi is member of IEEE, International Society of Research in Science and Technology (ISRST), USA and senior member of the International Association of Computer Science and Information Technology (IACSIT), Singapore. His research interest includes Parallel and Distributed Systems, Grid Computing, Mobile Computing.

Index